Internal Structure of Fault Zones

Edited by
Chi-yuen Wang

1986

Birkhäuser Verlag
Basel · Boston · Stuttgart

Reprinted from Pure and Applied Geophysics
(PAGEOPH), Volume 124 (1986), No. 1/2

Editor's address:

Chi-yuen Wang
Professor of Geophysics
Department of Geology and Geophysics
University of California
Berkeley, CA 94720
USA

CIP-Kurztitelaufnahme der Deutschen Bibliothek

Internal structure of fault zones / ed. by Chi-yuen Wang;
Basel; Boston; Stuttgart: Birkhäuser, 1986.
Aus: Pure and applied geophysics; Vol. 124, 1986
ISBN 3-7643-1845-7

NE: Wang, Chi-yuen [Hrsg.]

ISBN 3-7643-1845-7
ISBN 0-8176-1845-7

PURE AND APPLIED GEOPHYSICS
Vol. 124, No. 1/2, 1986

Contents

PAGEOPH, Vol. 124, Nos. 1/2 (1986)
0033-4553/86/020001-02$1.50+0.20/0

Introduction

Traditionally, faults are defined as fractures along which the opposite wall rocks have moved past each other; their function in transporting crustal material has long been recognized. In global tectonics, plate boundaries are marked by faults; movements on these faults and the resistance to such movements determine how the plates interact and what magnitude of force is required to drive their motions. Many faults are also the passageways for massive infiltrations of fluids and chemical elements. Thus a better understanding of the nature of fault zones may deepen our understanding of the dynamics of plate motion, the mechanics of faulting, the nature of the failure process and its predictability, and the evolution of the crust as a whole.

Faults are often regarded by geophysicists as dislocations without thickness, across which strain accumulates and elastic rebound occurs, giving rise to interplate and intraplate earthquakes. Geological observations of exhumed faults and geophysical soundings of subsurface faults, however, show that in nature faults occur as zones with substantial thicknesses, within which are complex structures that are the product of intense and repeated deformation. The study of the internal structures of fault zones may reveal the mechanisms controlling the evolution and the behavior of faults.

This special issue of *Pure and Applied Geophysics* puts together sixteen papers devoted to in-depth investigations of the internal structures of fault zones. Most of the faults investigated were natural; some were created artificially in controlled laboratory environments. The approaches of authors vary broadly, being geological, geophysical, and geochemical. It is the intention of this special issue to make available to the readers an integrated and balanced presentation of the subject. An advantage of such a presentation is that the conceptual models developed in the various approaches may be compared 'side by side'.

Rutter and his co-workers, Moore and colleagues, and Sammis examine the microstructures and grain-size distributions in natural and in experimentally produced fault gouges and show that there may be a scale-invariant similarity among some of the structures. Chester and Logan, Wallace and Morris, and Wang and co-workers focus on the overall structures of fault zones; the results from both geological observation and geophysical modeling point to a picture of fault zones (of different scales and types) containing chiefly fault gouge and bordered on the sides by extensively fractured wall rocks. Mooney and Ginzburg review the seismic velocities in fault zones, which appear to support this model. Sibson, Davies and Pollard, and Deng and his co-workers investigate the processes of brecciation and faulting and their interrelation, special attention being given to the interaction among subsidiary faults within the main fault zones; models of the generation and evolution of fault zones are proposed. Several investigators provide evidence of the active involvement of fluids in the evolution of fault zones; in particular, Kerrich focusses on the geochemical evidence of fluid infiltration in several major fault zones. Simpson, Stel, Hobbs and co-workers and Xu and co-workers explore the important topic of the brittle–ductile transition in shear zones and its role in earthquake faulting; a most interesting observation is that the transition may be cyclic (Stel, Hobbs et al.) and that brittle fracturing may predate the development of ductile shear zones (Simpson). Finally, Watterson presents an empirical relation between fault displacement and fault width, and he develops from this relation a model of fault growth.

Chi-yuen Wang
Guest Editor

Vol. 124, Nos. 1/2 (1986) 0033-4553/86/020003-28$1.50 + 0.20/0

Comparative Microstructures of Natural and Experimentally Produced Clay-Bearing Fault Gouges

E. H. RUTTER,[1] R. H. MADDOCK,[1] S. H. HALL[1] and S. H. WHITE[1]

Abstract—Fault rocks formed in phyllosilicate-bearing rocks formed over a wide range of environmental conditions within the Earth's crust are characterised by similar structural and microstructural features. The most striking of these are (a) P foliation, defined by the preferred alignment of phyllosilicates in a plane oblique to the direction of shear and (b) small-scale shear zones either parallel to the shear direction (Y shears) or oblique to the direction of shear but with the opposite sense of obliquity relative to the P foliation (Riedel shears, R_1). The minor shear zones have the same sense of displacement as the host shear zone.

The occurrence of these and other structures in clay-rich fault gouges from exceptionally well-exposed fault zones in southeastern Spain is described. The pervasive development of these flow structures throughout large volumes of fault gouge permits fault-displacement vectors to be inferred. For the region studied the movement picture is relatively simple and is superposed on a complex network of variably oriented fault zones.

The naturally produced fault-gouge structures are compared with fault gouges produced experimentally by shearing kaolinite–quartz mixtures between intact blocks over a wide range of experimental conditions. Good correspondence between their respective microstructural features was observed.

Finally, attention is drawn to the fact that natural clay-bearing fault gouges are the products of deformation accompanied by very low-grade retrogressive metamorphism, and that part of the microstructure of these rocks may be ascribed to crystallization under stress. Microstructures are described that are from long-duration experimental runs, (5 months at high temperature and in the presence of water) which go some way towards simulating these effects.

Key words: Structural Geology, faults, S.E. Spain, experimental rock mechanics.

Introduction

In recent years several experimental studies of the mechanical properties and, to a lesser extent, the microstructural evolution of clay-bearing fault rocks (gouge, in the loose sense) have been made (e.g., MORGENSTERN and TCHALENKO, 1967; LOGAN et al., 1979; LOGAN et al., 1981; SUMMERS and BYERLEE, 1977; BYERLEE and SUMMERS, 1976; BYERLEE et al., 1978; RUTTER, 1979; WANG et al., 1980; MORROW et al., 1981;

[1] Department of Geology, Imperial College, London SW7, Great Britain. S. H. Hall is now at British Petroleum, London.

GAMOND, 1983). By comparison, relatively few studies of the structural features of naturally produced fault rocks have been made, and even fewer of their characteristics at the microstructural level. This is perhaps due largely to the difficulty of collecting and thin-sectioning such materials, which are often extremely friable (but see, for example, ENGELDER, 1974; ANDERSON et al., 1983; STEL, 1981; HOUSE and GRAY, 1982; CHESTER et al., 1985).

We have recently been engaged on a detailed study of the regional setting, movement history, structure, and microstructure of fault rocks of Neogene age from southeastern Spain (HALL and RUTTER, 1986; HALL, 1983). In parallel with this, we have been studying the mechanical properties and microstructural evolution with progressive strain of synthetic clay–silt fault gouges by means of experiments in which gouge samples are sheared between intact rock pieces.

In this paper we report a comparison of structural and microstructural features of experimentally and naturally produced fault gouge, as a basis for a discussion of the extent to which the movement picture and mechanical characteristics of fault zones may be inferred from their structure.

Characteristic Geometric Features of Shear Zones

The formation of localized zones of intense shear deformation (faults) is a characteristic feature of rocks deformed over a surprisingly wide range of physical conditions. Shear faults involving localized intense *cataclastic* deformation are characteristic of rock deformation both near the Earth's surface in nature and under low temperatures and confining pressures in laboratory experiments. On the other hand, shear faults involving localized intense *plastic* deformation of mineral grains are quite common in rocks deformed even at the highest grades of regional metamorphism. In such shear zones intense plastic deformation often results in dynamic recrystallization, leading to reduced grain size and the possibility of a consequent deformation mechanism change to grain-boundary sliding accommodated by solid-state diffusive mass transfer. In either case the localization of the deformation depends upon mechanical weakening accompanying the progressive accumulation of strain.

What is very remarkable is that geometrically similar structural features are frequently found in fault rocks deformed under radically different physical conditions and by different deformation mechanisms. The similarities are often most striking in rocks rich in phyllosilicate minerals, whether they be clay-rich fault gouge, phyllonites, or mica schists. Many of these structural features can be used in deducing fault-displacement vectors. Especially with respect to higher-grade, mylonitic rocks, Simpson and Schmid (1983) have reviewed the microstructural indicators of shear zone movement sense. Figure 1 shows schematically the commonly observed features with particular reference to fault gouges, and these are described below. The

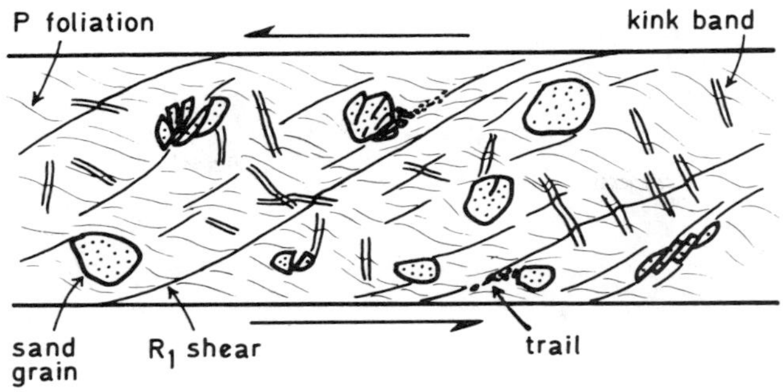

Figure 1
Schematic illustration of the interrelations between the commonly observed structural features of clay-bearing fault gouges. The sense of shear is the same on R_1 and on the fault zone, generally, as revealed by the deflection of the P foliation and the shearing of included clasts.

attachment of labels, such as 'P,' 'R_1,' 'Y,' to microstructural features is intended to imply orientation only with respect to the shear zone and direction, and not to mechanism of formation.

1. A preferred alignment of platy minerals generally develops in an orientation between 135° and 180° to the shear direction (see Figure 1 for orientation). Deformable particles or mineral grains, or both, are usually flattened in this plane and tend to be drawn out in a direction parallel to the trace of the shear direction. In the terminology of LOGAN et al. (1979) this is the P orientation. Though originally coined to describe a fracture preferred orientation, it is now clear that in clay-rich gouges an intense preferred orientation of clay platelets forms in the P orientation, and this may guide fracturing. LOGAN et al. (1981) also suggested that heavily microcracked fragments in the gouge may be drawn out in the P orientation by combined slippage on R_1 and R_2 Riedel shear sets (see below) to form 'ductile stringers' ('trails' in our terminology; see Figs. 1 and 2). Such features may equally arise through shearing parallel to the P fabric after it has formed. In the terminology of, for example, BERTHE et al. (1979), which is often applied to well-foliated mylonitic rocks, this orientation is called the S foliation. In this instance at least, it is held to correspond to the principal flattening plane of the finite strain ellipsoid. Localized sliding and folding may occur on a P-oriented foliation after it has formed and suffered rotation.

2. The foliated fault rock is often transected by a more or less pervasively developed set of smaller-scale shear zones, having the same sense of movement as the host shear zone but lying in an antithetic orientation with respect to the P foliation up to 45° to the shear direction. The sense of movement of these is often obvious by virtue of the offset of marker bands or transected particles, or because of the sense of

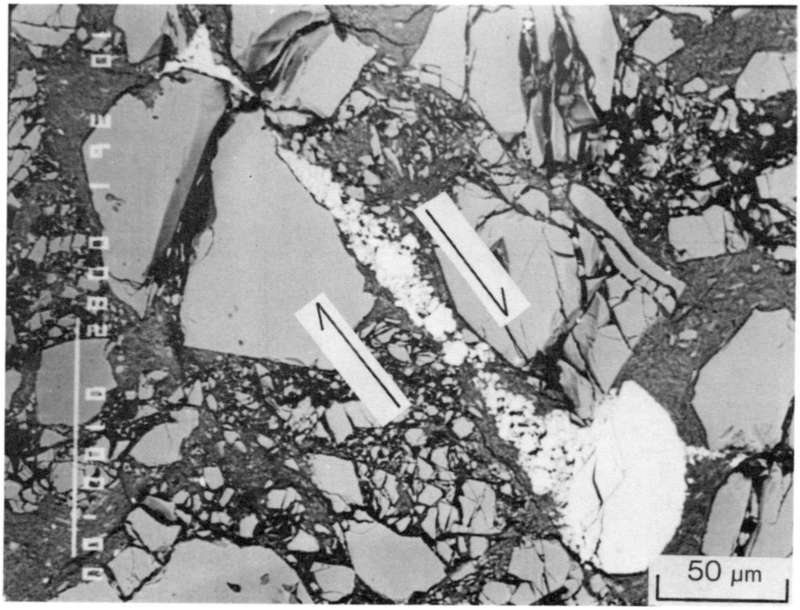

Figure 2
Backscattered electron micrograph showing trail development in experimentally deformed clay–silt gouge, through the intersection of an R_1 shear with an included rutile particle (white).

deflection of the P foliation. These are R_1 Riedel shears in the terminology of LOGAN et al. (1979), applied to fault gouges, and correspond geometrically to the 'shear bands' of WHITE et al. (1980) or 'extensional crenulation cleavage' of PLATT and VISSERS (1980), as frequently applied to higher-grade, schistose metamorphic rocks. The latter features are probably the most reliable indicators of sense of shear in foliated rocks.

3. The foliated fault rock may also be transected by a set of minor shear zones exactly parallel to the host shear zone and possessing the same sense of movement. Like R_1-oriented minor shear zones, these may deflect the P foliation and offset and distort transected particles of known original shape. In the terminology of LOGAN et al. (1979) these are known as Y surfaces. In experimental studies (e.g., LOGAN et al., 1981; MOORE and BYERLEE, 1985) marked intensification of slip in the same orientation has often been reported to develop preferentially along the interface of gouge and intact rock.

In the terminology of BERTHE et al. (1979) they correspond to C surfaces. Unlike R_1 surfaces which, because of their orientation, can only accommodate limited amounts of slip, Y-oriented surfaces in natural fault gouges can accommodate large finite displacements.

4. A number of smaller-scale features are commonly found in addition to those mentioned above.

Trails. These are produced when deformable particles included in the fault zone are transected by local slippage surfaces such as R_1, Y, or P surfaces (Figs. 1 and 2). In fault gouges, included fragments of widely varying sizes often are heavily traversed by cracks, so that they flow in a ductile manner like loose sand when sheared. The asymmetric trails produced by the transection of such particles by small shear zones gives a very clear indication of sense of movement in natural fault gouges. The drawing out of fragmented clasts in the P foliation contributes significantly to a foliated appearance even to clay-poor fault gouges (cf. CHESTER et al., 1985). LOGAN et al. (1981) show spectacular examples of trail development in experimentally deformed samples of San Andreas fault gouge. We describe further experimental and natural examples below.

Slickensides and wear grooves. Linear features on fault surfaces have often been used for inferring fault-movement vectors. The term *slickenside* is frequently applied equally to linear fibre growth features produced in dilatant areas of a sliding fault surface (e.g., Elliott, 1976) and to grooves and striations produced by both frictional ploughing and pressure-solution indentation by asperities on the opposing fault surface. Genuine wear grooves may also be characterised by asymmetric plucking and indentation cracking of the substrate material (e.g., HAMILTON and GOODMAN, 1966; ENGELDER, 1977), and these features can sometimes be used in inferring shear sense. Slickensides as regional fault-movement direction indicators are only reliable if the thickness of ductile gouge is small. In wide (hundreds of metres) gouge-bearing fault zones with included, relatively intact blocks of a wide range of sizes (metres to tens or hundreds of metres) local slippage of the gouge against the block faces and slippage along cracks developed within the block can occur in directions not necessarily equal to the slip vector of the fault zone as a whole. Small fault displacements (of the order of centimetres only are required to produce slickensides, and it is not unusual for more than one set of slickenside orientations to be developed on one fault surface. The planar features of the flowing gouge referred to in the paragraphs numbered 1, 2 and 3 above lead to more consistent movement-direction indications.

Kink bands and R_2*-oriented faults.* Perhaps less frequently than the features described above, kink bands deflecting the P foliation or small-scale antithetic faults, or both can occur in gouge zones at a high angle to the main fault zone. This is the R_2 Riedel shear orientation, in the terminology of LOGAN et al. (1979). Kink bands can be localized apparently by the strain heterogeneities associated with the edges of included clasts (see Fig. 3). Lineations produced by the intersection of kink bands with P or R_1 surfaces can easily be mistaken for wear grooves in phyllosilicate-rich natural fault gouge.

Extension-crack arrays. These well-known features often occur in the host rock adjacent to gouge-bearing fault zones or within higher-grade fault zones. In ex-

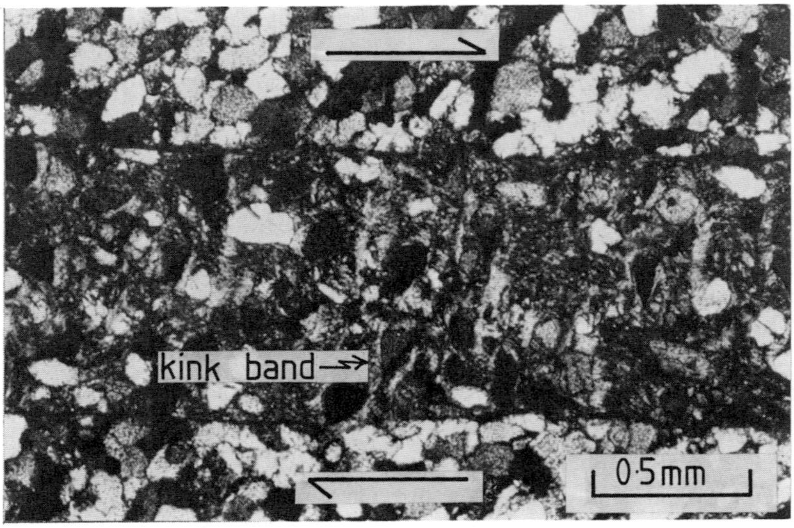

Figure 3
Optical micrograph (crossed polars) showing kink-band development at a high-angle to the sheared zone in experimentally deformed clay-rich gouge (67% kaolinite). The kink bands appear light in colour, whilst the differently oriented host-clay matrix is relatively dark.

perimental rock-deformation studies they are the principal sources of dilatant strain associated with rock failure. In nature they often occur in en-echelon arrays associated with localized zones of shearing. They are often filled with quartz or other minerals when found in nature, and they are very reliable indicators of shear sense and direction.

The Structure and Microstructure of Fault Gouges from Southern Spain

The Betic Zone of southeastern Spain is characterised by a series of uplifted blocks composed largely of low-to-medium-grade metamorphic rocks, whose orogenic history was largely terminated by the beginning of the Miocene period. These mountain ranges are dissected by a number of intermontane sedimentary basins containing postorogenic sedimentary sequences ranging in age from lower Miocene to Recent. The intermontane basins owe their origin largely to fault movements that led to the development of graben, half-graben, and pull-apart features which depended on the strike-slip nature of some of the faulting. In the area we have studied, which lies entirely within the Betic Zone in the strict sense, vertical fault-displacement components may range up to 6 km, and strike-slip components up to perhaps 30 km. A description of the rock products of faulting in the context of the regional geology and the evolution of the sedimentary basins forms the subject of a

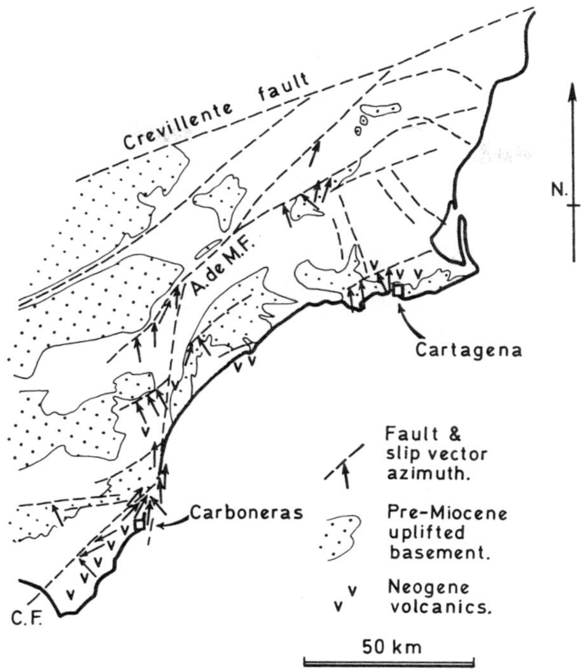

Figure 4
Map showing the disposition of the main fault zones of southeastern Spain. The unshaded areas represent the Neogene sedimentary basins, which are flanked by uplifted mountain ranges composed largely of metamorphic rocks formed during the Alpine orogeny. Legends: A. de M. F., Alhama de Murcia fault; C. F., Carboneras fault. Azimuths of fault slip vectors determined from fault-gouge structures are shown. Most faults show a combination of reversed dip-slip plus left-lateral strike-slip movement, inferred to have developed mainly in the middle and upper Miocene but extending up to the Quaternary.

separate paper (Hall and Rutter, 1986; but see also Hall, 1983).

Unlike fault zones in more temperate regions, the semi-arid climate and lack of vegetation in the region had led to excellent exposure of the fault rocks, from which the high degree of internal structuring within the fault zones is readily perceived in the field.

Figure 4 shows a simplified map of the fault–basin–uplifted-block configuration in the region studied, together with the trace of fault displacement vectors inferred from the internal structural features (see below) of the fault zones. Figure 5 shows fault plane orientations from the entire region, together with the associated slip vector. It can be seen that despite the wide range of orientations of fault segments there are indications of a remarkable simplicity and consistency in the Neogene to Quaternary movement picture. Slip vectors tend to lie in a vertical great circle, striking approximately NNE-SSW, they display mostly combinations of reverse dip-slip plus left-lateral strike-slip movements, and they are suggestive of compressional

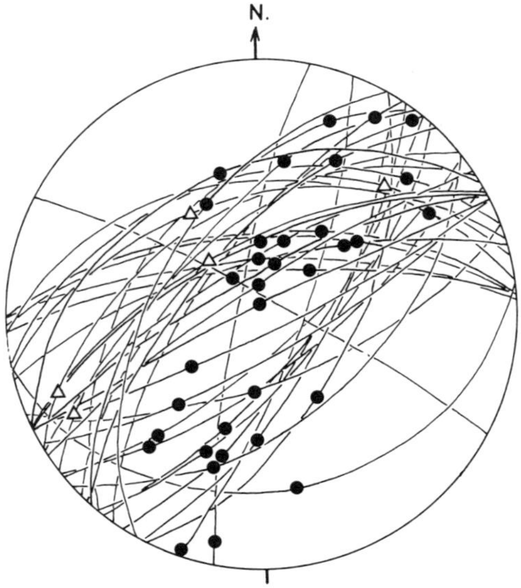

Figure 5
Equal-area lower-hemisphere projection showing fault plane orientations and slip vectors for the region shown in Figure 4. The tendency of slip vectors to lie in a broad great-circle girdle is suggestive of NNE-SSW-directed compression, activating slip on fault planes of varied orientations.

displacements applied in the NNE-SSW direction.

Individual fault movements can often be dated by reference to displacement of, and unconformities within, associated basin sediments of known age. Most of the movements indicated in Figure 5 relate to middle to upper Miocene events, when activity was most intense, but some relate to Pliocene and Quaternary events. In broad terms the movement picture seems to have remained fairly constant throughout this time span in the region considered; hence for the purpose of this paper we have not indicated the age of each movement shown in the figure. The fault orientations were probably established before the Miocene and reflect deep-seated crustal heterogeneities. There is no basis whatever for trying to use the *orientations* of faults in the regional pattern to deduce the regional movement picture, and less so to deduce regional 'principal stress orientations.' The regional movements have simply been accommodated by renewed slippage on weak, preëxisting zones in a variety of orientations, leading to the development of dip-slip, strike-slip, and oblique-slip zones.

The generally northward-directed regional compressive movement during the middle Miocene has been recognized previously (e.g., ARMIJO et al., 1977; SANZ DE GALDEANO, 1983), but we are unaware of any previous use of the internal structure of fault gouges, as discussed above, for inferring regional displacement pictures on

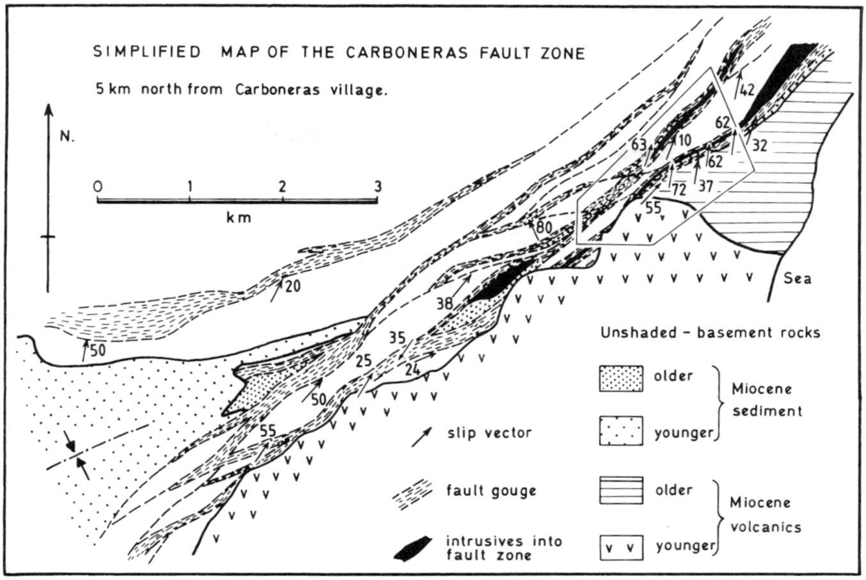

Figure 6
Simplified map showing the braided nature of the fault-gouge zones in the Carboneras fault zone about 5 km north from Carboneras village. Most individual fault-gouge zones dip steeply to the northwest. The fault zone bounds the uplifted basement block of the Sierra Cabrera to the north and the Carboneras volcanic complex (Miocene) to the south. To the south-west the fault zone branches, to bound the Nijar Miocene sedimentary basin. The oldest rocks of the Nijar basin have been incorporated into the fault zone. The marked area in the north-east is shown enlarged in Figure 7.

any systematic basis.

In our studies special attention has been given to the Alhama de Murcia and Carboneras fault systems (Fig. 4). Movement on these faults has involved both dip-slip and left-lateral strike-slip components; therefore the fault rock products might be expected to be derived at least in part from some kilometres depth below the present surface and to have suffered substantial total displacement (of the order of 10 km). Both fault zones at depth below the present surface are suspected of being involved in the present-day diffuse seismicity of the Betic Cordilleras (UDIAS et al., 1976). Most of the description below is specifically of the Carboneras fault zone 5 km north of the village of Carboneras, but the general characteristics of the fault rocks apply equally to all of the well-exposed fault zones in the region.

Fault zone structure

Figures 6 and 7 show simplified geological maps of part of the Carboneras fault zone on different scales, to illustrate the internal structuring of the zone and, in

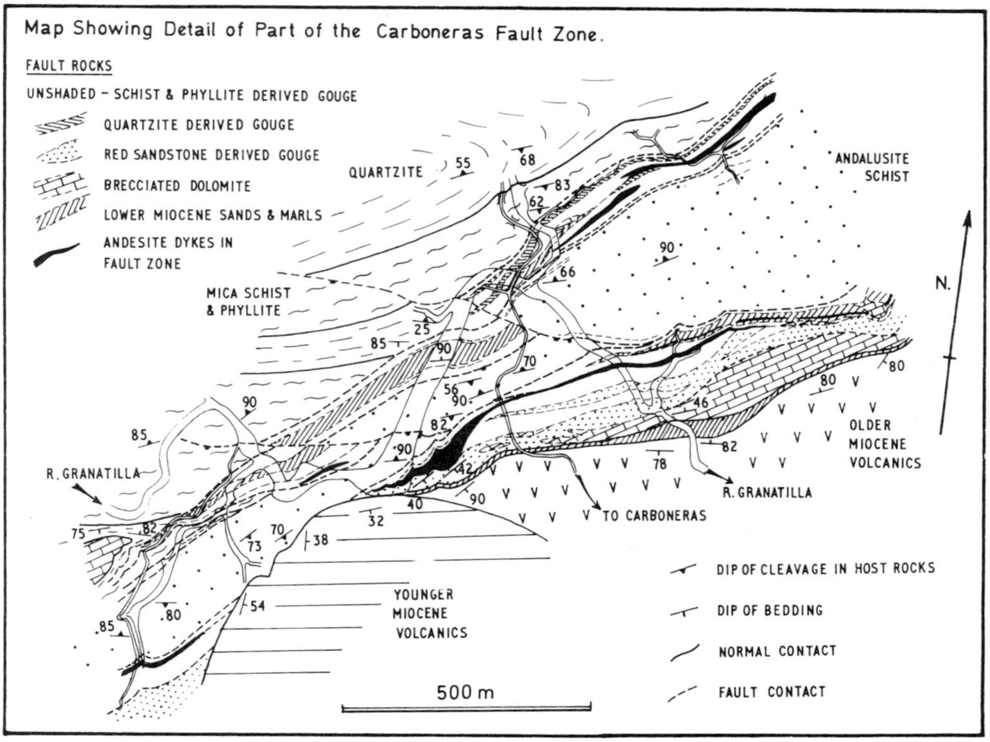

Figure 7
Detailed map of the inset area shown in Figure 6, showing the structuring of the gouge zones in the two main branches of the Carboneras fault. Each Y-oriented sliver is derived from a particular member from the regional stratigraphic sequence. The sandstones, quartzites, and dolomites show varying degrees of cataclasis, whilst the schist- and phyllite-derived fault rocks tend to be heavily degraded to clay-rich gouges with included clasts of various sizes. Andesitic dykes (feeders to the Carboneras volcanics?) were injected at various stages during the movement history and show varying degrees of mechanical and chemical transformation.

particular, the way in which the main fault strands are composed of long slivers of fault gouge and cataclasite, typically a few metres in thickness and hundreds of metres in length. Figure 8 shows this characteristic is readily perceived in the field. Each sliver typically represents one of the regional rock types. During its movement history the fault has sampled and incorporated within itself slivers of country rock derived from units spanning at least 3 km of stratigraphic separation, and including andalusite-and sillimanite-bearing schists of the Nevado-Filabride basement complex, on the one hand, to nonmetamorphic, lower Miocene sediments of the nearby Nijar basin on the other. The fault surfaces separating these slivers are extremely sharp (<1 cm in contact width) and must be regarded morphologically as Y shears. These are the surfaces upon which most of the net fault displacement must have

Figure 8
View looking north-west, showing the southern arm of the Carboneras fault zone where it intersects the Granatilla river valley on the eastern side of the area shown in Figure 7. The NNW-dipping hanging wall of the fault is clearly visible, and the Y slivers of gouges of different host-rock type can be perceived.

been accommodated, and if the fault displacements seen at the present erosion level were developed seismically, then these Y shears must represent the ancient seismogenic surfaces.

No evidence of frictional melting associated with movement on these surfaces has yet been found. On the basis of the syntectonic mineral assemblages of these fault rocks (see below), a maximum temperature during formation of the rocks of about 180 °C has been inferred, which may correspond to a depth of origin of about 6 km. It must be borne in mind, however, that fault movements on the Carboneras zone were accompanied by volcanic activity (HALL, 1983), including the injection of magma into some parts of the fault zone.

The fault rocks within each of the Y slivers referred to above vary in their degree of mechanical and chemical degradation, according to their initial mineralogical composition and, presumably, the time since their incorporation into the fault zone.

Cataclastic rock

Slivers initially composed of fine-grained dolomite, quartzite, or sandstone show no mineralogical changes but are often pervasively cracked and microcracked, the cracks becoming infilled with hydrothermal cement of the same composition as the host rock. The cracks often exhibit a discernible preferred orientation and are thought to be analogous to axially oriented extension cracks, which are a characteristic feature of the experimental deformation of many brittle rocks. These rocks therefore appear to be carried along as passive blocks within the more deformable fault rocks, suffering internal pervasive cracking, a cementation of the cracks keeping pace with crack growth, so that the block tends to retain its mechanical integrity. These blocks are usually heavily slickensided where they contact neighbouring fault rocks. They range in size from millimetres up to 1 km in length and 150 m in width, and fragments of such rocks are often seen incorporated into Y slivers of the more deformable fault rocks.

Clay-bearing gouges

Host rocks that originally contained feldspars, micas, and other aluminous silicates tend to undergo retrogressive chemical metamorphism as well as mechanical degradation during faulting. Thus preëxisting schists, phyllites, and volcanic rocks are transformed into Y slivers, each of which is dominated by clay-bearing fault gouge containing included host-rock fragments in various stages of breakdown together with included fragments of quartzites, sandstones, and dolomites, as mentioned above.

Clay-bearing fault gouges are usually very friable in the field. Specimens of these rocks were collected by mixing of a small quantity of epoxy resin inside a polythene specimen bag. The oriented specimen was then gently lowered into the bag and the resin smoothed around the rock. The coated specimen inside its bag was then allowed to harden. Thin sections of gouges were prepared by dry-cutting with a bandsaw and then impregnating the specimen with resin under vacuum to support the face that was to be thin-sectioned.

In studies of fault gouge from the San Andreas fault zone, California, much attention has been given to the composition of the clay fraction (e.g., WU, 1978; LIECHTI and ZOBACK, 1979) and to its role in the mechanical properties of fault gouges in general (e.g., WANG et al., 1980; SUMMERS and BYERLEE, 1977; LOGAN et al., 1981). Table 1 shows the mineralogical composition of several clay-rich fault gouges collected from southeastern Spain during the course of the present study, determined by the x-ray diffraction technique. The gouges have higher phyllosilicate contents, with more alteration products, than their host rocks. Montmorillonite is a significant component only of volcanic rock and Neogene sediment and their derived gouges. Mixed-layer clays are rare in the fault gouges. Particular segments of particular faults appear to be characterised by the dominance of certain clay minerals. This is probably a reflection largely of the host-rock composition and the temperature of formation. From a regional point of view, kaolinite, illite, and chlorite are the dominant clay minerals in the fault gouges.

Macroscopic structures in the clay-bearing fault gouges correspond precisely to

Table 1

Mineralogy of Gouges from the Granatilla Section through the Carboneras Fault Zone

Rock type (host)	Marl	Sandstone	Schist	Sandstone–schist	Phyllite–schist	Andesite	Quartzite
Illite–montmorillonite	—	—	—	—	—	T	—
Montmorillonite	M	—	T	—	T	T	—
Kaolinite	—	—	T	M	—	M	—
Chlorite	M	M	M	T	M	M	M
Illite	M	M	T	T	M	T	—
Paragonite	—	—	T	—	T	—	—
Muscovite	M	M	M	M	M	—	M
Dolomite	T	—	T	—	T	T	T
Calcite	M	—	—	—	T	T	—
Gypsum	T	—	T	—	T	—	T
Hornblende	—	—	—	—	—	T	—
Pyroxene	—	—	—	—	—	T	—
Andesine	—	—	—	—	—	M	—
Albite	T	T	T	—	T	—	T
Quartz	M	M	M	M	T	M	M
Rutile	—	—	T	—	T	—	—
Hematite	—	—	—	—	M	T	—
Goethite	—	M	—	T	—	—	—
Lepidocrosite	T	—	T	—	—	—	—

* M > 10%, T < 10% of volume.

the general characteristics suggested earlier. Clay-rich fault gouges always exhibit a marked P foliation, defined by the preferred orientation of clays and cataclastic mica fragments, at 150° to 170° to the fault walls, by R_1 shears spaced a few centimetres apart, and Y shears. Included fragments are often intensely microfractured and drawn into trails along these surfaces. The spacing of the R_1 shears often appears to be controlled by the dimensions of the included clasts. Figures 9 and 10 show examples of these features as seen in the field, and Figure 11 shows orientation data from a 300 m long section of the Alhama de Murcia fault near the town of Lorca, which demonstrates the angular relations generally observed between P, Y, and R_1 surfaces, and the slip vector.

The intensity of development the P foliation and R_1 shears is generally uniform throughout the volume of the fault gouge; thus the gouge can be said to have flowed fairly homogeneously within each gouge sliver. It is the pervasive development of these features throughout the flowing gouge mass which makes them reliable indicators of fault movement sense and direction. Strain hardening processes tend to result in the spreading of deformation throughout a flowing mass and, by analogy with the results of experimental studies discussed below, we believe that the observed tendency of the deformation to be uniformly spread throughout the gouge zones

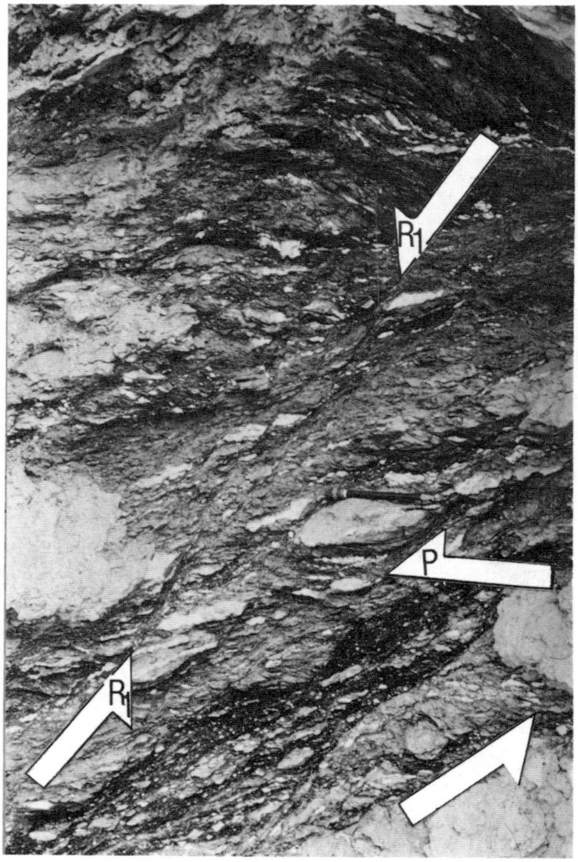

Figure 9
Typical field appearance of clay-rich fault gouge. The orientation of the P foliation, the discrete R_1 shears, and the overall fault-zone orientation (bottom right arrow) are indicated, the movement sense is left lateral in the plane of the photograph. Deflection of the P foliation, displacement of included clasts, and formation of trails along the R_1 shears can be seen.

may indicate that strain hardening processes were operative.

Under optical microscopy the same sets of structural features as those described above can be perceived, though on a smaller scale (see Fig. 12). With the help of transmission electron microscopy (Fig. 13) the nature of the preferred orientation of clay particles and clastic mica fragments in the P foliation can be seen. Beards of oriented clay plates develop on the mica clasts as a result of the chemical breakdown of micas and feldspars. The orientation of the mica plates must arise entirely from mechanical rotations, but some of the clay orientation probably arises from oriented growth controlled by the preoriented micas. This point is discussed below in connection with experimental studies.

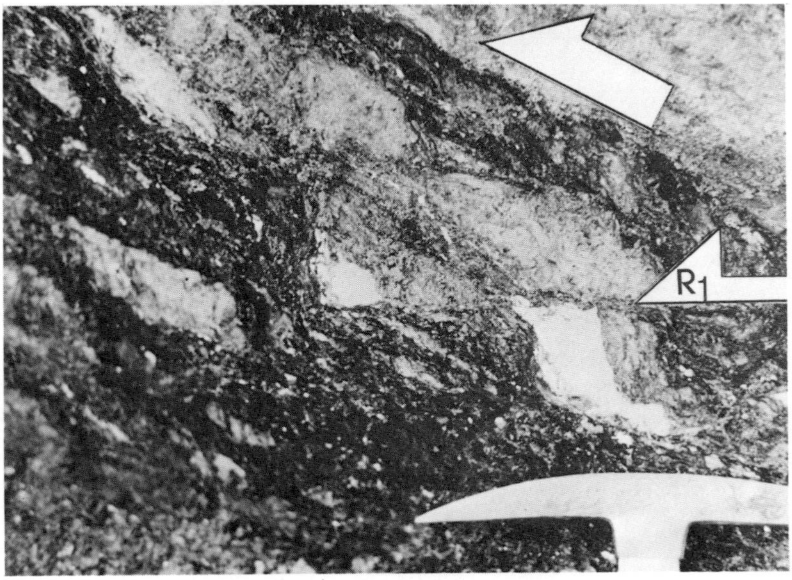

Figure 10
Natural fault gouge showing the development of P foliation, R_1 shears, and trails. The upper arrow is parallel to a Y shear, which in turn is parallel to the fault-zone boundary. The sense of shear is left lateral in the plane of the photograph.

Microstructure of Experimentally Produced Fault Rocks

In this section we attempt to make some broad generalisations about the microstructural features of experimentally produced fault rocks, based on experiments performed both by others and by ourselves. In the descriptions of microstructures emphasis will be placed on persistent features which develop over a wide range of environmental conditions. No attempt will be made to describe systematically the mechanical properties observed in our tests, the results of which have not yet been published; only those features which we consider most pertinent to the interpretation of natural fault gouges will be mentioned.

Experimental configurations

This discussion is based on the results of more than one hundred experiments, mainly on 'synthetic' fault gouge, prepared by mixing kaolinite and silt-grade quartz in various proportions (10, 33, 50, 67, and 90 percent clay) and shearing it between intact rock pieces, under various conditions of strain rate, confining pressures up to 300 MPa, pore-fluid pressures up to 175 MPa, and temperatures up to 600 °C.

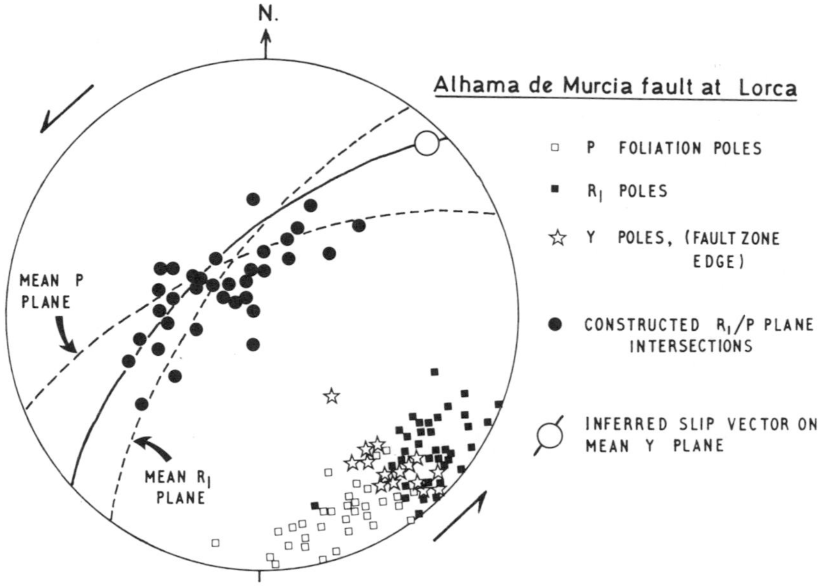

Figure 11
Equal-area lower-hemisphere projection of P, Y (fault-zone wall), and R_1 surfaces from 300 m of exposure of clay-bearing gouge parallel to the trace of the Alhama de Murcia fault zone near the town of Lorca, approximately midway along the fault outcrop. The clay-gouge zone-outcrop width is about 20 m. Constructed intersections between individual pairs of P and R_1 measurements are also shown. The inferred slip vector then lies in the fault plane, normal to the P–R_1 intersection.

In most of the experiments a gouge layer 1 mm thick was prepared by smearing a stiff paste made from the mixture together with a few drops of water onto the surface of a 35 ° saw cut made in an intact rock cyclinder. The assembly was dried at 90 °C, then deformed in a conventional fluid-medium triaxial-deformation apparatus. In experiments requiring pressurized pore fluid an access hole 0.8 mm in diameter was drilled in the upper rock piece. For room-temperature experiments either serpentinite or Tennessee sandstone was used for the intact rock pieces, whereas for high-temperature experiments sintered alumina end pieces were used.

A few experiments have been performed to date on synthetic gouge samples deformed at room temperature in a direct shear apparatus, for studies of the development of microstructure at shear strains up to about 16.

We also report here microstructural observations made on Tennessee sandstone gouge sheared in very long-duration tests (up to 5 months) at 300 and 400 °C under controlled pore-water pressure conditions. These observations are pertinent to the question of the role of mineral reactions and hydrothermal cementation in the development of fault-gouge microstructures.

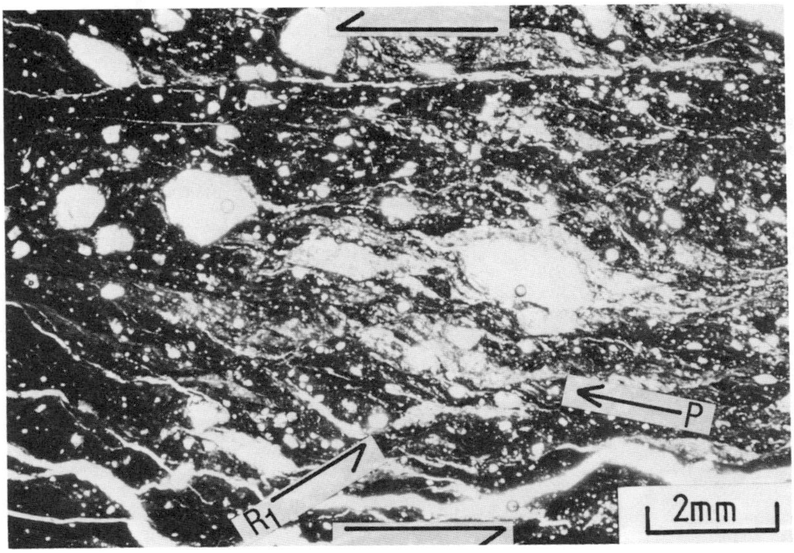

Figure 12
Optical micrograph of thin section of natural clay-rich fault gouge, showing the P foliation, which is deflected by R_1 shears.

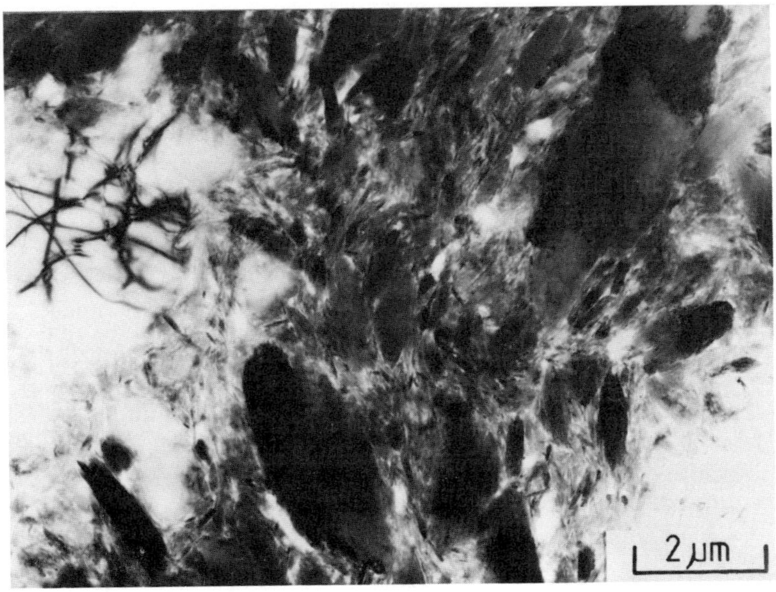

Figure 13
High-voltage transmission electron micrograph of kaolinite-rich natural fault gouge, showing oriented and rounded clasts of mica produced by mechanical granulation of the host mica schist. These are held in a matrix of new growths of similarly oriented, but much more finely grained, kaolinite.

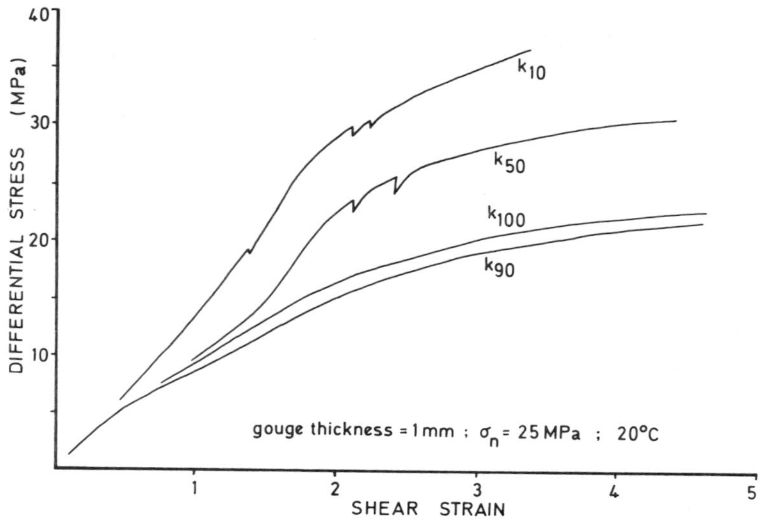

Figure 14
Stress–strain curves for dry, synthetic kaolinite–quartz fault gouge. Percent kaolinite by weight is given with each curve. The increased strength with decreasing kaolinite content is believed to be attributable at least in part to the increased shear strain in the remaining strain hardening kaolinite.

Mechanical properties

Our experience mirrors exactly that of other workers studying the mechanical properties of clay-bearing gouges (e.g., SUMMERS and BYERLEE, 1977; LOGAN et al., 1981). Under comparable conditions of effective confining pressure the frictional strength of kaolinite-bearing fault gouge is commensurate with that of many other crustal rocks; the clay is not anomalously weak. The strength is not very sensitive to temperature and tends to increase when temperature becomes sufficiently high to initiate dehydration of the clay (>300 °C).

Under all environmental conditions clay-bearing gouges exhibit strain hardening. For a given thickness of gouge we observe that the higher the proportion of quartz (particle size ranging up to 200 μm) the stronger the gouge at a given amount of strain. We interpret this as follows. At a given imposed shear strain, higher local shear strains are concentrated into a decreasing total amount of clay as the clay fraction is reduced. Because even a pure kaolinite layer strain-hardens, the apparent strength at a given imposed strain is increased. It is also inferred that the same effect is responsible for the apparent weakening of pure kaolinite gouge with increasing thickness of the gouge layer (RUTTER, 1979). These features are illustrated in Figure 14.

At present it is difficult to make generalizations regarding the conditions favouring stick-slip behaviour in fault gouges (MOORE et al., 1983). In our high-

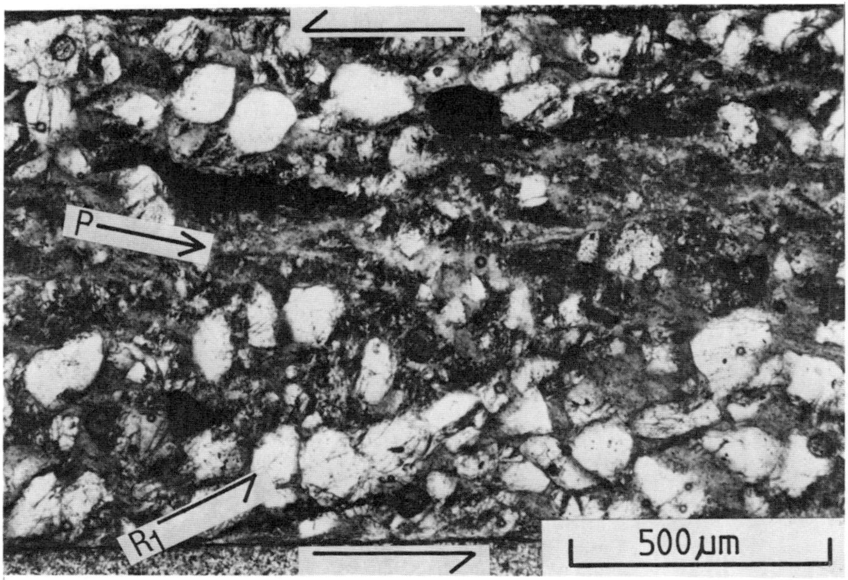

Figure 15
Optical micrograph (plane-polarised light) of experimentally deformed mixed clay–silt gouge (67% quartz) showing development of P foliation defined by the kaolinite orientation, oriented microcracks in the quartz grains, and R_1 shears.

strain-rate experiments, with effective confining pressures up to 175 MPa and temperatures up to 400 °C, none of the synthetic quartz–kaolinite gouges exhibited systematic stick-slip behaviour.

Microstructures

Experiments with High strain rate. All experiments on clay-bearing gouges lasting no more than a few days, whatever the imposed physical conditions, resulted in the formation of a P foliation early in the deformation history (see Fig. 15). The initial preparation and assembly of the sample resulted in the formation of a flattening fabric in the clay parallel to the saw-cut surfaces. Within a few percent of shear strain, this is reoriented to a planar fabric about 150° to the shear direction. The preferred orientation is easily perceived in optical studies, because the clay tends to go into optical extinction at a constant angle and, in the scanning electron microscope, it is evident by virtue of the dimensional orientation of the clay platelets (see Fig. 16).

This fabric does not appear to be substantially modified by further large increases in shear strain. Although we do not yet have quantitative data on its development, it

Figure 16
Backscattered electron micrograph of experimentally deformed clay–silt fault gouge (33% kaolinite) showing the well-developed P foliation defined by the phyllosilicate orientation, together with oriented microcracking of the quartz grains resulting from impingement.

does not appear to track the finite extension direction corresponding to the imposed shear strain. It is not understood how this fabric develops, because the relative roles of intracrystalline plasticity, frictional sliding, and dilatancy in the deformation of the clay aggregates are not yet known.

Closely following the development of the P fabric, R_1 microshears begin to form. In 100% clay gouge, these are extremely closely spaced, and some of the clay is reoriented into the R_1 orientation in the microshear zones. Thus, between crossed polars with the optical microscope, two dominant extinction angles may be seen, one corresponding to the P foliation, the other to closely spaced R_1 shear zones. In samples with higher proportions of silt grains, larger-displacement, discrete, more widely spaced R_1 shears become apparent (Figs. 2 and 15). At the boundaries of the sheared zones these characteristically curve into the gouge–intact-rock boundary. Thus the discrete R_1 shears behave as transfer faults, transferring slippage from one of the shear zone walls across to the other. These shears often appear to be localized by heterogeneities on the gouge–intact-rock boundary or in connection with single silt particles or clusters of them.

The formation of R_1-oriented shears occurs at all scales (e.g., Tchalenko, 1970) within shear zones in all types of rock deformed over a wide range of physical conditions and, hence by different deformation mechanisms. Theories of their origin that depend on a particular deformation mechanism are therefore untenable. We suspect that the development of these features in some way depends upon the deformation-induced anisotropy which accompanies shearing, though the nature of the anisotropy is different in shear zones of different types.

One of the most significant attributes of the deformation of fault gouge by a combination of P-foliation development and R_1 shears is the fact that the deformation spreads throughout the volume of the gouge material. We interpret this in terms of the mutual interference of the processes involved in the development of each feature, leading to the strain hardening which is seen in the stress–strain behaviour. Large fault displacements cannot then be developed without further instabilities and localization of the deformation. In experiments this is seen in the tendency to localize slippage at the gouge–intact-rock boundary and in nature, by the formation of Y shears, which can accommodate very large shear displacements. Such localizations within the otherwise ductile deformation of fault gouge may correlate with stick-slip events in experiments. In nature the localization of Y faults may be associated with earthquakes.

In none of our experiments on kaolinite–silt gouges, even up to shear strains in excess of 10, have we yet observed the development of Y shears. The only recorded account of these in experiments on clay-bearing gouge seems to be that of LOGAN et al. (1981), who observed them in elevated-temperature experiments on reconstituted montmorillonite-bearing San Andreas fault gouge. Their formation may depend upon the successful suppression of frictional sliding along the gouge–intact-rock interface, perhaps in their case in association with the dehydration of montmorillonite.

Microfracturing of the quartz grains included in the gouge always occurred, though always more intensely as the percentage of included fragments was increased, and hence differential load on the gouge became more completely borne by impingements between the quartz fragments. Microfractures exhibited a preferred orientation subparallel to the greatest applied stress direction, as might be expected (Figs. 16 and 17). Microfractured included clasts were occasionally seen to be drawn into trails along P- and R_1-oriented microshears in much the same way as that often seen in natural fault gouges (Fig. 2). Very spectacular development of trails and 'ductile stringers' derived from included clasts was described by LOGAN et al. (1981) from their experiments on reconstituted San Andreas fault gouge. The microstructures they showed paralleled closely those we have observed in natural fault gouges from southeastern Spain.

Kink bands were generally observed to form in gouge experiments, being best developed in quartz-poor gouge compositions (e.g., 67% kaolinite) but not at all in pure clay gouge (but MORGENSTERN and TCHALENKO (1967) obtained kink bands in their experiments on pure clay blocks). They were invariably triggered through strain heterogeneities associated with the included quartz particles (Fig. 3) and developed at a high angle to the shear zone walls. Similar, though narrower, kink bands are

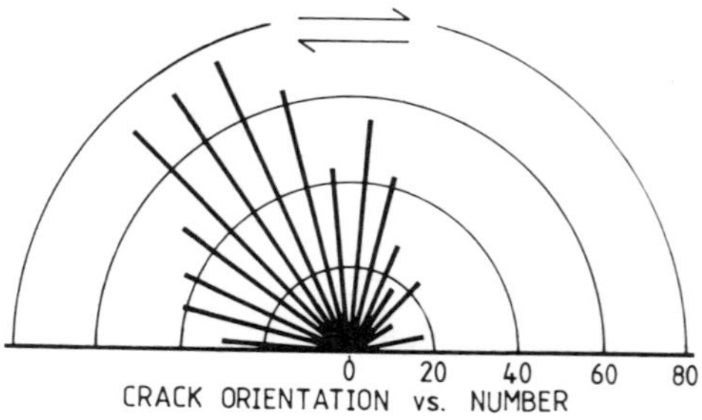

Figure 17
Typical histogram at 10° intervals, showing the preferred orientation of cracks (ranging from about 20 microns to 1 grain diameter in apparent length) relative to the saw cut orientation developed in quartz grains in quartz–kaolinite gouge deformed experimentally at 400 °C with elevated pore-fluid pressure. The saw cut was 35° to the axis of the cylindrical specimen. Cracks measured: 708, not weighted according to crack length.

evident in the photomicrographs from the experiments of LOGAN et al. (1981), who interpreted them as R_2 Riedel shears. In clay-free fault gouges deformed experimentally, R_2-oriented microfaults do occur. In clay-bearing-gouge experiments the kink bands are presumably the manifestation of the tendency for the same displacement patterns to be produced.

Slow-creep experiments at elevated temperature. RUTTER and WHITE (1979) described experiments on the creep of prefaulted specimens of Tennessee sandstone at temperatures of 300 and 400 °C, 150 MPa effective confining pressure, and 20 MPa pore-water pressure. Three of these experiments were run for time durations of up to 5 months. Microstructures from these experiments have not previously been described, and they cast light on the role of water-assisted redistribution of material and neomineralization during the evolution of fault gouge. In these experiments approximately linear viscous-creep behaviour was observed at low stresses and strain rates, and a transition in dominant deformation mechanism towards low strain rates from frictional and cataclastic processes to water-assisted grain-boundary diffusion creep was inferred.

The rock is 85% quartz, with the remainder dominated by illitic clay, together with kaolinite and hydrated iron oxides. Initial faulting of the rock yields a fault zone about 1 grain in diameter (100 μm) containing 'bands' of crushed quartz grains. Where the fault intersects pores containing large concentrations of diagenetic clay, elongated 'smears' of illite mixed with small concentrations of quartz fragments are formed. Thus the fault is composed of bands of gouge with varying concentrations of clay and quartz.

Microstructures seen by high-voltage transmission electron microscopy are shown in Figs. 18–21. In the quartz-rich areas cementation of the originally angular quartz fragments occurs, by means of optically continuous overgrowths of quartz

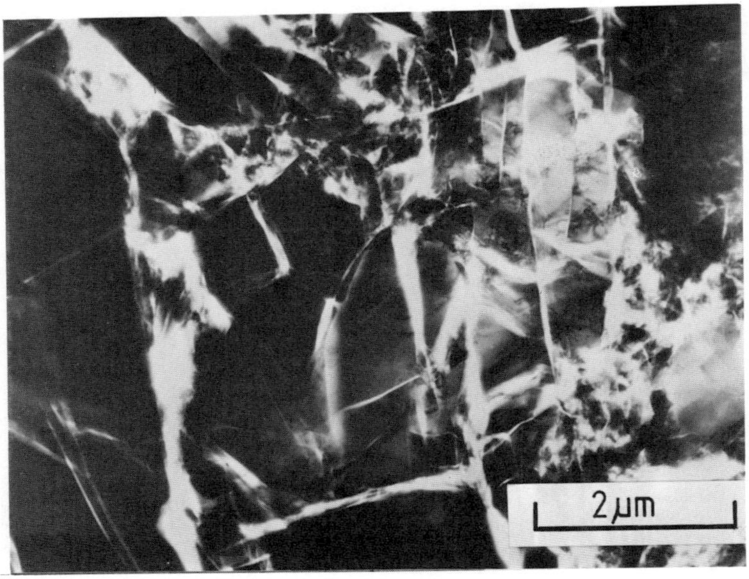

Figure 18
High-voltage transmission electron micrograph of quartz-rich area in gouge produced by experimental faulting of Tennessee sandstone. Note porosity and angular shapes of fragmented quartz.

(Figs. 18 and 19). In the clay-rich areas, diagenetic kaolinite disappears, the illite is completely recrystallized (probably to a mica) with concomitant grain-size reduction by 10 ×, and the included quartz fragments become rounded and embayed (Figs. 20 and 21). The recrystallized illite exhibits an intense preferred orientation subparallel to the shear zone. The development of oriented growth of phyllosilicate minerals to reëquilibrate with the newly imposed physical conditions during the deformation should be compared with the oriented growth of clays accompanying the breakdown of mica in the clay-bearing gouges from southeastern Spain (Fig. 13).

The cementation of the quartz-rich areas of the fault gouge is not surprising in view of the vigorous hydrothermal circulation of pore water which must occur in the sample, the clear corrosion of quartz in the mixed quartz–clay areas, and the likelihood of silica being released into the pore fluid in association with the recrystallization of clay into what is thought to be a less siliceous mica. The likelihood of a decreased pH of the pore fluid in the mica-rich regions, together with silica saturation in such regions through the equilibrium between the clay and the local pore fluid, makes it less easy to see why quartz should be preferentially corroded in such regions.

The following explanation is offered. In the clay-rich areas the quartz grains are loaded by the relatively deformable clay at an interfacial normal stress level commensurate with the mean stress in the rock. This raises the equilibrium activity of

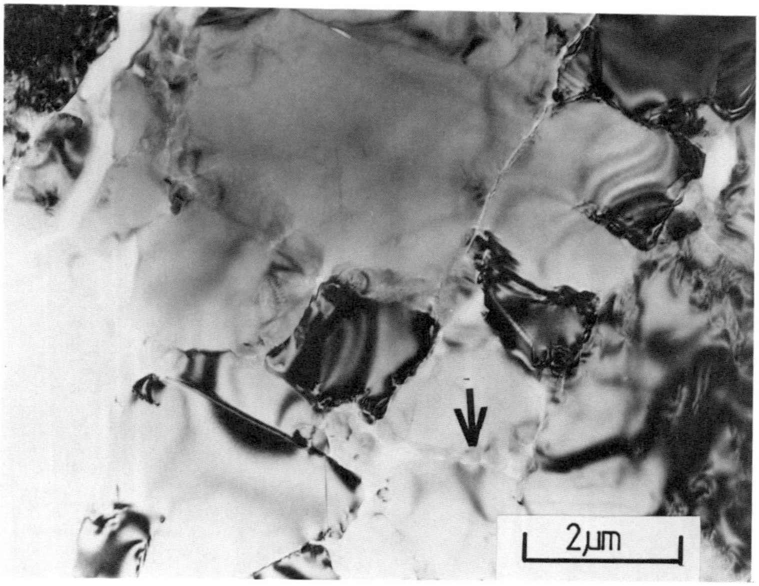

Figure 19
High-voltage transmission electron micrograph of quartz-rich area in gouge produced by experimental faulting of Tennessee sandstone after slow fault creep, at 400 °C, 150 MPa effective confining pressure, 25 MPa pore-water pressure, for 5 months. Note cementation of void spaces by overgrowths. Terminated quartz crystals can be seen growing into one void space (arrowed).

silica in the aqueous interstitial phase between the clay particles. From the commonly held view that interfacial diffusion within clay–quartz interfaces is much easier than within quartz–quartz interfaces, we infer that in the clay-poor areas the equilibrium activity of silica in the pore fluid is determined largely by the pore-fluid pressure, which in these experiments is more than one order of magnitude less than the mean rock pressure. Quartz should therefore undergo diffusive transfer from the clay-rich areas to the clay-poor areas.

A possible implication of these observations is that, once a compositional heterogeneity like the above has been established, further amplification of the intensity of the heterogeneity should occur spontaneously, even at a very low differential stress level, provided the pore pressure is less than the overburden pressure. The effect may contribute to the development of tectonically induced mineralogical banding, which is a common feature of metamorphic rocks. These experiments also indicate the rapidity with which hydrothermal cementation can occur in fault rocks (cf. also SMITH and EVANS, 1984; STEL, 1981). Such cementation might be expected to affect the strength and frictional characteristics of a fault zone, perhaps being a contributory factor in the determination of timing of repeated faulting events (e.g., RUTTER and WHITE, 1979; ANGEVINE et al., 1982).

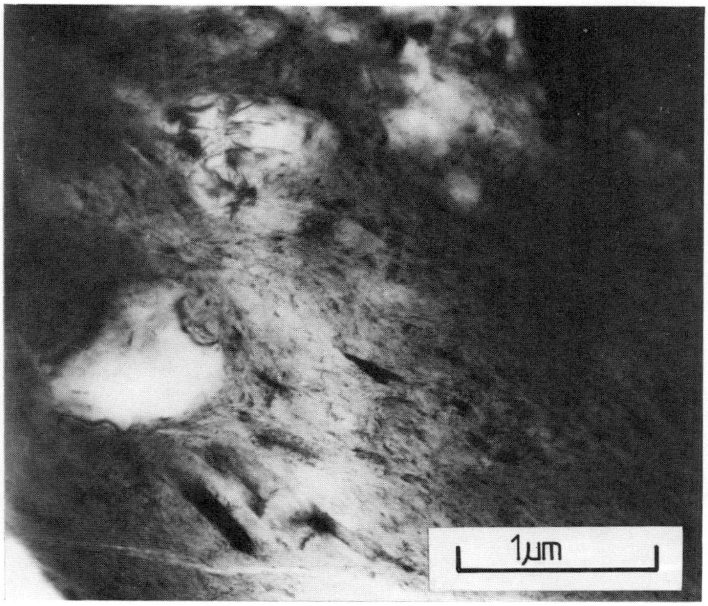

Figure 20
High-voltage transmission electron micrograph of clay-rich area of gouge in same sample as that figured in Figure 19. Original illite crystals several microns long are now replaced by a well-oriented matrix of submicron mica crystals. Included quartz fragments in such areas, initially angular, are rounded and embayed.

Conclusions

In this paper we have summarized the characteristic structural and microstructural features which tend to develop in both natural and experimentally produced fault gouge, emphasizing their remarkable geometric similarities. By reference to ongoing studies of fault zones in southeastern Spain it has been shown that the characteristic structural features of fault gouges can be used for inferring regional movement pictures. In describing the microstructural features of experimentally produced fault gouge we have highlighted those features which are common to gouges formed over a wide range of environmental conditions. We have here emphasized the 'comparative anatomy' approach to the study of natural and experimentally produced fault gouges, whilst recognizing that the mechanisms of formation of the P foliation and R_1 shears remain unclear.

It is pointed out also that clay-bearing fault gouges are metamorphic rocks which have suffered chemical metamorphism in addition to physical degradation. A great deal remains to be done in order to understand such effects in fault gouges.

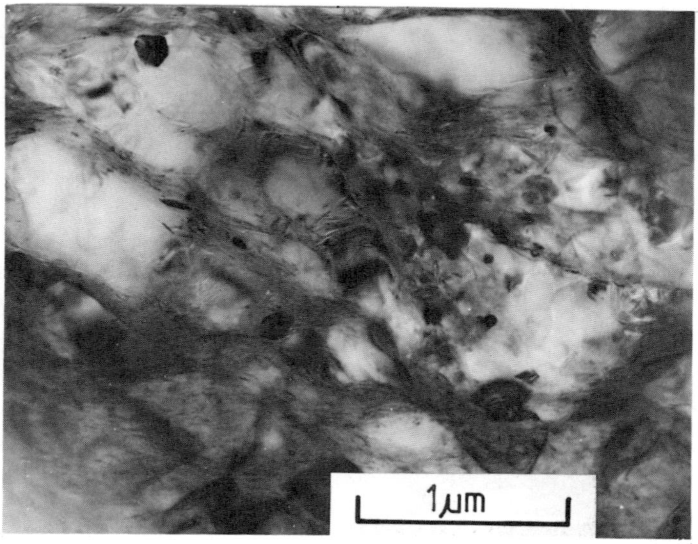

Figure 21
High-voltage transmission electron micrograph of clay-rich area of gouge in same sample as that figured in Figure 19. As in Figure 20, the clay is completely recrystallized and well oriented, but there is a higher proportion of quartz in this area, also showing evidence of rounding and embayment.

Acknowledgments

This work is currently supported by U.K. Natural Environment Research Council Grant GR3/5109. S. H. Hall acknowledges an N. E. R. C. research studentship. The Tennessee sandstone studies described here were performed with support from the U.S. National Earthquake Hazards Reduction Program, Contract 14-08-0001-17662. R. F. Holloway maintained the rock-deformation equipment used in this study. The microcomputer program used for crack-orientation analysis was developed by W. Shea. Work in Spain was carried out with the permission of the Spanish National Commission for Geology. Thanks are due P. Bush, H. Shaw, N. Harbury, and D. McDougall for helpful discussion. We appreciate the constructive criticisms and helpful comments made by reviewers J. Logan and D. Moore.

References

Anderson, J. L., Osborne, R. H. and Palmer, D. F. (1983), *Cataclastic rocks of the San Gabriel fault—An expression of deformation at deeper crustal levels in the San Andreas fault zone.* Tectonophys. *98*, 209–251.

Angevine, C. L. Turcotte, D. L. and Furnish, M. P. (1982), *Pressure solution lithification as a mechanism for the stick-slip behaviour of faults.* Tectonics *1*, 151–160.

ARMIJO, R., BENKHELIL, J., BOUSQUET, J. C., ESTEVEZ, A., GUIRAUD, R., MONTENAT, C., PAVILLON, M. J. PHILIP, H., SANZ DE GALDEANO, C. and VIGUIER, C. (1977), *L'histoire tectonique recent (Tortonien à Quaternaire) de l'Arc de Gibraltar et des bordures de la mer d'Alboran.* Bull. Geol. Soc. France *19*, (3) 575–614.

BERTHE, D., CHOUKROUNE, P. and JEGOUZO, P. (1979), *Orthogneiss, mylonite and non-coaxial deformation of granites: the example of the south Armorican shear zone.* J. Struct. Geol. *1*, 31–42.

BYERLEE, J. O. and SUMMERS, R. (1976), *A note on the effect of fault gouge thickness on fault stability.* Int. J. Rock Mech. Min. Sci. *13*, 35–36.

BYERLEE, J. D., MJACHKIN, V., SUMMERS, R. and VOEVODA, O. (1978), *Structures developed in fault gouge during stable sliding and stick-slip.* Tectonophys. *44*, 161–171.

CHESTER, F. M., FRIEDMAN, M. and LOGAN, J. M. (1985), *Foliated cataclasites.* Tectonophys. *111*, 139–146.

ELLIOT, D. (1976), *The energy balance and deformation mechanisms of thrust sheets.* Phil. Trans. R. Soc. *283A*, 289–312.

ENGELDER, T. (1974), *Catalasis and the generation of fault gouge.* Geol. Soc. Amer. Bull. *85*, 1515–1522.

ENGELDER, T. (1977), 'Aspects of asperity-surface interaction during frictional sliding,' in *Proceedings of Conference II, Experimental studies of rock friction with application to earthquake prediction.* U.S. Geol. Surv. 117–137.

GAMOND, J. F. (1983), *Displacement features associated with fault zones: A comparison between observed examples and experimental models.* J. Struct. Geol. *5*, 33–45.

HALL, S. H. (1983), *Post Alpine tectonic evolution of S.E. Spain and the structure of fault gouge.* Ph.D Dissertation, Univ. London.

HALL, S. H. and RUTTER, E. H. (1986), *Late orogenic faulting in S.E. Spain,* to be submitted to J. Geol. Soc. London.

HAMILTON, G. M. and GOODMAN, L. E. (1966), *The stress field created by a circular sliding contact.* J. Appl. Mech. *33*, 371–376.

HOUSE, W. M. and GRAY, D. R. (1982), *Cataclasites along the Saltville thrust, U.S.A., and their implications for thrust sheet emplacement.* J. Struct. Geol. *4*, 257–269.

LIECHTI, R. and ZOBACK, M. D. (1979), 'Preliminary analysis of clay gouge from a well in the San Andreas fault zone in central California,' in *Proc. Conf. VIII: Analysis of actual fault zones in bedrock.* U.S. Geol. Survey Open File Rep., 79–1239, 268–275.

LOGAN, J. M., FRIEDMAN, M., HIGGS, N. G., DENGO, C. and SHIMAMOTO, T. (1979), 'Experimental studies of simulated gouge and their application to studies of natural fault zones,' in *Proc. Conf. VIII, Analysis of actual fault zones in bedrock.* U.S. Geol. Survey Open File Rep. 79–1239, 305–343.

LOGAN, J. M., HIGGS, N. G. and FRIEDMAN, M. (1981), *Laboratory studies on natural gouge from the U.S. Geological Survey Dry Lake Valley No. 1 well, San Andreas fault zone.* Geophys. Monogr. 24, Amer. Geophys. Union 121–134.

MOORE, D. E., SUMMERS, R. and BYERLEE, J. D. (1983), *Strengths of clay and non-clay fault gouges at elevated temperatures and pressures.* Proc. 24th U.S. Symp. Rock Mech. 489–499.

MOORE, D. E. and BYERLEE, J. D. (1985), *Deformation textures developed in heated fault gouge.* EOS, Trans. Am. Geoph. Union. *66*, p. 1100.

MORGENSTERN, N. R. and TCHALENKO, J. S. (1967), *Microscopic structures in kaolin subjected to direct shear.* Geotechn. *17*, 309–328.

MORROW, C., SHI, L. Q. and BYERLEE, J. D. (1981), *Permeability and strength of San Andreas fault gouge under high pressure.* Geophys. Res. Lett. *8*, 325–328.

PLATT, J. P. and VISSERS, R. L. M. (1980), *Extensional structures in anisotropic rocks.* J. Struct. Geol. *2*, 397–410.

RUTTER, E. H. (1979), *The mechanical properties of kaolinite fault 'gouge' at moderate confining pressure, 20 °C.* Int. J. Rock Mech. Min. Sci. *16*, 407–410.

RUTTER, E. H. and WHITE, S. H. (1979), *The effects of water, temperature and time on the microstructural and mechanical properties of experimentally produced fault gouge.* Bull. Mineral. *102*, 93–101.

SANZ DE GALDEANO, C. (1983), *Los accidentes y fracturas principales de las Cordilleras Beticas.* Estudios Geol. *39*, 157–165.

SIMPSON, C. and SCHMID, S. (1983), *An evaluation of criteria to deduce the sense of movement in sheared rocks.* Geol. Soc. Am. Bull. *94*, 1281–1288.

SMITH, D. L. and EVANS, B. (1984), *Diffusional crack healing in quartz.* J. Geophys. Res. *89*, 4125–4135.
STEL, H. (1981), *Crystal growth in cataclasites: diagnostic microstructures and implications.* Tectonophys. *78*, 585–600.
SUMMERS, R. and BYERLEE, J. (1977), *A note on the effect of fault gouge composition on the stability of frictional sliding.* Int. J. Rock Mech. Min. Sci. *14*, 155–160.
TCHALENKO, J. S. (1970), *Similarities between shear zones of different magnitudes.* Geol. Soc. Am. Bull. *81*, 1625–1640.
UDIAS, A., LOPEZ ARROYO, A. and MEZCUA, J. (1976), *Seismotectonic of the Azores-Alboran region.* Tectonophys. *31*, 259–289.
WANG, C., MAO, N. and WU, F. T. (1980), *Mechanical properties of clays at high pressure.* J. Geophys. Res. *85*, 1462–1468.
WHITE, S. H., BURROWS, S. E., CARRERAS, J., SHAW, N. D. and HUMPHREYS, F. J. (1980), *On mylonites in ductile shear zones.* J. Struct. Geol. *2*, 175–188.
WU, F. T. (1978), *Mineralogy and physical nature of clay gouge.* Pure Appl. Geophys. *116*, 685–689.

(Received 15th August 1985, revised 3rd February 1986, accepted 7th February 1986)

PAGEOPH, Vol. 124, Nos. 1/2 (1986)
0033-4553/86/020031-22$1.50 + 0.20/0

The Effects of Sliding Velocity on the Frictional and Physical Properties of Heated Fault Gouge

D. E. Moore,[1] R. Summers[1] and J. D. Byerlee[1]

Abstract—The frictional properties of a crushed granite gouge and of gouges rich in montmorillonite, illite, and serpentine minerals have been investigated at temperatures as high as 600 °C, confining pressures as high as 2.5 kbar, a pore pressure of 30 bar, and sliding velocities of 4.8 and 4.8×10^{-2} μm/sec. The gouges showed nearly identical strength behaviors at the two sliding velocities; all four gouges, however, showed a greater tendency to stick-slip movement and somewhat higher stress drops in the experiments at 4.8×10^{-2} μm/sec. Varying the sliding velocity also had an effect on the mineral assemblages and deformation textures developed in the heated gouges. The principal mineralogical difference was that at 400 °C and 1 kbar confining pressure a serpentine breakdown reaction occurred in the experiments at 4.8×10^{-2} μm/sec but not in those at 4.8 μm/sec. The textures developed in the gouge layers were in part functions of the gouge type and the temperature, but changes in the sliding velocity affected, among other features, the degree of mineral deformation and the orientation of some fractures.

Key words: Fault gouge, friction, deformation textures.

Introduction

Earthquakes are generated to depths of about 20 km on transform faults such as the San Andreas (Brace and Byerlee, 1970), and many of the largest earthquakes are concentrated at the base of this seismogenic zone (Sibson, 1982). To help determine the controls on stable and unstable slip along faults at such depths, friction studies at elevated temperatures need to be conducted on appropriate rock materials. The sliding characteristics of a variety of faulted rocks have been investigated at elevated temperatures and pressures (Brace and Byerlee, 1970; Olsson, 1974; Engelder et al., 1975; Stesky et al., 1974; Stesky, 1978). Thus far, however, few such studies of heated gouge materials have been conducted (Logan et al., 1979, 1981; Moore et al., 1983).

We have begun a systematic investigation of the high-temperature friction strengths of a group of four gouges. An initial set of experiments, in which the temperature and confining pressure were varied at a constant pore pressure and sliding velocity, was described in Moore et al. (1983). This study considers the effects

[1] U.S. Geological Survey, 345 Middlefield Road, MS/977, Menlo Park, CA 94025.

Table 1

Mineralogy of Gouges Used in this Study

Gouge	Composition
Dry Lake Valley: 212–235 m depth interval of drill hole into the San Andreas near Hollister, Calif.	About 80% clays, 20% quartz + plagioclase. Approximate clay proportions: 50% montmorillonite + mixed-layer clays, 30% kaolinite, 10% chlorite, 10% illite
Serpentinite: from a trench in the San Andreas near San Carlos, Calif.	Clinochrysotile, lizardite, trace opaques
Illite: disaggregated illitic shale, from Fithian, Ill.	Approximately 70% illite, 20% quartz, 10% chlorite + kaolinite
Westerly granite: crushed sample from quarried material, R.I.	35% plagioclase, 25% quartz, 25% potash feldspar, 10% biotite + muscovite + chlorite, 5% opaques + alteration minerals (epidote, calcite, fluorite).

of changing the slip velocity on the strength and sliding stability of these four gouges. In addition, this paper contains the first petrographic description of the run products from both the earlier and present experiments. Along with temperature and gouge type, the slip rate was found to have some important controls on the deformation textures observed in these gouges. TCHALENKO (1970) and LOGAN et al. (1979), among others, have shown that many of the microscopic structures developed in laboratory experiments can be correlated with the regional structures of natural fault zones. Therefore, the relationships observed in this petrographic study may have application to the interpretation of fault-zone characteristics.

Experimental Procedure

Four gouges of differing composition were studied (Table 1); they were two clay-rich gouges, a serpentinite gouge, and a gouge composed of crushed Westerly granite. The montmorillonite-rich Dry Lake Valley gouge and the serpentinite gouge were natural fault-gouge materials collected from a drill hole and a trench, respectively, in the San Andreas fault. The granite gouge was representative of natural granitic and gneissic rock flours that have been found along the San Andreas fault zone in southern California (ANDERSON et al., 1980). The illite-rich sample provided a nonexpanding clay-bearing gouge for comparison with the montmorillonite-rich gouge. Identification of the clay minerals and estimation of their relative proportions in the Dry Lake Valley and illite-rich gouges (Table 1) were made by the procedure in SCHULTZ (1964). Mineral proportions in the granite gouge were determined from point counts of thin sections of the starting Westerly material. Serpentine mineral identifications were based principally on WHITTAKER and ZUSSMAN (1956), with additional reference to the indexed diffraction lines of serpentine minerals in NEMECZ (1981) and the lizardite x-ray traces in BRINDLEY and ZUSSMAN (1957).

The experimental assembly is described in detail in MOORE et al. (1983). Briefly, each sample consisted of a layer of gouge 0.65 mm in thickness that was sandwiched between 30° finely ground saw-cut surfaces in a Westerly granite cylinder 19.0 mm in diameter and 41.3 mm long. The gouge-filled cylinder and insulating pieces were placed within an annealed copper jacket; the jacketed assembly was then placed in a cylindrically shaped resistance heater. All the experiments were run at a pore pressure of 30 bar, with deionized water as the pore fluid. Temperatures of 200, 400, and 600 °C and confining pressures of 1 and 2.5 kbar were tested. The temperature measurements were accurate to about ± 5 °C. The experiments were run at strain rates of 10^{-4}/sec and 10^{-6}/sec, which correspond to average sliding velocities along the saw cut of 4.8 and 4.8×10^{-2} μm/sec, respectively. These velocities refer to the average rate at which one granite piece slides past the other along the saw cut. For samples which show stick slip, velocities within the gouge layer will differ significantly between the stick and the slip portions of each cycle.

Confining and pore pressures were applied first to the jacketed sample. After these pressures had stabilized, the temperature was raised. The sample was held at temperature and pressure for 1800 sec before the load was applied. After completion of the experiments several of the run products were set aside for x-ray powder diffraction analysis. Thin sections of the remaining samples were prepared for standard petrographic analysis; each section was cut parallel to the long axis of the cylindrical sample and perpendicular to the plane of the fault gouge.

Results

Strength and sliding behavior

Plots of differential stress versus axial compression for the gouges at each temperature and sliding velocity studied are shown in Figures 1 to 3. The reproducibility of the results was verified in repeat experiments; some of the experiments at 600 °C and 2.5 kbar had to be repeated several times because of sample failures at these extreme conditions. For clarity, only one curve for each gouge has been plotted. As seen in these figures, the results at the two sliding velocities show very similar trends relative to temperature and confining pressure. Several of the curves at the slower slip rate are at slightly higher differential stresses than the corresponding experiments at the faster slip rate, but the results are essentially the same within the experimental variation.

At a given temperature and confining pressure at either sliding velocity, the three sheet-silicate-rich gouges supported roughly the same stresses. They and the granite gouge all showed increases in strength with increasing confining pressure. At 200 °C (Fig. 1) the granite gouge supported considerably higher stresses than the clay and

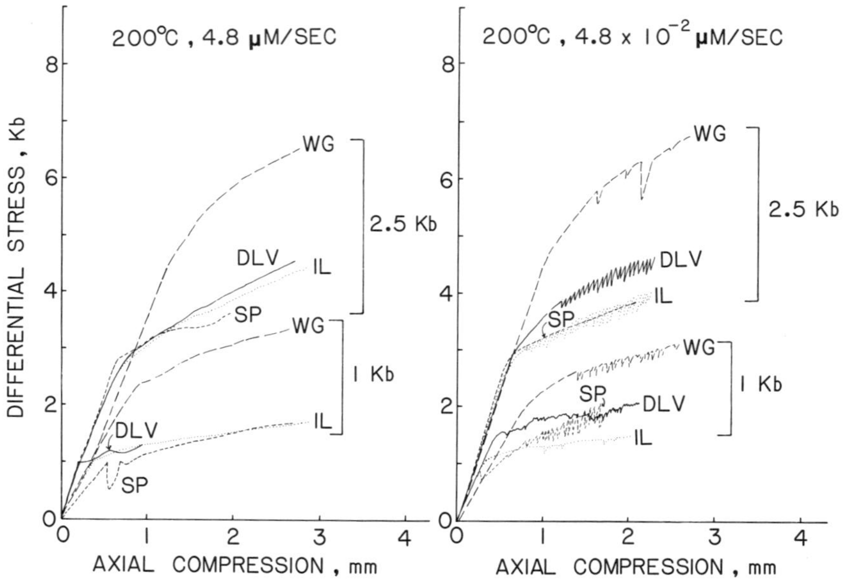

Figure 1
Plot of differential stress against axial compression for friction experiments at 200 °C, comparing results at 4.8 and 4.8×10^{-2} μm/sec sliding velocities: WG, Westerly granite gouge (long-dashed lines); DLV, Dry Lake Valley gouge (solid lines); IL, illite gouge (dotted lines); SP, serpentinite gouge (short-dashed lines).

serpentinite gouges at both confining pressures. The strength of the granite gouge, however, was unaffected by temperature, whereas the three sheet-silicate-rich gouges showed marked strength increases with temperature. At 400 °C (Fig. 2) the granite gouge was still stronger than the other gouges at 2.5 kbar, but all four gouges supported similar stresses at 1 kbar. At 600 °C (Fig. 3) the clay and serpentinite gouge strengths equalled that of the granite gouge at 2.5 kbar and surpassed it at 1 kbar.

In contrast to their strengths, the sliding behavior of the gouges did vary with the sliding velocity (Figs. 1 to 3). At the 4.8 μm/sec slip rate the granite gouge slid stably at all temperatures and pressures, and the three sheet-silicate-rich gouges slid stably at 200 °C. At higher temperatures the clay-rich and serpentinite gouges had a greater tendency to show stick slip, particularly at 1 kbar confining pressure. The magnitude of the stress drops tended to increase with temperature. Reducing the sliding velocity to 4.8×10^{-2} μm/sec expanded the field of stick-slip behavior for the sheet-silicate gouges to lower temperatures and higher confining pressures. In addition, the granite gouge slid unstably in some of the experiments at the 4.8×10^{-2} μm/sec slip rate. The differences in sliding stability at the two slip velocities were particularly marked at 200 °C (Fig. 1). The size of the stress drops was also slightly higher in the experiments at 4.8×10^{-2} μm/sec. The best correlation between slip rate and stress drop was seen in the 400 °C experiments (Fig. 2), but also at

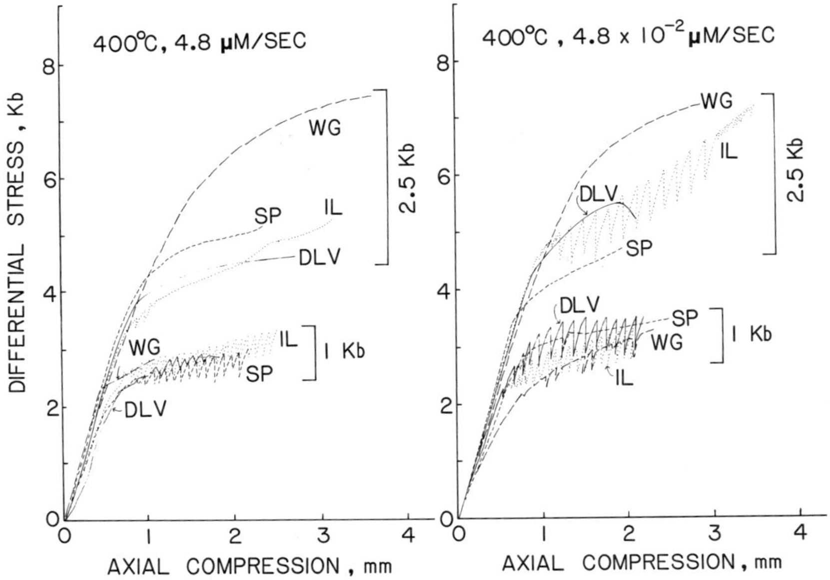

Figure 2
Frictional strengths of the four gouges at 400 °C and 4.8 and 4.8×10^{-2} μm/sec sliding velocities; same symbols as in Figure 1.

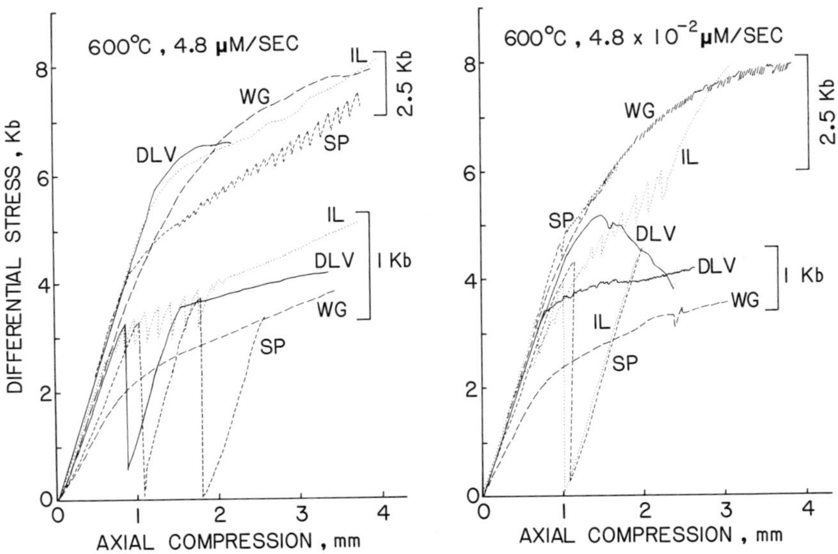

Figure 3
Strength plots, comparing results at 600 °C and 4.8 and 4.8×10^{-2} μm/sec sliding velocities; same symbols as in Figure 1.

600 °C the largest stress drops (approximately 4 kbar) were measured in the experiments at the slower displacement rate (Fig. 3).

Mineralogy

Samples of the starting material and run products of each gouge were prepared for x-ray diffraction analysis, by Cu K_α radiation, to establish the extent of mineral alteration during the experiments. The procedures are the same as described in MOORE et al. (1983); the results for the experiments at 4.8 μm/sec, along with representative diffraction patterns, also were described in MOORE et al. (1983). To summarize briefly the results at 4.8 μm/sec, the granite gouge showed almost no mineralogical changes with increasing temperature and pressure. The only visible effect was the gradual disappearance with temperature increase of the peaks for the minor alteration mineral chlorite. The x-ray peaks for clay minerals from the Dry Lake Valley and illite gouges generally decreased in intensity with temperature and pressure increase. Two exceptions were observed, however. The 10 Å illite peak was augmented at 600 °C because of the collapse of the dehydrated montmorillonite lattice from 14 to 10 Å. The 4.5 illite peak also increased in size, apparently because of modifications accompanying water loss from the illite structure.

Changing the sliding velocity to 4.8×10^{-2} μm/sec had no effect on the mineralogy of the granite gouge. The illite and Dry Lake Valley gouges did show some slight differences in the temperatures and pressures at which observable changes in the clay mineral peak intensities occurred. In addition, at 600 °C the intensities of the clay mineral peaks that decreased were somewhat lower in the experiments at the 4.8×10^{-2} μm/sec slip rate, and the intensities of the two peaks that increased were rather higher. No new mineral phases were observed in any of the run products of these three gouges.

Selected x-ray patterns at the two sliding velocities are shown for the serpentinite gouge in Figure 4. The serpentinite starting material was previously described (MOORE et al., 1983) as containing chrysotile as the sole serpentine mineral; the x-ray pattern, however, is more consistent with a combination of chrysotile and lizardite (Fig. 4), both of them low-temperature serpentine minerals. In contrast to the other gouges, the clinochrysotile and lizardite in some of the serpentinite gouge run products were replaced by new minerals; and the pressure and temperature conditions at which the reaction was observed varied somewhat with the slip rate. Every 600 °C sample at both sliding velocities contained forsteritic olivine instead of serpentine (Fig. 4). Poorly to well developed peaks attributable to talc also appeared at both confining pressures in the 600 °C x-ray traces, and some possible quartz peaks were present at 1 kbar. Serpentine was also partly or completely replaced by forsteritic olivine at a 400 °C, 1 kbar confining pressure and 4.8×10^{-2} μm/sec sliding velocity. The olivine peaks were less intense at 400 than at 600 °C, and no peaks appeared for either talc or quartz. The rest of the 400 °C samples showed no evidence of olivine growth, although the serpentine mineral diffraction lines from those samples did have reduced intensities.

Petrography

The four gouges of this study offer a wide range of initial textures for petrographic comparison. The granite gouge consists of a framework of quartzofeldspathic grains ranging in size up to approximately 0.15 mm in diameter, with minor amounts of fine-grained, interstitial matrix. The illite-rich gouge, on the other hand, is almost completely composed of clay-sized material, with only a few silt- or sand-sized grains (maximum of 0.07 mm in diameter) of quartz and opaque minerals. The Dry Lake Valley gouge is intermediate between these two extremes; it contains many more quartzofeldspathic grains (maximum of 0.12 mm in diameter) than the illite gouge, but not enough to form a framework-grain-supported material. The serpentinite gouge texturally resembles the granite gouge, in that it consists predominantly of rounded grains of serpentine up to about 0.12 mm in diameter, which may be pseudomorphs of the original ultramafic minerals. This gouge therefore has a framework of coarse grains, but the serpentine grains are much softer than the quartzofeldspathic grains of the granite gouge.

Because the emphasis of this special volume is on 'fault zone' structures, the deformation textures of the run products will be described in some detail. Several similar schemes have been put forward for labeling the various fracture sets that have been identified in natural fault zones and in the products of gouge friction and shear box experiments (e.g., SKEMPTON, 1966; TCHALENKO, 1968, 1970; TCHALENKO and AMBRASEYS, 1970; LOGAN et al., 1979). The terminology of LOGAN et al. (1979), which is illustrated in modified form in Figure 5, was adopted for this study because their scheme was the only one used for describing the run products of gouge friction experiments. The other schemes describe fracture sets with the same orientations and offset directions as shown in Figure 5, but they employ slightly different letter labels (e.g., the Y fractures in Figure 5 are labeled 'D' by TCHALENKO, 1968). For convenience in reporting, the letter scheme in Figure 5 has been used for describing not only the relative orientations of fracture and fault planes but also the orientations within the plane of the thin section of features such as kink bands and aligned minerals. This does not necessarily mean that the latter features will have the same orientation in the third dimension as the fault planes.

Granite gouge. Of the four gouges examined the crushed granite gouge shows the greatest variety of deformation textures. At 200 and 400 °C a series of low-angle R1 fractures that cross-cut the gouge layers are among the most prominent features in the samples. Individual grains that are split by the fractures show offsets of up to 0.05 mm (Fig. 6a). Opaque and phyllosilicate minerals are commonly smeared out along the fracture traces. At 600 °C the R1 traces are much more closely spaced, but

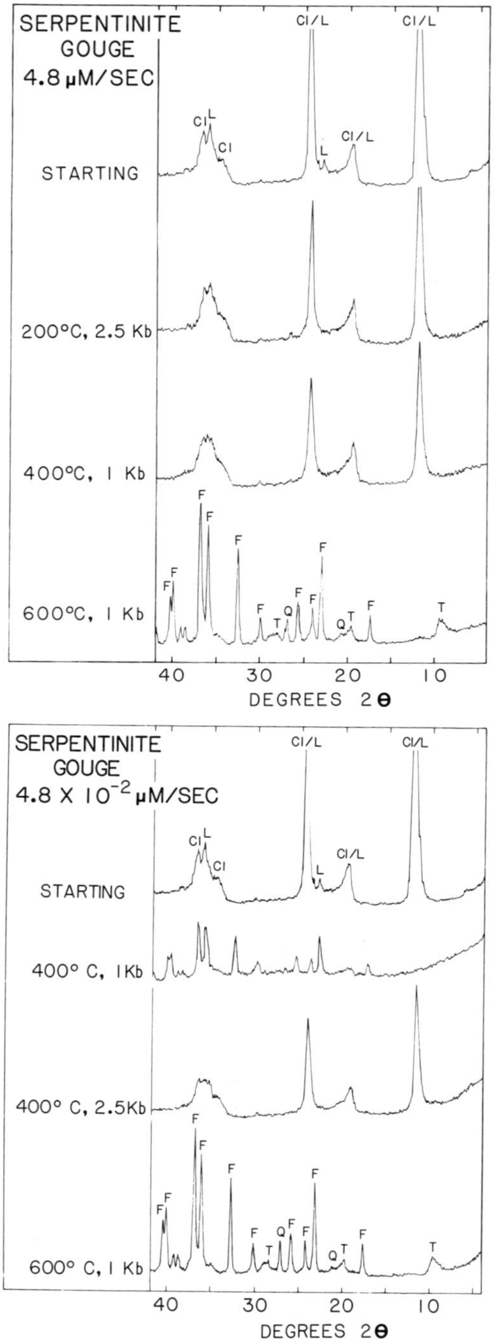

Figure 4

X-ray diffraction patterns of serpentinite gouge starting material and run products at 4.8 and 4.8×10^{-2} μm/sec sliding velocities: Cl, clinochrysotile; L, lizardite; F, forsteritic (Mg-rich) olivine; T, talc; Q, quartz. Cu K_α radiation.

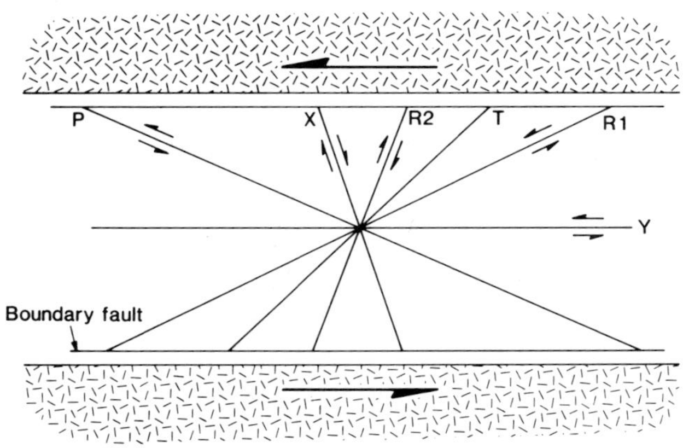

Figure 5
Labeling scheme of LOGAN et al. (1979, p. 325–6) for fracture sets developed in gouge layers deformed in triaxial friction experiments. The terminology of LOGAN et al. is used in this paper. The orientation of this diagram is the same at that of the examined thin sections, that is, parallel to the long axis of the cylindrical sample and perpendicular to the plane of the fault gouge.

individual planes are less well defined than at lower temperatures. The angle between the R1 fractures and the edge of the gouge layer ranges from 6° to 19°; the range is 6° to 11° for the samples with 4.8 μm/sec sliding velocity and 9° to 19° for the samples at 4.8×10^{-2} μm/sec. Cross-cutting fracture sets with different orientations were not observed, though a few individual quartz and feldspar grains in the gouge show fractures and offsets consistent with the R2 fractures.

The R1 fractures terminate at a fracture that is oriented parallel to the edge of the gouge layer and that is located at or within about 0.1 mm of the boundary (Fig. 6b). These boundary fractures commonly are filled with a very fine-grained grayish-brown material; in places, smeared-out phyllosilicate or opaque grains can be identified in the fracture filling. Some of the smeared grains were plucked from the adjoining granite cylinder. A boundary fracture is developed to at least some extent on both sides of every granite gouge sample, but it is generally more continuous on one side than on the other. The principal boundary fracture is usually straight, whereas the subsidiary one may have a jagged trend, with some of the segments extending into R1 fractures (Fig. 6b). In the samples at 600 °C and a 2.5 kbar confining pressure, short segments of the boundary fractures cut through the rock cylinder rather than the gouge. The slices of rock trapped within these segments have been partly crushed. In the 600 °C samples run at the faster sliding velocity, the granite cylinder adjacent to the gouge layer contains a number of short fractures oriented at angles between 0° and 45° to the gouge–rock boundary.

The granite gouge shows increasing degrees of grain-size reduction with increasing temperature and confining pressure. At 200 and 400 °C the zones of most

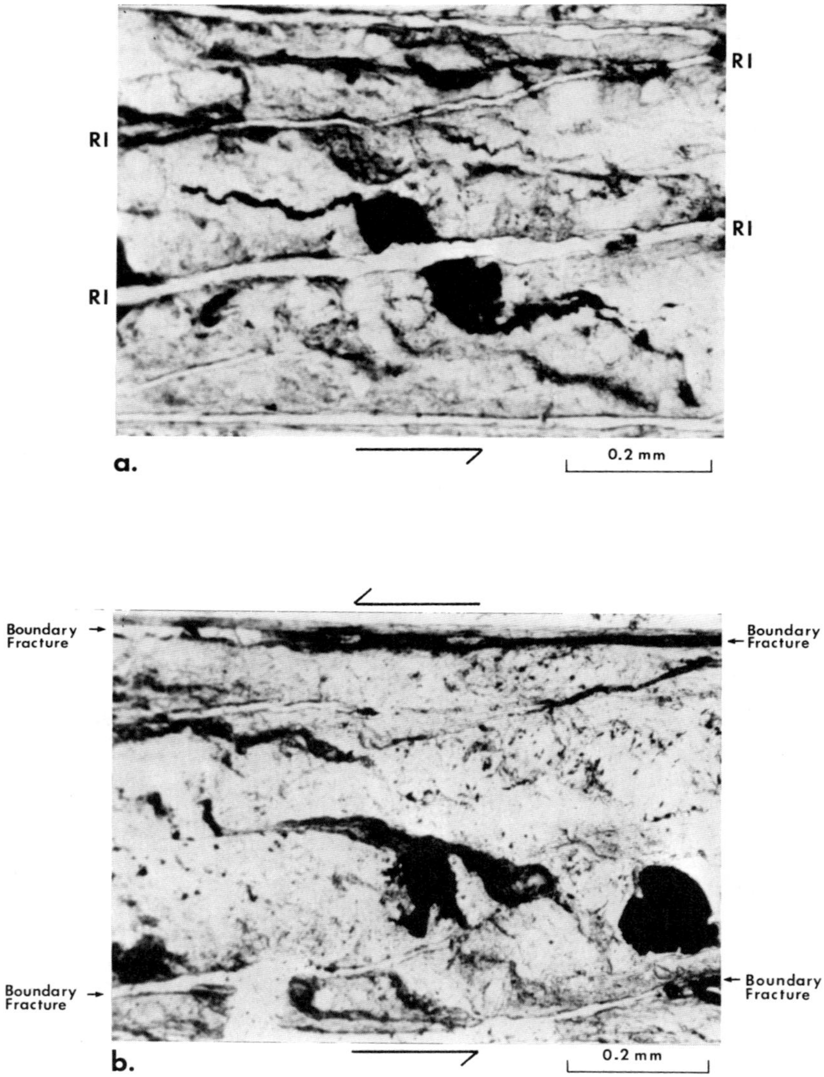

Figure 6
Textures of heated granite gouge. (a) Opaque grain offset along an R1 fracture. Note the tails of opaque material extending from each half of the opaque grain. The shear direction is as indicated in the figure (sample at 400 °C and 1 kbar confining pressure) (b) Examples of boundary fractures. The fracture at the top of the gouge layer is relatively straight and filled with fine-grained, dark material. The fracture at the base consists of a set of short segments that terminate in R1 fractures. Much of the fine-grained filling of the lower boundary fracture was lost during the thin-sectioning process (400 °C, 1 kbar). (c) An otherwise well-rounded quartz clast with flattened surfaces along R1 fractures. This grain also contains internal fractures aligned roughly in the R1 direction (400 °C, 1 Kbar). (d) Opaque and phyllosilicate grains deformed in a stepwise manner in the R1 and X directions (600 °C, 2.5 kbar). All four photos are in plane-polarized light; all four show the entire width of the gouge layer.

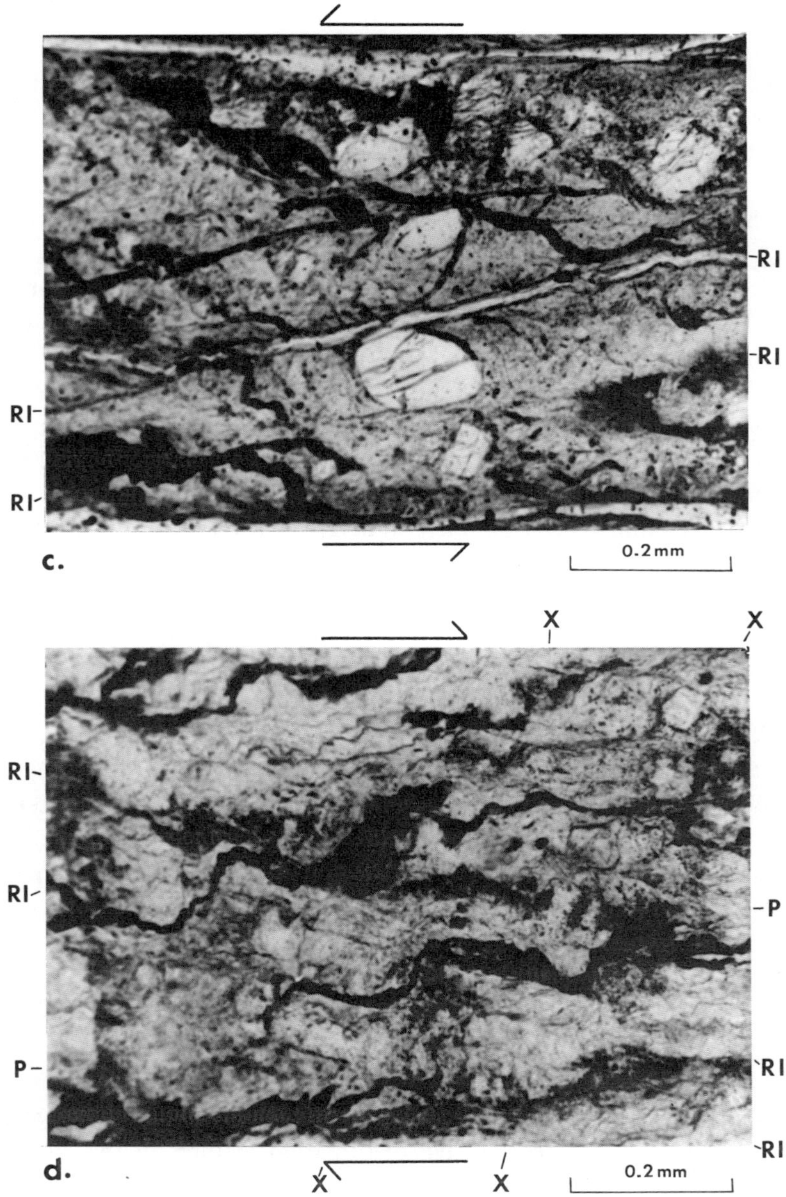

intense grain crushing show no particular relationship to other deformation textures in the gouge, with the exception of otherwise rounded grains that have flattened surfaces adjacent to the R1 fractures (Fig. 6c). However, the 600 °C gouge samples contain alternating bands of coarser and more finely grained material that are bounded by the R1 planes.

The abundant dark-colored biotite and opaque minerals in the granite gouge provide good indicators of grain deformation. These grains characteristically occur as 'ductile stringers' (LOGAN et al., 1981, p. 130) that have an overall orientation in the P direction (Figure 6d). In detail, however, the deformed grains consist of long segments on adjacent R1 planes that are connected by shorter segments parallel to X. At lower temperatures this stepwise, segmented appearance is well defined, but in the more highly deformed, 600 °C samples the changes in orientation are less abrupt and the deformed opaques and phyllosilicates have an undulating appearance. The number and amount of strain of these deformed grains increases with temperature; at 600 °C this mineral fabric is much more pronounced in the samples run at the faster sliding velocity.

Dry Lake Valley gouge. In general, the Dry Lake Valley gouge contains the same types of deformation texture as the granite gouge, but they are developed to a lesser degree. Because of this the Dry Lake Valley gouge has a relatively massive appearance. The principal difference from the granite gouge is the general lack of grain-size reduction, resulting from the isolation of the larger grains of quartz and feldspar in soft, clayey matrix. Some quartz and feldspar grains, however, are offset along breaks that have the R2 fracture orientation. Some of the larger grains are deformed in the same stepwise manner as that described for the granite gouge; such strained grains are more common in the samples at the 4.8 μm/sec sliding velocity.

The Dry Lake Valley gouge also contains low-angle R1 fractures and boundary faults. In a 600 °C sample the boundary fault along one side of the gouge layer is discontinuous, terminating where it intersects an R1 fracture. As was observed with the granite gouge, the angle between the R1 fractures and the edge of the gouge layer varies with the sliding velocity: at 4.8 μm/sec the angle ranges from 8° to 15°, and at 4.8×10^{-2} μm/sec it is between 14° and 23°.

Illite gouge. The illite gouge contains a rather different set of deformation textures from the granite and Dry Lake Valley gouges. For example, the 200 °C illite samples contain cross-cutting fractures with a variety of orientations. Fractures with the T orientation (40–55°) predominate, and R2 fractures (65–70°) are also common; R1 and X fractures are less abundant. The gouge adjacent to the fractures generally is compacted. Fractures are the principal deformation features of the 200 °C gouge samples; at higher temperatures, however, they occur in lesser numbers and are generally oriented in the R1 direction. The R1 fracture angles are 6° to 15° for the samples at a sliding velocity of 4.8 μm/sec and 11° to 33° for those at 4.8×10^{-2} μm/sec.

The 400 and 600 °C illite samples contain kink bands that are oriented approximately parallel to the R1 fractures (10° to 17°). The clay minerals in the narrow, kinked portions are aligned parallel to the edge of the gouge layer. The clays in the wider zones between the kinks are aligned parallel to P. In some samples the kink

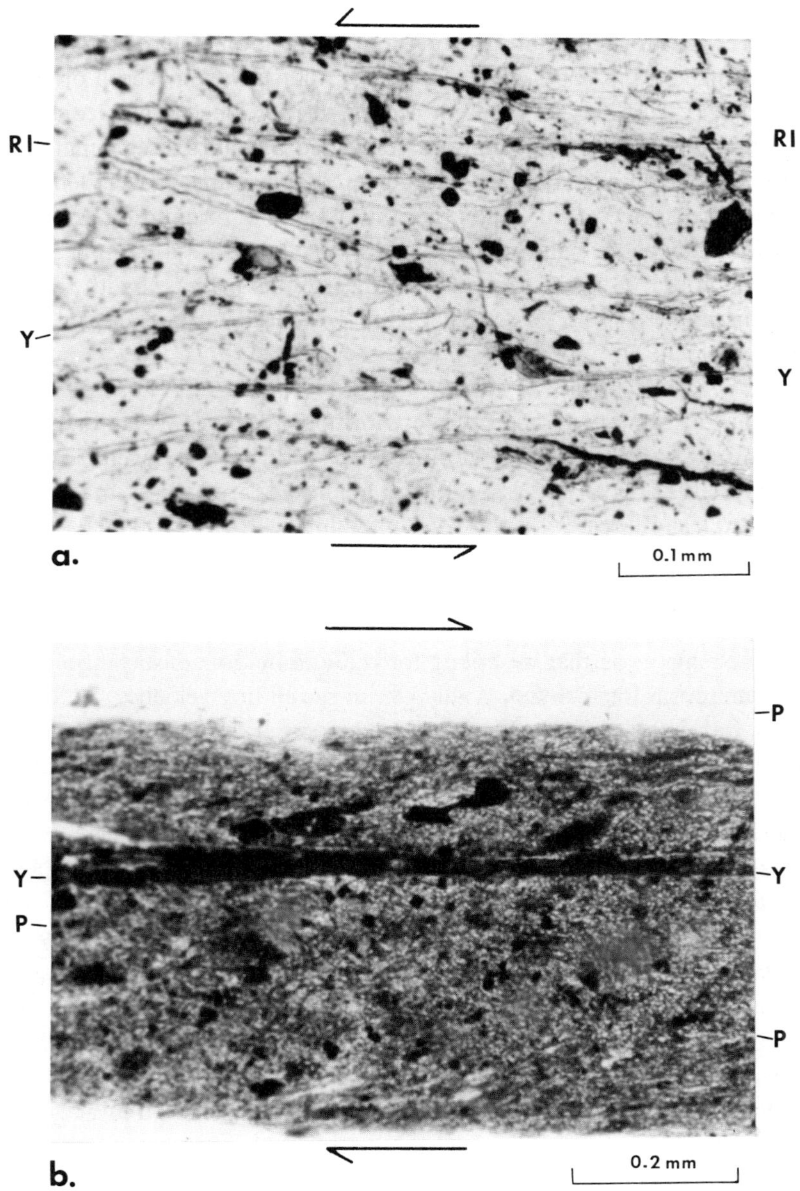

Figure 7

Textures of heated serpentinite gouge, in plane-polarized light. (a) Serpentine-bearing sample, at 200 °C and 2.5 kbar, showing Y and R1 fractures. In the center a serpentine grain is offset along a Y fracture. (b) Olivine-bearing sample at 600 °C and 2.5 kbar. The gouge consists principally of small, rounded crystals of brownish-colored olivine; the lighter-colored interstitial areas are talc. A fracture filled with dark, fine-grained material lies roughly parallel to the boundary of the gouge layer but well within the gouge. The upper third of the gouge layer contains some relict opaque grains stretched in the P direction. In the lower right-hand corner a faint lamination, also in the P direction, is visible.

bands extend only partway across the gouge layer from one side, whereas in others they transect the entire gouge layer.

Opaque grains in the 600 °C illite samples have been strained; most of them are elongate parallel to P, although a few are nearly parallel to the gouge–rock boundary. The illite gouge samples also contain the boundary faults filled with isotropic material that were observed in the other two gouges. One 600 °C sample has a boundary fault on only one side, and a second 600 °C sample has two, closely spaced boundary faults along part of one side.

Serpentinite gouge. The 200 and 400 °C serpentine-bearing samples contain intersecting sets of Y and R1 fractures (Fig. 7a). The Y fractures characteristically have a wavy appearance and are discontinuous; some serpentine grains cut by the Y fractures are slightly offset in accordance with the shear direction (Fig. 7a). Many other serpentine grains have been crushed and smeared out along the fractures. Fine-grained serpentine in the zones between fractures shows a faint extinction alignment parallel to X. Some of the remaining large serpentine grains contain kink bands that also are oriented in the X direction. Some red-colored phyllosilicate grains and opaque minerals that are present in the starting serpentinite gouge have been strained parallel to P. Boundary fractures filled with fine-grained material are not as common in the serpentine-bearing gouges as in the other gouges examined; some samples have a fracture along only one side and others none at all.

The 400 and 600 °C olivine-bearing gouge samples show a combination of deformation and crystallization textures. The 400 °C olivine-bearing sample preserves the original serpentine fabric to a large degree, but the edges of the grains are poorly defined. These relict clasts have a dusty appearance, not seen in the serpentine-bearing samples, that may be very fine-grained olivine. In one 600 °C sample olivine occurs as a series of small, round, brownish-colored crystals that form around the edges of relict serpentine grain shapes. With increasing amounts of olivine growth the relict grain shapes gradually disappear, and at the extreme case the gouge consists of a mat of small, round, brownish-colored olivine crystals with minor amounts of interstitial talc and quartz (Fig. 7b).

The olivine-bearing gouges principally contain R1 fractures (14° to 22°) and boundary fractures, although some T fractures (40°) were observed in one sample. In place of boundary fractures one 600 °C sample contains a single Y fracture that is located about a third of the way across the layer (Fig. 7b). Some of the relict serpentine grain shapes in the 400 °C olivine-bearing sample are offset along R1 fractures, whereas others are flattened in the Y planes. The 400 °C sample also contains somewhat wavy kink bands that are oriented roughly perpendicular to the gouge–rock boundary. One 600 °C sample contains laminae parallel to P that extend partway into the gouge from the edge (Fig. 7b). Strained opaque grains also are elongate in the P direction.

Discussion

Strength and sliding stability

The effects of temperature and confining pressure on the frictional properties of these four gouges were considered in MOORE et al. (1983); this discussion will concentrate principally on the effects of changing the sliding velocity. Few studies of rock or gouge friction have dealt with changing displacement rates. Of these, OLSSON (1974), ENGELDER et al. (1975), and TEUFEL and LOGAN (1978) found no dependence of friction strength on slip rate. By contrast, DIETERICH (1981) varied the sliding velocity of the same sample of granite gouge during direct shear experiments, and he found that decreasing the sliding velocity produced a slight increase in the residual strength of the gouge. Dieterich's results are consistent with the observation in our experiments that many of the curves at the slower slip rate plotted at slightly higher differential stresses. TEUFEL and LOGAN (1978) observed a relationship between sliding velocity and sliding stability similar to that reported here; during room-temperature friction experiments on cut specimens of Tennessee sandstone they reported a change from stable sliding to stick slip as the sliding velocity was lowered. ENGELDER et al. (1975) found that at slower sliding rates the size of the stress drops was larger for samples of quartz gouge showing stick slip; this also corresponds to the relationships in Figures 1 to 3.

The correlation between slip rate and stick slip in room-temperature friction experiments on gouge has been explained as a result of the interlocking of grains in the gouge (SUMMERS and BYERLEE, 1977). As a gouge becomes more highly compacted, sliding must take place by fracturing through grains rather than over them. MOORE et al. (1983) proposed a similar mechanism for the increase in the occurrence of stick slip in the sheet-silicate-rich gouges with increasing temperature. The clay-rich and serpentinite gouges showed greater degrees of lithification at higher temperatures, and sliding would then have occurred by fracturing through the hardened mineral aggregate. Decreasing the sliding velocity would also promote lithification and thereby stick-slip behavior in these three gouges, because of the concomitant increase in heating time.

In the experiments at 4.8 μm/sec the granite gouge was compacted but not lithified (MOORE et al., 1983). Because of this, the occurrence of stick slip during some of the granite gouge experiments at 4.8×10^{-2} μm/sec may be due simply to increased compaction of this gouge at the slower deformation rate; but lithification may also be a factor in the lower-velocity experiments. Given enough time, cementation lithification of quartz-bearing gouges could occur through solution and redeposition of silica from pore fluids (ANGEVINE et al., 1982). This process may have occurred to some extent in the experiments at the slower displacement rate with their longer reaction times. Cementation of the granite gouge probably was not very extensive even at the slower slip rate, however, because less than half of the granite samples showed stick slip at that displacement rate and the run products were also not visibly lithified.

Mineralogical changes

As described previously, the x-ray patterns of quartz and feldspar minerals in the gouges were unaffected by changes in the sliding velocity. The effects on the clay mineral x-ray intensities were also not particularly significant. Many clay minerals begin to lose some of their loosely bound water at temperatures between 100 and 200 °C (DEER et al., 1962), so that the temperature conditions of all the friction experiments of this study were within the field of clay mineral dehydration. At any given temperature and pressure the dehydration reactions were more thorough at the slower sliding velocity, but this difference is readily explained by the corresponding increase in the amount of heating time (e.g., BRINDLEY and ZUSSMAN, 1957).

In contrast to the clays, the lowest temperature at which an olivine-forming reaction in the serpentinite gouge was observed did vary with the sliding velocity (Fig. 4). On the basis of the mineral assemblages observed in the x-ray patterns (Fig. 4) and in thin section, the lizardite and chrysotile of the serpentinite gouge reacted at 600 °C as follows:

$$\underset{\text{serpentine}}{5\,Mg_3Si_2O_5(OH)_4} = \underset{\text{olivine}}{6\,Mg_2SiO_4} + \underset{\text{talc}}{1\,Mg_3Si_4O_{10}(OH)_2} + 9\,H_2O \qquad (1)$$

$$\underset{\text{serpentine}}{2\,Mg_3Si_2O_5(OH)_4} = \underset{\text{olivine}}{3\,Mg_2SiO_4} + \underset{\text{quartz}}{1\,SiO_2} + 4\,H_2O \qquad (2)$$

One of these two reactions is probably also responsible for the appearance of olivine in the slower sliding-velocity experiments at 400 °C, even though neither quartz nor talc appeared with olivine in the 400 °C x-ray traces. As seen in Figure 4b, the x-ray peaks for olivine are much weaker at 400 than at 600 °C. A similar reduction in the talc or quartz x-ray intensities at 400 °C could reduce them to the level of the background noise. Typically, a mineral must be at least 5 to 10 percent of a sample to appear on an x-ray pattern.

Of the two, reaction (1) is the more likely to have occurred at 400 °C. Reaction (2) takes place in air at temperatures near 600 °C (DEER et al., 1962), and at the higher confining pressures of the friction experiments the reaction would probably shift to slightly higher rather than to lower temperatures. In the phase diagram for the system $MgO–SiO_2–H_2O$ of EVANS et al. (1976), part of which is shown in Figure 8, reaction (1) is located at about 430 °C at 1 kbar of confining pressure and at about 450 °C at 2.5 kbar. Such temperatures are also slightly too high to explain the observed 400 °C results for the serpentinite gouge. EVANS et al. (1976), however, described some uncertainties in applying the existing experimental data for reaction (1) to their diagram, which they resolved by consideration of field evidence. Therefore, the placement of reaction (1) in their diagram is not completely certain; it may occur at slightly lower temperatures than shown. In addition, the phase diagram of EVANS et al. (1976) was constructed for conditions of $P_{conf} = P_{fl}$ whereas in the

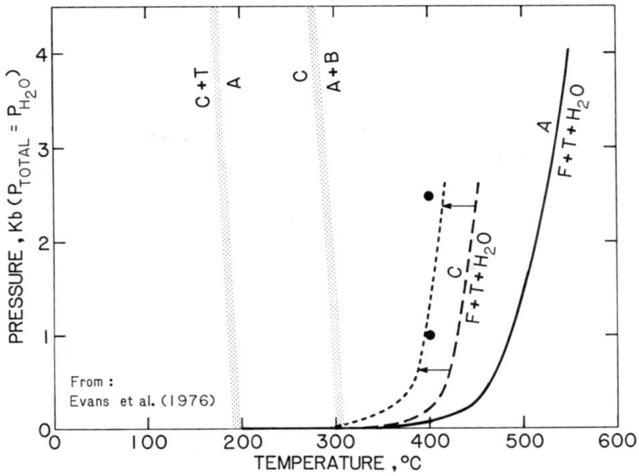

Figure 8
A portion of the phase diagram of EVANS et al. (1976) for the system $MgO–SiO_2–H_2O$: C, chrysotile; T, talc; A, antigorite; B, brucite; F, forsterite. Reaction (1) is metastable in this system, and reaction C = A + B would be the stable reaction for the 400 °C serpentinite gouge experiments. This reaction, however, is sluggish both in nature and in the laboratory. Reaction (1)′ indicates a possible position for Reaction (1) at low fluid pressures, which would allow olivine to form in the gouge experiments at 400 °C and 1 kbar confining pressure.

friction experiments $P_{conf} \gg P_{fl}$. At lower fluid pressures relative to confining pressure, the reaction should be displaced to slightly lower temperatures. A shift of Reaction (1) to Reaction (1)′ in Figure 8 would be sufficient to make the serpentine-forming reaction possible at 400 °C and a 1 kbar confining pressure.

Assuming that the placement of reaction (1)′ in Figure 8 is approximately correct, the restricted occurrence of the olivine-forming reaction to the slower-velocity experiments at 400 °C and a 1 kbar confining pressure may be due to the combined effects of deformation and increased reaction time. Although the reaction theoretically could occur at those temperature–pressure conditions, sluggish reaction kinetics at such relatively low metamorphic temperatures might inhibit the breakdown of serpentine even at relatively long times during hydrostatic experiments. This was tested by heating of a serpentinite gouge sample for 66 hours at 400 °C and a confining pressure of 1 kbar, without an axial load being applied. Although the amount of time in this experiment was considerably greater than that for an experiment at a sliding velocity of 4.8×10^{-2} μm/sec, no evidence of reaction was found. An increase in the strain and surface energy of the serpentine grains during deformation may have promoted the reaction to the extent that it is observable in the longer experiments.

Gouge textures

Many of the textures observed in the gouges are functions of the initial physical character of the gouges. The relative amounts of hard and soft grains, in particular, control the development of cataclastic textures and kink bands in the gouges. A relationship also exists between gouge type and fracture development, for example, the restriction of Y fractures to the serpentine-bearing and olivine-bearing gouges, and the many different fracture orientations in the low-temperature illite gouge samples. Temperature and confining pressure also have important effects on gouge texture, including the amount of grain-size reduction and the degree of induration of a given gouge.

Of particular interest in this discussion are the relationships between sliding velocity and gouge textures. Perhaps the most significant correlation is the increase in the R1 angles with decreasing displacement rate, which was observed for three of the gouges. The physical significance of the R1 fracture orientations has been considered by a number of workers, particularly those involved in soil mechanics research (e.g., RIEDEL, 1929; HANSEN, 1961; TCHALENKO, 1970; see discussion in MORGENSTERN and TCHALENKO, 1967). The results of many of the shear box tests and Riedel experiments have been interpreted on the assumption that the deformation is of the simple shear type; in such a case the following relationship should hold:

$$\theta' = \varphi/2 \tag{3}$$

where θ' is the angle between the R1 fracture and the direction of shear, and φ is the angle of internal friction (HANSEN, 1961; TCHALENKO, 1970). Using this equation, TCHALENKO (1970) has reported that the fracture angles developed during shear box tests on kaolin, bentonite, and London Clay yielded angles of internal friction consistent with the experimentally determined values for those clays. Equation (3) also is valid for room-temperature triaxial friction tests. For example, LOGAN et al. (1979) found a close agreement between coefficients of friction of pure quartz gouge obtained from equation (3) and from strength measurements. BYERLEE et al. (1978) reported R1 angles for room-temperature experiments on quartz gouge that were consistent with those of LOGAN et al. (1979). The two sets of results for quartz gouge covered a range of sliding velocities, which indicates the lack of a velocity relationship at room temperature. Equation (3) also appears to work reasonably well for the subsidiary R1 fractures developed in surface sediments during a recent earthquake in Iran (TCHALENKO and AMBRASEYS, 1970).

On the basis of these results, the observed correlation between sliding velocity and R1 angle in the heated gouge samples would not be predicted from the strength measurements. Instead, all the granite gouge samples would be expected to have the same R1 angles, whereas the clay-rich and serpentinite gouges should show a change in R1 angle with temperature because of the pronounced temperature-related changes in the coefficients of friction for these three gouges (MOORE et al., 1983, Fig. 8, p. 495). In contrast to our results, LOGAN et al. (1979) did find a decrease with increasing temperature in both the strength and the R1 angles of calcite gouge.

The origin of the sliding-velocity relationship in the experiments of this study is as yet unknown. The increase in R1 angle with decreasing slip velocity is consistent with observations of DIETERICH (1981) in room-temperature experiments, that gouge strength was inversely related to sliding velocity. However, a velocity relationship to strength could not definitely be proved in the experiments of this study. One possible explanation of the high-temperature R1 angles is that at least one of the assumptions used in analyzing the room-temperature shear box and friction experiment data is not applicable at high temperatures. Alternatively, some additional factors may have to be considered for the heated gouge samples, such as changes in the degree of cohesion of the clay and serpentinite gouges at elevated temperatures. Further work needs to be done to elucidate the origin of the fracture-angle variations and to identify any other controlling factors.

Because stick-slip motion in triaxial friction tests has been correlated with earthquakes in nature (BYERLEE, 1970), it is important in earthquake-prediction programs to identify the causes of stick-slip behavior. On the basis of room-temperature friction studies of quartz gouge ENGELDER et al. (1975) and BYERLEE et al. (1978) proposed that the onset of stick slip corresponds to a shift in slip from the cross-cutting (R1) faults to the boundary faults. For the four heated gouges of this study, however, both types of fracture were observed in samples that slid stably as well as in ones that showed stick slip. SHIMAMOTO (1977) also found that slip was concentrated along the gouge–rock interface for room-temperature quartz and dolomite gouges that slid stably. Therefore, the development of boundary faults cannot be correlated with stick slip. In this study a decrease in sliding velocity has been related to an increase in the frequency of stick slip and also to an increase in R1 fracture angle. From this one might predict an additional correlation between high R1 angle and the occurrence of stick-slip motion. At the present time, however, this additional relationship is not confirmed, because stably and unstably sliding samples of a particular gouge at a given displacement rate had similar R1 angles. The number of samples examined petrographically for this study, however, was relatively small. Many more thin sections from other experiments are at present being prepared for further tests for this relationship and for other physical manifestations of stick slip.

The observed relationship between sliding velocity and fracture angle may have a potential use in estimations of the relative slip rates of fault zones. Fracture sets equivalent to the experimental R1 fractures have frequently been observed in natural fault zones (e.g., TCHALENKO and AMBRASEYS, 1970; LOGAN et al., 1979). Because the change in fracture angle occurs for a variety of compositions of heated gouge, this tool may be applicable at least to those portions of a fault that have been subjected to elevated temperatures.

Another potential slip-rate indicator for heated gouges is the amount of grain deformation. In the granite and Dry Lake Valley gouge samples examined, strained and elongated grains were more common and had higher aspect ratios in the experiments at a sliding velocity of 4.8 μm/sec. The results of LOGAN et al. (1981) for strength experiments at 300 °C with the Dry Lake Valley gouge are also consistent with those reported here. Quartz and feldspar clasts in their run products, which were deformed at a 10^{-2}/sec strain rate, or roughly 5.1×10^2 μm/sec displacement rate along the saw cut, were considerably stretched (see Fig. 15 of LOGAN et al., 1981, p. 132). In this study the samples of Dry Lake Valley gouge tested at the 4.8 μm/sec sliding velocity contained only a few slightly strained quartz and feldspar grains, although some opaque and phyllosilicate grains were relatively strongly deformed. At 4.8×10^{-2} μm/sec almost none of the grains in the Dry Lake Valley gouge were deformed.

The mineral fabric may prove to be a relatively poor indicator of slip rates in high-temperature fault zones, because many factors besides displacement rate may probably affect the degree of grain deformation. The age of the fault and the total amount of displacement along it may be particularly significant controls on the deformation texture. Fluid pressure may also be important (LOGAN et al., 1981). Additional experiments need to be conducted for assessment of the significance of these other factors.

Conclusions

Changing the sliding velocity has been found to have several effects on the frictional and physical properties of heated fault gouge. At the slower slip rate (a) the gouges showed a greater tendency to stick-slip motion, and individual stress drops tended to be larger, (b) observable replacement of serpentine minerals by olivine was found at a lower temperature, (c) the angle between low-angle Riedel (R1) fractures and the boundary of the gouge zone was higher, and (d) grain deformation was less pronounced at a given temperature and pressure.

The mineralogical effect may have resulted from a combination of deformation and increased reaction time at the slower slip rate. The observed relationships between sliding velocity and sliding stability may have important applications in determinations of the conditions under which earthquakes occur at depth in fault zones. The textual variations relative to the sliding velocity may have potential application in determinations of the relative slip rates of natural fault zones.

References

Anderson, J. L., Osborne, R. H. and Palmer, D. F. (1980), *Petrogenesis of cataclastic rocks within the San Andreas fault zone of southern California.* Tectonophys. *67*, 221–249.

Angevine, C. L., Turcotte, D. L. and Furnish, M. D. (1982), *Pressure solution lithification as a mechanism for the stick-slip behavior of faults.* Tectonics. *1*, 151–160.

Brace, W. F. and Byerlee, J. D. (1970), *California earthquakes: Why only shallow focus?.* Science. *168*, 1573–1575.

Brindley, G. W. and Zussman, J. (1957), *A structural study of the thermal transformation of serpentine minerals to forsterite.* Amer. Min. *42*, 461–474.

Byerlee, J. D. (1970), *The mechanics of stick-slip.* Tectonophys. *9*, 475–486.

Byerlee, J., Mjachkin, V., Summers, R. and Voevoda, O. (1978), *Structures developed in fault gouge during stable sliding and stick-slip.* Tectonophys. *44*, 161–171.

Deer, W. A., Howie, R. A. and Zussman, J., *Rock-forming minerals, Vol. 3: Sheet Silicates.* John Wiley, New York, 1962, p. 170–190.

Dieterich, J. H. (1981), 'Constitutive properties of faults with simulated gouge', in N. L. Carter, M. Friedman, J. M. Logan and D. W. Stearns (eds.), *Mechanical Behavior of Crustal Rocks.* Amer. Geophys. Union Geophys. Mon. *24*, 103–120.

Engelder, J. T., Logan, J. M. and Handin, J. (1975), *The sliding characteristics of sandstone on quartz fault gouge.* Pure Appl. Geophys. *113*, 69–86.

Evans, B. W., Johannes, W., Oterdoom, H. and Trommsdorff, V. (1976), *Stability of chrysotile and antigorite in the serpentinite multisystem.* Schweiz. Min. Petrogr. Mitt. *56*, 79–93.

Hansen, B. (1961), *Shear box tests on sand.* Proc. Fifth Internat. Conf. Soil Mech. *1*, 127–131.

Logan, J. M., Friedman, M., Higgs, N., Dengo, C. and Shimamoto, T. (1979), *Experimental studies of simulated gouge and their application to studies of natural fault zones.* Proc. Conf. VIII, Analysis of Actual Fault Zones in Bedrock, U.S. Geol. Surv. Open-File Rept. *79–1239*, 305–343.

Logan, J. M., Higgs, N. G. and Friedman, M. (1981), '*Laboratory studies on natural gouge from the U.S. Geological Survey Dry Lake Valley No. 1 well, San Andreas fault zone*', in N. L. Carter, M. Friedman, J. M. Logan and D. W. Stearns (eds.), *Mechanical Behavior of Crustal Rocks.* Amer. Geophys. Union Geophys. Mon. *24*, 121–134.

Moore, D. E., Summers, R. and Byerlee, J. D. (1983), *Strengths of clay and non-clay fault gouge at elevated temperatures and pressures.* Proc. 24th U.S. Symp. Rock Mech., 489–500.

Morgenstern, M. R. and Tchalenko, J. S. (1967), *Microscopic structures in kaolin subjected to direct shear.* Geotechn. *17*, 309–328.

Nemecz, E., *Clay Minerals,* Akademiai Kiạdo, Budapest, Hungary, 1981, p. 119–123.

Olsson, W. A. (1974), *Effects of temperature, pressure, and displacement rate on the frictional characteristics of a limestone.* Int. J. Rock Mech. Mining Sci. Geomech. Abstr. *11*, 267–278.

Riedel, W. (1929), *Zur mechanik geologischer Brucherscheinungen.* Zentralbl. Mineral Geol. Palaeont. *1929B*, 354–368.

Schultz, L. G. (1964), '*Quantitative interpretation of mineralogical composition from X-rays and chemical data for the Pierre shale*', in *Analytical Methods in Geochemical Investigation of the Pierre Shale.* U.S. Geol. Surv. Prof. Paper *391-C*, C1–C31.

Shimamoto, T. (1977), *Effects of fault-gouge on the frictional properties of rocks: An experimental study.* Ph.D. Dissertation, Texas A and M Univ., p. 138–167.

Sibson, R. H. (1982), *Fault zone models, heat flow, and the depth distribution of earthquakes in the continental crust of the United States.* Seism. Soc. Amer. Bull. *72*, 151–163.

Skempton, A. W. (1966), *Some observations on tectonic shear zones.* Proc. 1st Congr. Internat. Soc. Rock Mech., Lisbon *1*, 329–335.

Stesky, R. M. (1978), *Rock friction—effect of confining pressure, temperature and pore pressure.* Pure Appl. Geophys. *116*, 690–704.

Stesky, R. M., Brace, W. F., Riley, D. K. and Robin, P.-Y. F. (1974), *Friction in faulted rock at high temperature and pressure.* Tectonophys. *23*, 177–203.

Summers, R. and Byerlee, J. (1977), *A note on the effect of fault gouge composition on the stability of frictional sliding.* Int. J. Rock Mech. Mining Sci. Geomech. Abstr. *14*, 155–160.

TCHALENKO, J. S. (1968), *The evolution of kink-bands and the development of compression textures in sheared clays.* Tectonophys. *6*, 159–174.

TCHALENKO, J. S. (1970), *Similarities between shear zones of different magnitudes.* Geol. Soc. Amer. Bull. *81*, 1625–1640.

TCHALENKO, J. S. and AMBRASEYS, N. N. (1970), *Structural analysis of the Dasht-e Bayaz (Iran) earthquake fractures.* Geol. Soc. Amer. Bull. *81*, 41–60.

TEUFEL, L. W. and LOGAN, J. M. (1978), *Effect of displacement rate on the real area of contact and temperatures generated during frictional sliding of Tennessee sandstone.* Pure Appl. Geophys. *116*, 840–865.

WHITTAKER, E. J. W. and ZUSSMAN, J. (1956), *The characterization of serpentine minerals by X-ray diffraction.* Mineral Mag. *31*, 107–126.

(Received 25th September 1985, revised 18th January 1986, accepted 24th January 1986)

PAGEOPH, Vol. 124, Nos. 1/2 (1986)

0033-4553/86/020053-26$1.50+0.20/0

Self-Similar Cataclasis in the Formation of Fault Gouge

CHARLES G. SAMMIS,[1] ROBERT H. OSBORNE,[1] J. LAWFORD ANDERSON,[1] MAVONWE BANERDT[1] and PATRICIA WHITE[1]

Abstract—Particle-size distributions have been determined for gouge formed by the fresh fracture of granodiorite from the Sierra Nevada batholith, for Pelona schist from the San Andreas fault zone in southern California, and for Berea sandstone from Berea, Ohio, under a variety of triaxial stress states. The finer fractions of the gouge derived from granodiorite and schist are consistent with either a self-similar or a logarithmic normal distribution, whereas the gouge from sandstone is not. Sandstone gouges are texturally similar to the disaggregated protolith, with comminution limited to the polycrystalline fragments and dominantly calcite cement. All three rock types produced significantly less gouge at higher confining pressures, but only the granodiorite showed a significant reduction in particle size with increased confining pressure. Comparison with natural gouges showed that gouges in crystalline rocks from the San Andreas fault zone also tend to be described by either a self-similar or log-normal particle distribution, with a significant reduction in particle size with increased confining pressure (depth). Natural gouges formed in porous sandstone do not follow either a self-similar or a log-normal distribution. Rather, these are represented by mixed log-normal distributions. These textural characteristics are interpreted in terms of the suppression of axial microfracturing by confining pressure and the accommodation of finite strain by scale-independent comminution.

Key words: Cataclasis, gouge, self-similar, fractal, fracture, faulting.

Introduction

One of the results of brittle failure in rock is the production of cataclastic material in the fault zone. The type of material formed depends primarily upon the conditions of deformation. Temperature and pressure conditions associated with shallow faulting tend to produce microbreccia and gouge, somewhat deeper conditions produce extremely fine-grained, indurated cataclasites, whereas still higher temperatures and pressures in the lower crust produce mylonites.

The physical properties of these cataclastic materials have been extensively studied because it is evident that the frictional strength and mode of failure as well as the hydrological state of an existing fault zone may depend on the nature of the cataclastic material. Recent papers include MORROW et al. (1981, 1982, 1984), WANG

[1] Department of Geological Sciences, University of Southern California, Los Angeles, CA, 90089-0741.

et al. (1980), Byerlee (1978), Summers and Byerlee (1977), Wu et al. (1975), Engelder et al. (1975), and Chu et al. (1981). Because fault gouge can be extremely fine-grained, the surface energy involved in its creation can be a significant part of an earthquake's energy budget. This problem has been addressed recently by Olgaard and Brace (1983).

In this paper we focus on the micromechanics by which gouge is generated. Specifically, we wish to address the question of how the state of stress at the time of faulting and the total fault displacement might be reflected in the textural characteristics of the resultant gouge. Quantifiable textural characteristics include the total width of the gouge zone, the particle-size distribution function within this zone (e.g., normal, log-normal, log-extreme, exponential, etc.), a measure of the mean gouge particle size, particle shape, and mineralogical composition.

Some intriguing clues have emerged from textural studies in the field, which suggest that natural gouge may contain useful information concerning the fracture process. First, gouge produced in fully dense crystalline rock tends to have a log-normal particle-size distribution. This has been documented by Anderson et al. (1980) for granodiorite gouge sampled from several locations along the San Andreas fault in southern California and by Olgaard and Brace (1983) for the finer particle fractions of quartzite gouge from mining-induced shear fractures in South Africa. In each case the gouge was a rock flour of the source rock with no signs of significant chemical alteration. It may be significant, however, that natural sandstone gouges studied by Engelder (1974) did not appear to follow a log-normal distribution.

A second textural trend is a tendency of the mean particle size of the gouge material to decrease with increasing depth (Anderson et al., 1980). At midcrustal levels more than 80% of the intrafault particles have been comminuted to a crystalline cataclasite groundmass with grain size on the order of 5 μm (Anderson et al., 1983; Phillips, 1982).

There also may be information in the width of the gouge zone. Engelder (1974) characterized the quartz gouge and cataclastic sandstone from four dip-slip fault zones which intersected massive sandstone formations. He made two important observations: (a) the mean particle size and sorting decrease as one approaches a major slip plane within the fault zone, presumably because of the associated increase in shear strain, and (b) the number of subfractures associated with a given fault strand is positively correlated with total displacement. Assuming each subfracture generates gouge, Engelder pointed out that the width of the gouge zone can be inferred to increase with displacement. Robertson (1982) has presented evidence for a direct proportionality between displacement and the width of the cataclastic zone which spans more than six orders of magnitude. Robertson limited his study to normal and high-angle reverse faults.

There is also some evidence that the width of the zone decreases with depth. On the basis of their study of cataclastic rocks in the San Andreas and San Gabriel fault

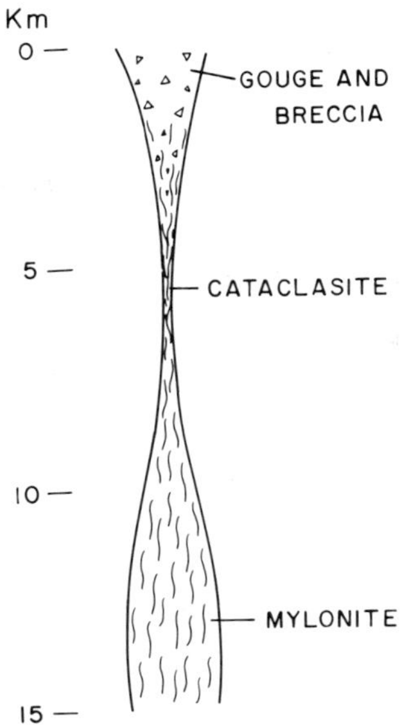

Figure 1
Hypothetical cross-section of the San Andreas fault, showing inferred changes in zone width and cataclastic rock type with depth. From ANDERSON et al. (1983).

zones, ANDERSON et al. (1983) present the schematic cross-section of the San Andreas fault reproduced as Figure 1. As discussed below, there is reason to expect the width of a fault zone to decrease in conjunction with the decreased particle size in the brittle field. Below the brittle–ductile transition the zone again broadens where mylonites predominate.

To help interpret these observations, we present textural data on gouge produced by the fresh fracture of granodiorite, schist, and sandstone samples under triaxial loading, and compare the resultant particle distributions with those observed in the field. We interpret these observations in terms of (a) the progressive brittle microfracture which leads to macroscopic failure on the fault plane and (b) further self-similar commination associated with fault displacement.

Gouge Generated by Brittle Failure in Triaxial Experiments

Cylindrical samples of granodiorite from the Sierra Nevada batholith, Pelona schist, and Berea sandstone were loaded to failure in the triaxial cell shown in Figure 2. Following each experiment the particle-size distribution was determined for all particulate material having a diameter less than 4 mm.

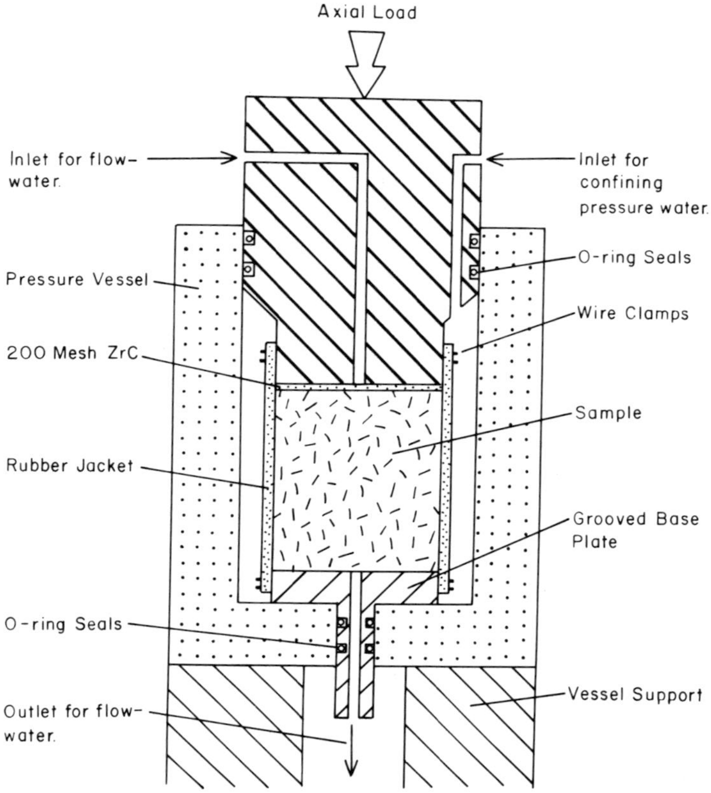

Figure 2
Schematic diagram of triaxial test cell used to create gouge in the laboratory by the fresh fracture of cylindrical samples. The pore-pressure capability was not used; all samples were failed in the 'dry' state. MoS_2 lubricant was used at the top and bottom of the sample in place of the ZrC powder.

Table 1
Granodiorite Gouge Data

Sample	σ_3 (MPa)	σ_1 (MPa)	Gouge wt. per sample wt. (%)	$\langle d \rangle$ (mm)
425-4B	0	62.3	14.4	0.81
425-3C	0	65.9	7.5	0.72
425-3AB	0	65.9	12.2	0.78
425-4D	3.5	147.7	5.6	0.52
425-3E	6.9	192.3	8.0	0.37
435-2B	6.9	271.2	1.7	0.37
435-4A	6.9	280.0	8.7	0.36
435-5D	10.3	328.5	1.4	0.24
425-1DA	10.7	234.5	4.6	0.25
425-2D	13.8	335.8	1.5	0.23

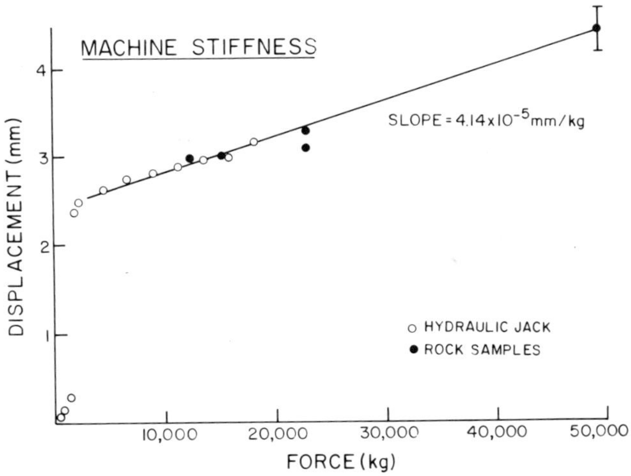

Figure 3
Machine-stiffness measurements for Rheile hydraulic testing machine. Displacements for confined specimens were smaller owing to support of the upper piston by the confining fluid (see Tables 2 and 3).

Sample selection and preparation

A 6 m drill core 5 cm in diameter from the depth interval of 67 to 73 m in the Sierra Nevada batholith was provided by the geothermal-energy group of the U.S. Geological Survey from their Jackass Creek site. The rock is a grey, medium-to-coarse-grained hornblende–biotite–sphene granodiorite. Cylindrical test specimens 10 cm long were prepared by grinding the ends flat and parallel to a precision limit of 0.025 mm.

Blocks of Pelona schist were removed from an outcrop in the San Francisquito Canyon area located in the San Gabriel Mountains of southern California. This material was chosen as a typical, foliated metamorphic rock which is a well-known rock unit that occurs adjacent to the San Andreas fault zone in southern California. The Pelona schist is a greenschist-grade meta-graywacke consisting predominantly of albite, quartz, and muscovite. Common accessory minerals include epidote, chlorite, calcite, sphene, pyrite, and graphite. Albite occurs as porphyroblasts and is poikilitic with graphite inclusions. Muscovite defines the well-developed schistosity. The blocks were cored to produce cylindrical samples 5 cm in diameter and 10 cm long. Cores were cut at angles of 0 and 90 degrees to the foliation. Sample ends were ground flat and parallel to within 0.025 mm.

The Berea sandstone cores were provided by Union Oil Research Center and were obtained from Berea, Ohio. The Berea sandstone sample is a medium-grained sandstone which is 66% monocrystalline quartz, 22% polycrystalline quartz, 5% feldspar, 3.5% calcite cement, 1.6% muscovite, 1.3% opaques, and 0.6% rock fragments (modal composition based on 800 point counts on two thin

Table 2

Pelona Schist Data

Sample	Foliation*	σ_3 (MPa)	σ_1 (MPa)	Disp.† (mm)	Gouge wt. per sample wt. (%)	$\langle d \rangle$ (mm)
PD1R	‖	0	98.8	0.17	4.7	1.70
PSC	⊥	0	130.9	0.13	1.1	0.70
PE2L	⊥	0	127.9	—	2.6	0.95
PG1	‖	4.8	140.6	0.31	0.2	0.48
PSD	⊥	4.8	148.8	1.30	0.9	0.56
PSA	‖	5.2	156.0	0.03	0.3	0.87
JLP	‖	7.9	102.2	0.25	2.2	1.40
PSB	⊥	8.3	164.2	—	1.1	0.44
PD1L	‖	8.3	192.4	0.39	1.0	1.10
PE2R	⊥	11.0	169.8	0.52	1.1	0.65
PSE	⊥	12.4	196.5	1.59	1.4	0.95

* Foliation parallel to the cylinder axis (σ_1 direction), ‖; Foliation perpendicular to the cylinder axis (σ_1 direction), ⊥.

† Post-failure displacement.

Table 3

Berea Sandstone Data

Sample	σ_3 (MPa)	σ_1 (MPa)	Displ.* (mm)	Gouge wt. per sample wt. (%)	$\langle d \rangle$ (mm)
SSD	0	37.7	—	10.6	0.18
SSK	1.4	37.7	1.02	8.3	0.19
SSI	2.1	42.7	0.14	3.1	0.19
SSB	6.9	79.8	—	5.1	0.18
SSG	6.9	75.3	0.11	2.2	0.18
SSC	14.5	94.7	—	2.4	0.15
SSE	20.7	118.9	0.09	3.6	0.17
SSF	20.7	120.1	0.30	1.2	0.14
SSH	34.5	136.8	—	24.3	0.17

* Post-failure displacement.

sections). The average porosity is 18%, but ranges from 17% to 22% within a few cubic inches (HANDIN et al., 1963). The core, 3.8 cm in diameter, was cut into samples 7.6 cm long, and the ends were ground flat on parallel to within 0.025 mm.

Apparatus and testing procedure

A schematic diagram of the triaxial loading cell is shown in Figure 2. Axial load, σ_1, was applied by an LH-160 Rheile hydraulic testing machine. An empirical

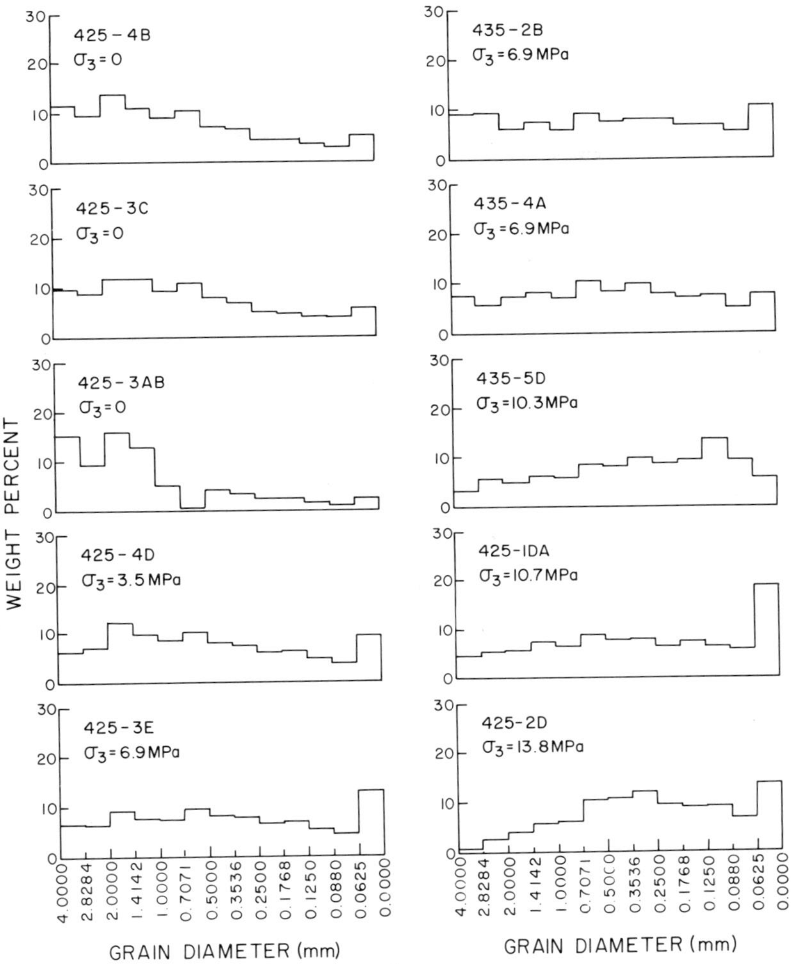

Figure 4

Histograms showing the grain-size distributions for the Sierran granodiorite gouge specimens. Note the shift toward finer particles sizes with increased confining pressure, σ_3.

machine-stiffness curve is given in Figure 3. The initial 2.5 mm of displacement is due to backlash in the crosshead screw. Displacements were measured by a displacement transducer mounted between the piston and pressure vessel.

The sample core assembly was prepared first with a MoS_2 lubricant applied to the sample ends and then with a rubber jacket (bicycle inner tube), fitted over the sample, the base plate, and the end of the loading piston, as shown in Figure 2. For the Berea sandstone a thin copper-foil disk was placed on the sample ends before the lubricant was applied. The jacket was secured to the piston and base plate by electrical tape and steel wire.

Samples were stressed to failure at constant confining pressure at a loading rate

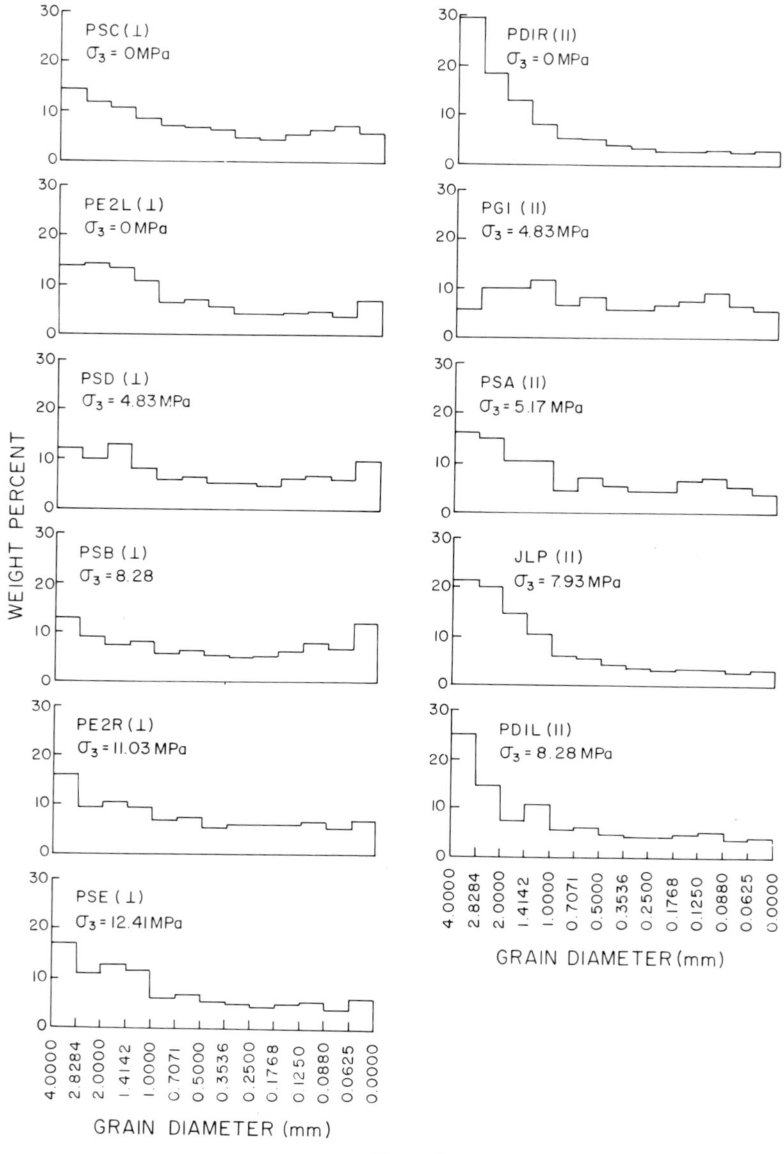

Figure 5
Histograms showing the grain-size distributions for the Pelona schist gouge specimens. Note that samples loaded with σ_1 parallel to the foliation (∥) tend to have more particles in the largest fractions than samples loaded perpendicular to the foliation (⊥).

of about 200 N/sec. Mechanical testing data for the Sierra granodiorite samples are given in Table 1, for the Pelona schist samples in Table 2, and for the Berea sandstone samples in Table 3.

Immediately after failure the axial load was removed, followed by the removal of

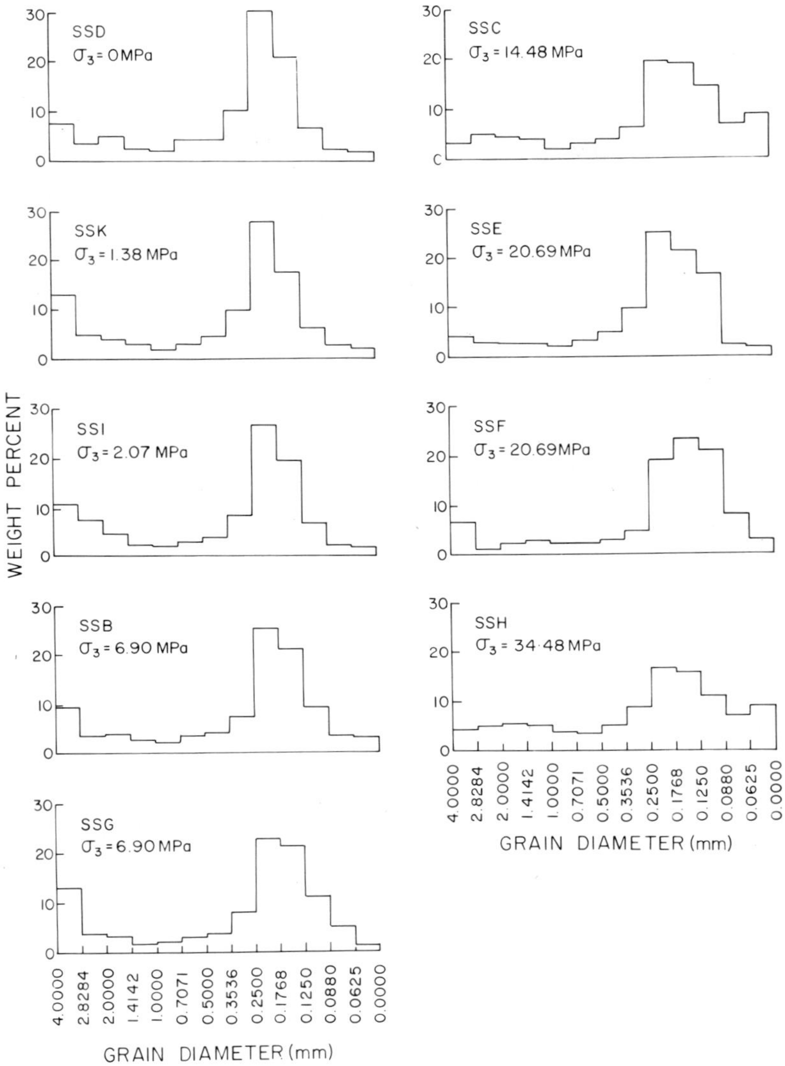

Figure 6
Histograms showing the grain-size distribution for the Berea sandstone gouge specimens.

confining pressure. This minimized post-test displacement; since the upper part of the piston is larger than the sample, the confining fluid increased the effective machine stiffness. As a result, the measured post-failure displacements of confined samples ranged from 0.3 to 1.6 mm, significantly less than predicted by Figure 3. The sample was removed from the core assembly at the piston and base plate by cutting the steel wires and tape. The sample jacket was removed and the gouge material collected and saved for textural analysis. All loose gouge was brushed from the fracture surfaces.

Textural analysis

Each gouge was sieved for 15 min on a Fisher-Wheeler sieve shaker. Using one half-Phi unit sieve intervals, the sieves were stacked in decreasing order from 4.0 mm (−2.0 Phi) at the top to 0.0625 mm (4.0 Phi) at the bottom. Particles finer than 0.0625 mm were collected in a pan at the bottom of the set of sieves. Each sieve fraction, and the pan fraction, were weighed to a precision of 0.01 gm. These data were used for calculating the weight percent in each sieve interval. Particle-size distributions by weight are shown in Figures 4, 5, and 6 for the granodiorite, schist, and sandstone, respectively. A general shift toward the finer particle fractions with increased confining pressure is evident in the granodiorite data in Figure 4. In Figure 5 there is a general tendency of schist samples loaded parallel to the foliation to produce coarser gouge than those oriented perpendicular. In Figure 6 no trends are apparent for the sandstone distributions. Comparison with the disaggregated Berea sample in Figure 7 shows that little if any comminution beyond disaggregation occurs at fracture. Microscopic examination of quartz grains in the finest gouge fractions failed to show any evidence of fresh breakage. The slight shift toward finer fractions (in Figure 6 relative to Figure 7) apparently is due to a breakup of polycrystalline quartz fragments and cement.

The logarithmic normal probability plots for these distributions are given in Figures 8, 9, and 10. In Figure 8 representative low, intermediate, and high pressure distributions are given. All other distributions are sub parallel to, and bounded by, these curves. Similarly, in Figure 9 all the sandstone distributions are subparallel to, and bounded by, the given distributions. Note that the smallest weight fractions of the granodiorite and schist gouges tend to follow a log-normal distribution, whereas the gouge distributions from the porous sandstone are not log-normal. Examination of such gouge indicates the presence of mixed log-normal distribution, perhaps representing an inherited log-normal component representing the framework grains of the Berea sandstone, admixed with a second log-normal component representing grains of broken calcite cement.

Table 4

*Least-Squares Linear Fits to Gouge Data**

Rock type	Table ref.	m	d_0	M	G_0
Sierran granodiorite (lab)	1	-0.045 ± 0.005	0.73 ± 0.03	-0.72 ± 0.20	10.8 ± 1.5
Pelona schist, ∥ (lab)	2	-0.047 ± 0.077	1.36 ± 0.47	-0.36 ± 0.25	3.6 ± 1.5
Pelona schist, ⊥ (lab)	2	-0.0040 ± 0.019	0.73 ± 0.15	-0.048 ± 0.053	1.7 ± 1.4
Berea sandstone (lab)	3	-0.002 ± 0.002	0.19 ± 0.005	-0.45 ± 0.23	7.7 ± 1.7
San Andreas gouge (natural)	4 and 5	-0.025 ± 0.013	0.32 ± 0.07		

* Here $\langle d \rangle = m\sigma_3 + d_0$ and $G = M\sigma_3 + G_0$, where d is in millimetres, σ_3 is in megapascals, and G is in weight percent.

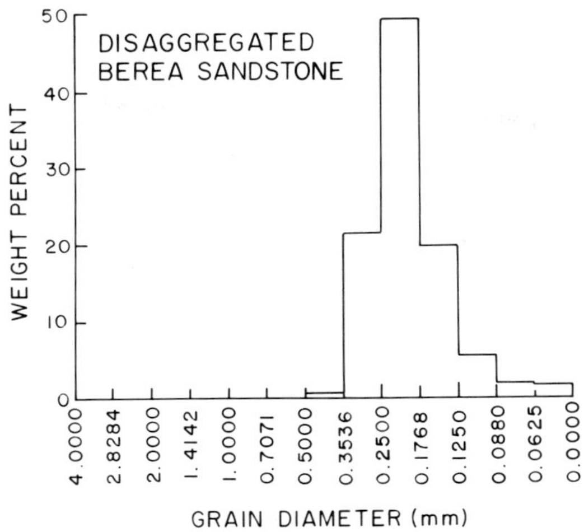

Figure 7
Histogram showing the grain-size distribution of a disaggregated Berea sandstone sample. Note the similarities to gouge specimens in Figure 6.

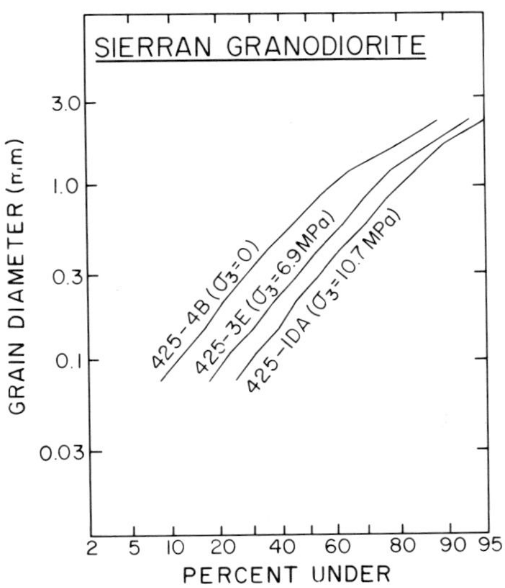

Figure 8
Cumulative probability curves for representative granodiorite samples. Distributions for all other specimens are subparallel to, and bounded by these curves. A straight line on this plot indicates log-normal distribution.

Two parameters were assigned to each distribution: (a) the total weight percent of the original laboratory sample, which was reduced to grain diameter $d \leqslant 4.0$ mm, G, and (b) the median grain diameter $\langle d \rangle$, which is given by the diameter at 50% on the abscissa in Figures 8, 9, and 10. These two parameters are given for each lab-

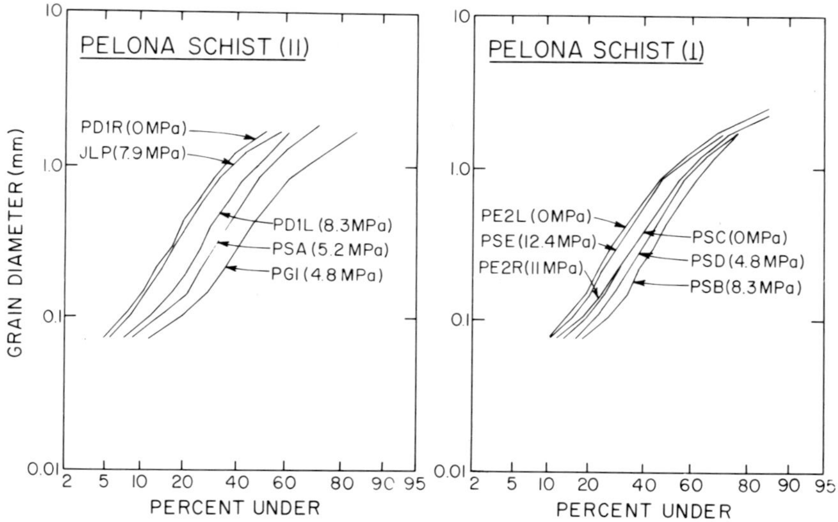

Figure 9
Cumulative probability curves for Pelona schist gouge specimens. A straight line indicates a log-normal distribution.

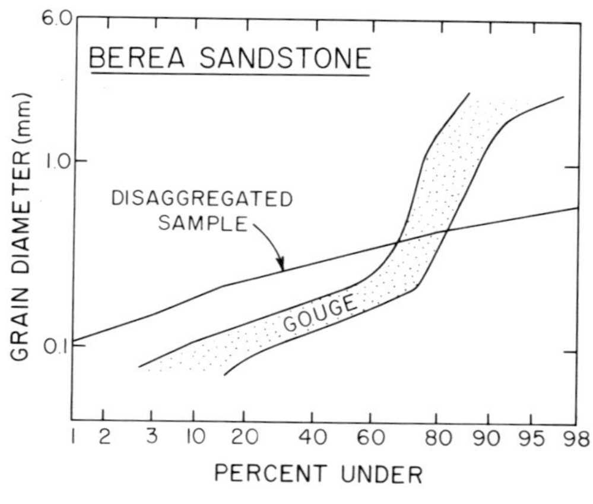

Figure 10
Cumulative probability curves for Berea sandstone gouge specimens and the disaggregated sample material. The two gouge curves shown bound the curves for the other seven specimens. The marked nonlinearity of the gouge curves indicates significant deviation from a log-normal distribution.

oratory sample in Tables 1, 2, and 3, and plotted as a function of the confining pressure σ_3 in Figures 11 and 12. Each set of data was fitted to a straight line by least-squares analysis. The resultant slopes and intercepts with their associated standard deviations are given in Table 4. Note that only the granodiorite gouge samples showed a statistically significant decrease in particle size with increasing

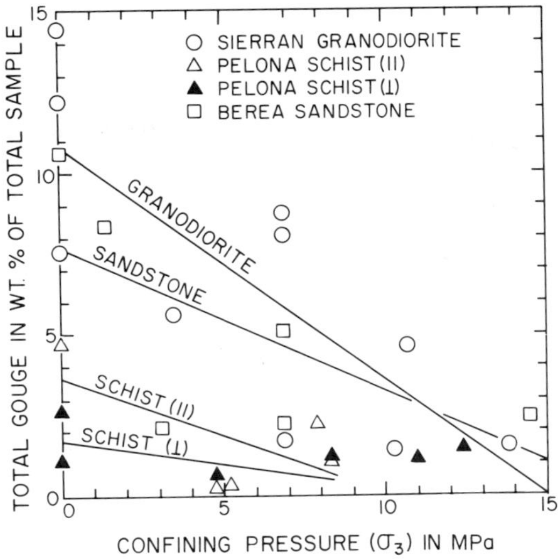

Figure 11
Total gouge produced, in weight percent of the sample, as a function of the confining pressure. Lines are linear least-squares fits to the data. The slopes and intercepts and their standard deviations are given in Table 6.

confining pressure. This linear fit for the gouge derived from granodiorite is given by the solid line in Figure 12. The gouge samples dervied from the granodiorite, the schist loaded parallel to foliation, and the sandstone, all showed a statistically significant decrease in total gouge produced with increasing confining pressure.

Naturally Occurring Fault Gouge

ANDERSON et al. (1980) have characterized gouge material from the San Andreas fault zone in the vicinity of Lake Hughes, California. The samples were collected from a section of drill core which intersected seven fault strands in the depth range of 213 to 373 m, as detailed in Table 5. Each of the seven associated gouge zones was petrologically and texturally characterized. No significant chemical alteration was observed; rather, the gouge consisted of a rock flour having mineralogy identical with the wall rock. Logarithmic normal plots of the particle distribution in the range $0 < d < 4$ mm for these seven zones are given in Figure 13, where they may be compared with the laboratory granodiorite gouge distributions. The median particle sizes are given in Table 5 and compared with the laboratory gouges in Figure 12. The confining pressure for the field samples was estimated from their depth by assuming a lithostatic pressure gradient of 25 MPa/km. Linear least-squares analysis (Table 4) suggests a statistically significant decrease in particle size with depth.

We have subsequently characterized gouge from a depth of 165 m in 'Robert's well,' which intersects the San Andreas fault zone (SAMMIS and OSBORNE, 1982). In

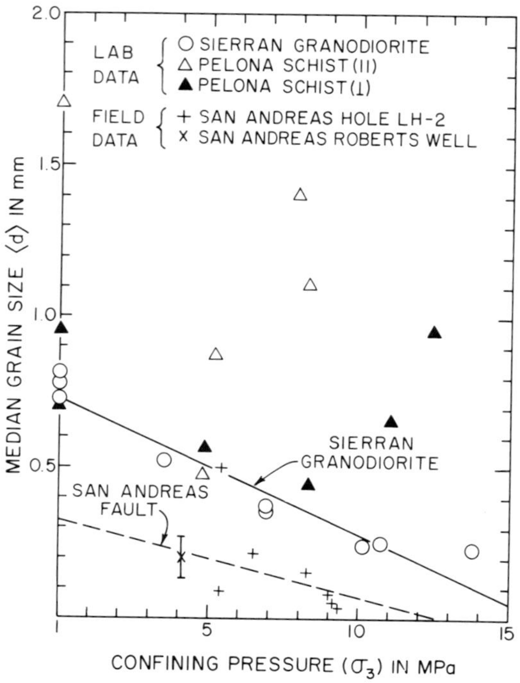

Figure 12
Median grain size as a function of the confining pressure for laboratory and field gouges. The solid line is the linear least-squares fit to the Sierran granodiorite data, while the dashed line is the fit to natural gouges from the San Andreas fault zone in southern California. Other laboratory gouges did not show a statistically significant decrease in particle size with confining pressure (see Table 6).

Table 5

*Summary of Textural Data for Granodiorite Gouge from Core LH-2, Site 4, from the San Andreas Fault Zone near Lake Hughes, California**

Sample	Depth (m)	σ_3 (MPa)	$\langle d \rangle$ (mm)
2-12T	213	5.3	0.09
2-13BT	215	5.4	0.50
2-27T	260	6.5	0.21
2-46T	331	8.3	0.16
2-52CT	361	9.0	0.08
2-53BT	368	9.2	0.050
2-54AT	373	9.3	0.042

* ANDERSON et al., 1980.

order to assess the homogeneity of natural gouge samples, one meter of core was divided into eleven identical samples which were sieved as described above; bounding

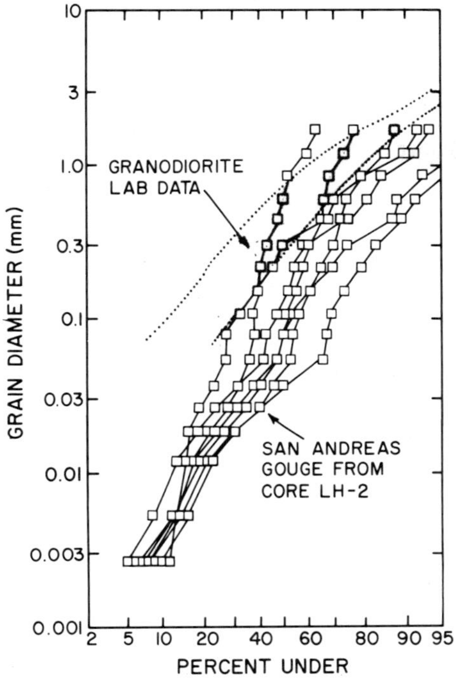

Figure 13
Cumulative probability curves for San Andreas gouge from hole LH-2. The envelope of distributions for the granodiorite gouge formed in the laboratory is also given for comparison.

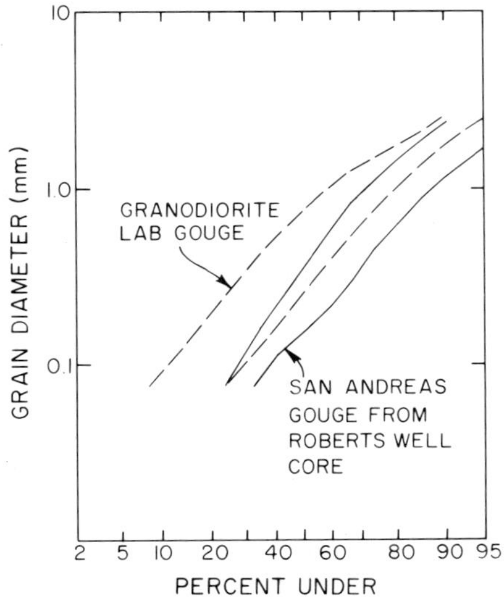

Figure 14
Cumulative probability curves for San Andreas gouge from Roberts Well. Ten of the eleven samples are bounded by the two solid curves. The envelope of granodiorite lab gouges is given by the broken curves.

log-normal distributions are given in Figure 14 and the median grain sizes in Table 6. The mean and standard deviation of these 11 samples are plotted in Figure 12. Petrographic modal analysis revealed no significant alteration of the granitic-gneiss source rock. A slight enrichment in mica relative to quartz and feldspar was noted in the finest-sized fractions of the gouge.

Table 6

Summary of Textural Data for Gouge Derived from Granitic Gneiss from Roberts Well in the San Andreas Fault Zone, San Francisquito Canyon, San Gabriel Mountains, California

Sample	Depth (m)	σ_3 (MPa)	$\langle d \rangle$ (mm)
RO-1	165	4.1	0.20
RO-2	165	4.1	0.08
RO-3	165	4.1	0.16
RO-4	165	4.1	0.33
RO-5	165	4.1	0.20
RO-6	165	4.1	0.16
RO-7	165	4.1	0.20
RO-8	165	4.1	0.17
RO-9	165	4.1	0.14
RO-10	165	4.1	0.30
RO-11	165	4.1	0.31

Mean $\pm$ S.D.: 0.20 $\pm$ 0.07

Discussion

In order to understand the observed differences in the particle-size distributions of gouge derived from granite, schist, and sandstone, as well as the mechanisms which result in a reduced particle size with increasing confining pressure (depth), we must first address the micro-mechanics of the brittle failure of rock under compressive loading.

Failure by progressive microfracture under compressive loading

Brittle solids fail in compression by progressive microfracture. Acoustic emission studies by SCHOLZ (1968) show that microfracturing begins at a stress of approximately one half the ultimate failure stress. Near failure the density of microfractures increases rapidly, and the sample volume increases, even though the stress state is a compressive one. Because of this dilatancy the volume at fracture exceeds the unstressed sample volume. Microscopic examination shows that most of the micro-

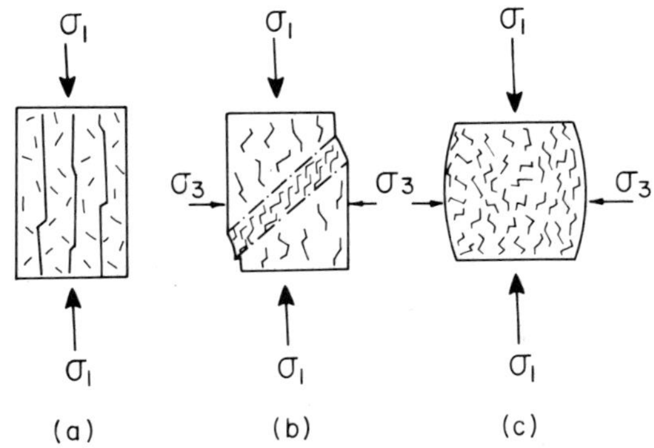

Figure 15
Change in the mode of macroscopic failure with increased confining pressure. After JAEGER and COOK, 1979.

fractures have propagated in a direction that is more or less parallel to the largest principal stress axis.

The mode of macroscopic failure depends upon the confining pressure (JAEGER and COOK, 1979). In a typical triaxial experiment, unconfined samples tend to fail by 'vertical slabbing'; the microcracks grow or combine to form continuous vertical failure planes. At low and intermediate confining pressures, failure occurs in a shear band. At high confining pressures a single failure surface does not form. Instead, the sample deforms in a pseudo-ductile mode with large-scale deformation being taken up on many short, homogeneously distributed shear bands. This change in the mode of macroscopic failure with increasing confining pressure (Fig. 15) and the observed decrease in the particle size of the resultant gouge can be understood, at least qualitatively, in terms of the effect of confining pressure on the propagation of axial microcracks.

Since the work of GRIFFITH (1924) it has been known that macroscopic failure of brittle solids is the result of the propagation of one or more pre-existing cracks. In tensile modes of loading a single flaw propagates unstably. Macroscopic failure occurs by the unstable propagation of the most dangerous (the longest and most favorably oriented) flaw; knowledge of the critical-stress intensity factor and information about the distribution of intrinsic flaws allows calculation of the macroscopic failure stress.

The mechanics of compressive brittle fracture are less well developed. Most attempts to model compressive fracture start from the stress field of an elliptical flaw and calculate (with or without allowance for friction across the crack face) the conditions under which a crack will first initiate at some point on its surface (INGLIS, 1913; MCCLINTOCK and WALSH, 1962; MURRELL, 1963, 1965). The propagation of axial microcracks is a stable process, and they propagate in such a way that they do not readily link (NEMAT-NASSER and HORII, 1982; ASHBY and HALLAM, 1986; SAMMIS

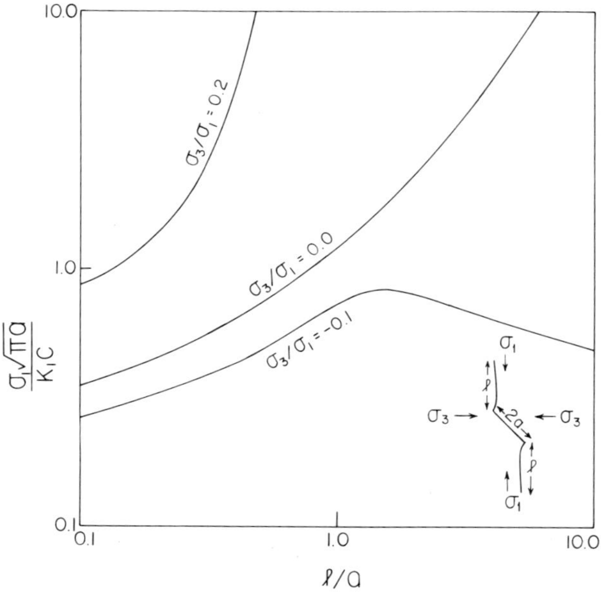

Figure 16

Normalized stress $\sigma_1 \sqrt{\pi a}/K_{1C}$ required to extend an axial crack to normalized length l/a. Note that the extension is sensitive to the confining pressure σ_3. After Nemat-Nasser and Horii, 1982; Ashby and Hallam, 1986.

and ASHBY, 1986). The cooperative process by which these microfractures produce macroscopic failure has remained one of the more elusive problems of modern fracture mechanics.

In the case of unconfined specimens, PENG and JOHNSON (1972) have shown that the use of a lubricant on the ends of the sample actually puts part of the specimen in slight radial tension. NEMAT-NASSER and HORII (1982) and ASHBY and HALLAM (1986) have investigated the axial extension of inclined cracks as a function of the triaxial stress state, while SAMMIS and ASHBY (1986) have analyzed the growth of cracks from spherical pores. The results of these studies are summarized in Figure 16. A slight radial tension results in an extremely long extension of the initial flaw. When such fractures grow or combine to intersect the ends of the sample, failure by vertical slabbing occurs (Fig. 15a).

Confining pressure is extremely effective in suppressing the axial extension (Fig. 16). Even under a mild confining pressure the fractures do not grow long enough to intersect the ends of the sample. Rather, they interact to form a shear band. It shall be assumed for this discussion that the axial fractures combine to form weak clusters which nucleate the ultimate shear failure. Such a hypothetical cluster is shown in Figure 17. It is assumed that the cluster is critical when a certain fracture density has been exceeded in a region of some critical length L. The critical cluster now forms a 'mega-defect' which leads to the macroscopic failure.

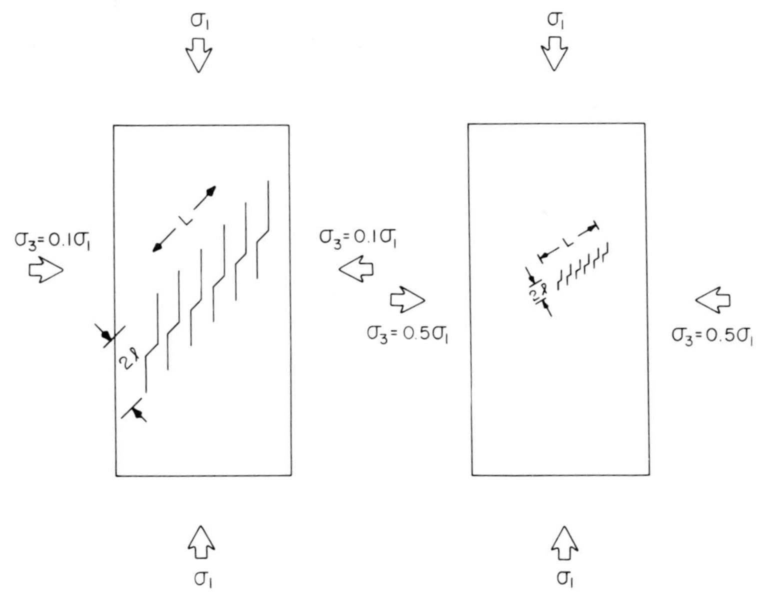

Figure 17
Schematic showing how the supression of the axial extension of microfractures, l, by an increased confining pressure, σ_3, results in a change in the scale of hypothetical critical cluster size, L.

Although the statistical and mechanical modelling of such 'mega-defects' is beyond the scope of this paper, the obvious way such a cluster would scale down with an increase in confining pressure gives some insight into both the change in failure mode and the reduction in gouge particle size. As the axial extension of individual microfractures is suppressed by increased confining pressure, the length L and critical spacing also are scaled down. Hence, at higher confining pressures the probability of there being more than one critical cluster in sample of a given size increases (as L decreases relative to the sample size). At very high confining pressures there will be many shear bands, all active, resulting in the observed pseudo-ductile deformation.

The reduction in gouge particle size (as well as the reduction in total gouge) can be understood in terms of the decrease in the spacing of the axial microcracks as the critical cluster is scaled down with increasing confining pressure. In this view it is the columns between the vertical microfractures which 'set the scale' of the resultant fault gouge. This model is consistent with PENG and JOHNSON's observations (1972) that Barre granite fails by the buckling of columns between axial fractures (Fig. 18). SAMMIS and ASHBY (1986) and ASHBY and HALLAM (1986) have shown that the bending of such columns between growing microfractures dominates the crack-driving force in dense arrays. They suggest that the principal interaction between growing microcracks is through such mutual growth enhancement, rather than through any type of crack linkage. Mutual growth enhancement also leads to clus-

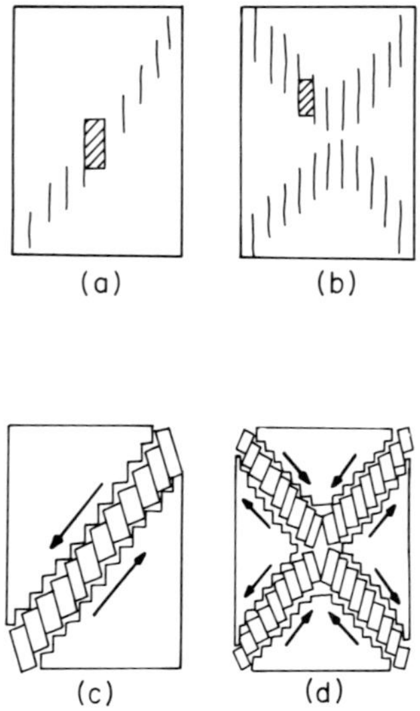

Figure 18
Idealized concept of the development of shear zones by the buckling of rock columns between cracks. After PENG and JOHNSON, 1972.

tering, as observed by SONDERGELD and ESTEY (1981) in acoustic emission studies and hypothesized above. In this context, the tendency of Pelona schist samples loaded parallel to the foliation to produce more gouge with a larger median particle size (see Fig. 11 and 12) can be interpreted as resulting from the propagation of longer cracks in this favorable orientation.

Macroscopic failure results from the cooperative buckling and collapse of columns in one or more shear zones which traverse the sample. The mechanism by which the buckled columns are further comminuted by displacement in the shear zone to produce the observed gouge particle-size distribution is discussed below.

Self-similar cataclasis and the particle-size distribution

One of the most interesting textural characteristics of natural gouge formed in crystalline rock is the tendency of the finer fractions to follow a logarithmic normal distribution (ANDERSON et al., 1980; SAMMIS and OSBORNE, 1982; OLGAARD and BRACE, 1983), whereas natural gouge formed from sandstone does not follow a logarithmic normal distribution (ENGELDER, 1974). Similarly, the finer gouge fractions formed by the fracture of granodiorite and schist in the laboratory (discussed above) tend to follow a logarithmic distribution, whereas those produced by the fresh fracture of sandstone clearly do not.

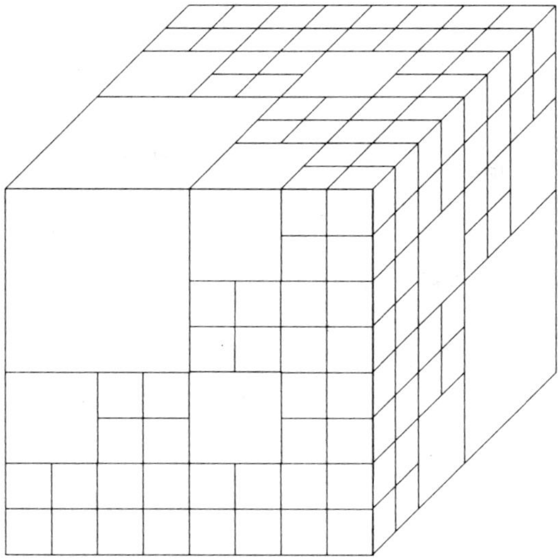

Figure 19
Self-similar gouge of cubic particles. Note that each time the particle size is halved, there are six times as many particles. Hence, the particle distribution follows a power law. The fractal dimension for this array is $D = (\log 6)/(\log 2) = 2.58$.

The production of a logarithmic normal particle-size distribution appears to be a direct consequence of the accommodation of finite strain in the shear zone by self-similar (scale-independent) fracturing. The failure of sandstone to produce a loga-

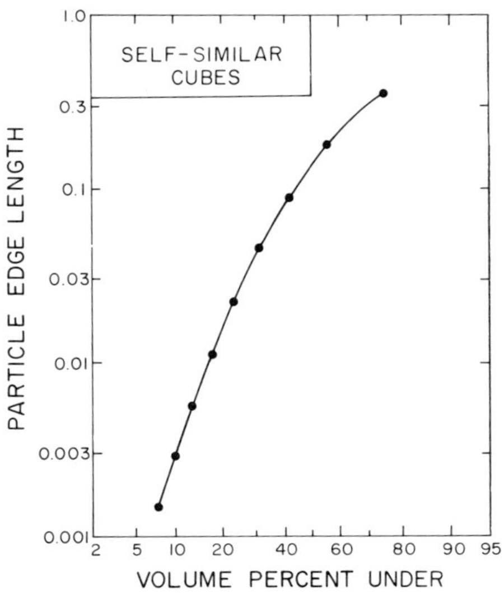

Figure 20
The power-law volume distribution for the array in Figure 19 plotted on log-normal axes. Note that the smaller fractions are approximately linear and could be mistaken for log-normal ones.

rithmic normal gouge may be interpreted as a breakdown of self-similar microfracturing at the scale of the pores.

KING (1983) applied the 'fractal geometry' of MANDLEBROT (1982) to the problem of regional strain and seismicity. He has shown the following.

1. A three-dimensional system of self-similar faults is capable of accommodating finite strain.
2. The self-similar system accommodates the strain with a minimum surface area.
3. Simple assumptions regarding the displacement on, and activation of, faults within this system lead directly to the logarithmic frequency–magnitude relation observed for earthquakes and even predicts the observed *b* value of 1.

The observation that acoustic emissions in laboratory specimens also follow a logarithmic frequency–magnitude relation (SCHOLZ, 1968) suggests that failure on the laboratory scale also might involve strain accommodation on a self-similar system of fractures.

The connection between scale-independent comminution and a logarithmic normal particle-size distribution was made by Epstein (1947), who proved that any initial distribution of particle sizes will be comminuted to a logarithmic normal distribution if the following obtains.

1. The probability of the breakage of any piece during any step of the process is a constant, independent of the size of the piece, the presence of other pieces, or the number of prior breakage events.
2. The distribution of pieces obtained from the application of a single breakage event to a given piece is independent of the dimension of the piece broken.

It is important to note that while this *process* is scale-independent, the resultant particle distribution is *not* self-similar; it is log-normal. A self-similar particle distribution follows a power law, as can be seen from the self-similar gouge of cubic particles in Figure 19. For this array of cubes, each decrease in particle size by a factor of 1/2 corresponds to an increase in the number of particles by a factor of 6 (giving a fractal dimension of $D = (\log 6)/(\log 2) = 2.58$).

When the cumulative volume fraction for this self-similar distribution is plotted on log-normal probability paper, as in Figure 20, it is evident that the finer fractions might be interpreted as log-normal (i.e., a straight line). In fact, Figure 20 looks disturbingly similar to many of the data curves presented above. We conclude that our data do not cover a sufficiently wide range of particle sizes to distinguish between a log-normal and a self-similar distribution.

In a series of classic papers Gilvarry and his co-workers (GILVARRY, 1961; GILVARRY, 1962a; GILVARRY, 1962b; GILVARRY and BERGSTROM, 1961; GILVARRY and BERGSTROM, 1962) demonstrated that the repeated crushing of particles does not produce a logarithmic normal distribution but, rather, a Rosin-Rammler distribution. They showed that this distribution is determined by the Poissonian distribution of starter flaws in the original material. In this case the fracture probability is not independent of the particle size. It is important to note, however, that there is a fundamental difference between crushing a collection of particles and accommodating strain in a fully dense solid. In a loose aggregate of particles, strain can be accommodated by motion of the particles. In a fully dense solid, strain can be accommodated only by the extension of the fracture 'fractal' pattern to smaller scales.

How far down in scale the self-similar fracturing extends depends upon the total strain to be accommodated and the 'inner fractal limit.' For small strains, fractures only need extend down to the scale at which elastic strains may cope. The 'lower fractal limit' is set by the scale at which the solid is no longer fully dense but can accommodate strain by the displacement of particulate material into voids. In the Berea sandstone samples the pore size is on the order of 10 μm (HANDIN et al., 1963). At this size scale the logarithmic normal distribution is expected to break down, and this may explain why the particle-size distributions of gouge produced by the fresh fracture of the Berea sandstone are not logarithmic normal at the scale being investigated. Since single-crystal quartz particles make up less than 60% of the Berea samples by volume (when porosity is included in the total volume), they are below 'critical packing' (see BRADY and BOSSIS, 1985) and shear strain can be accommodated by the calcite cement without fracturing the quartz, as observed above. From this point of view, gouge experiments involving saw-cut specimens may be misleading, since the gouge, artificially introduced into the saw cut, has a porosity on the scale of the particle size.

The 'upper fractal limit' is set by the spacing of axial fractures in the critical cluster. This spacing should decrease with increasing confining pressure. It is the decrease in this initial fracture spacing which may produce the observed decrease in particle size with increasing confining pressure. In this view, increased confining pressure scales down the entire comminution process within the localized shear zone to shorter scale lengths.

Comparison of laboratory and field gouge samples

Both the laboratory and field gouges show finer particle size at higher confining pressure, yet the gouges from the San Andreas fault zone clearly are finer than those produced in the laboratory at the same confining pressure (Fig. 12). Two possible explanations are that (a) the San Andreas gouges were formed at greater depth and have subsequently been tectonically transported to their present depth (which was used to calculate σ_3), or that (b) the San Andreas gouges have been comminuted to a finer particle size by larger strains, or both. To assess the first possibility, we can ask what depth of faulting would be required to make the laboratory and field gouges colinear in Figure 12. If the San Andreas gouges were formed at a 7 MPa greater pressure (280 m), then the field data would be colinear with the lab data. The second possibility is difficult to assess. Indeed, a quantitative relationship between strain and the progress of self-similar comminution is the missing link in a scheme by which stress and displacement might be read from the texture of fault gouge.

In sum:

1. Gouges formed in crystalline rock may be described by either log-normal or self-similar particle distributions, especially for the fractions of size less than about 1 mm.
2. Particle size tends to decrease with increased confining pressure.
3. Gouges derived from porous sandstone do not form either log-normal or self-similar particle distributions, but reflect mixed log-normal components probably inherited from the grain-size distribution of framework gains and an admixture of broken calcite cement.

These observations may be explained by the suppression of axial microcracks by confining pressure and the accommodation of strain in macroscopic shear zones by scale-independent cataclasis.

Acknowledgments

We would like to thank Geoffrey King for many stimulating discussions. This work was supported by National Science Foundation Grant EAR-82-12753.

References

ANDERSON, J. L., OSBORNE, R. H. and PALMER, D. F. (1980), *Petrogenesis of cataclastic rocks within the San Andreas fault zone of southern California.* Tectonophys. *67*, 221–249.

ANDERSON, J. L., OSBORNE, R. H. and PALMER, D. F. (1983), *Cataclastic rock of the San Gabriel fault—An expression of deformation at deeper crustal levels in the San Andreas fault zone.* Tectonophys. *98*, 209–251.

ASHBY, M. F. and HALLAM, S. D. (1986), *The failure of brittle solids containing small cracks under compressive stress states.* Acta Metall. *34*, 497–510.

BRADY, J. F. and BOSSIS, G. (1985), *The rheology of concentrated suspensions of spheres in simple shear flow by numerical simulation.* J. Fluid Mech. *155*, 105–129.

BYERLEE, J. D. (1978), *Friction of Rocks.* Pure Appl. Geophys. *116*, 615–626.

CHU, C. L., WANG, C. Y. and LIN, W. (1981), *Permeability and frictional properties of San Andreas fault gouges.* Geophys. Res. Lett. *8*, 565–568.

ENGELDER, J. T. (1974), *Cataclasis and the generation of fault gouge.* G.S.A. Bull. *85*, 1515–1522.

ENGELDER, J., LOGAN, J. M. and HANDIN, J. (1975), *The sliding characteristics of sandstone on quartz fault-gouge.* Pure Appl. Geophys. *113*, 69–86.

EPSTEIN, B. (1947), *The mathematical description of certain breakage mechanisms leading to the logarithmic-normal distribution.* J. Franklin Inst. *244*, 471–477.

GILVARRY, J. J. (1961), *Fracture of brittle solids. I. Distribution function for fragment size in single fracture (theoretical)*. J. Appl. Phys. *32*, 391–399.

GILVARRY, J. J. (1962a), *Fracture of brittle solids. IV. Two-dimensional distribution function for fragment size in single fracture (theoretical)*. J. Appl. Phys. *33*, 3214–3217.

GILVARRY, J. J. (1962b), *Fracture of brittle solids. V. Two-dimensional distribution function for fragment size in single fracture (experimental)*. J. Appl. Phys. *33*, 3218–3224.

GILVARRY, J. J. and BERGSTROM, B. H. (1961), *Fracture of brittle solids. II. Distribution function for fragment size in single fracture (experimental)*. J. Appl. Phys. *32*, 400–410.

GILVARRY, J. J. and BERGSTROM, B. H. (1962), *Fracture of brittle solids. III. Experimental results on the distribution of fragment size in single fracture*. J. Appl. Phys. *33*, 3211–3213.

GRIFFITH, A. A. (1924), *Theory of rupture*. Proc. First Internat. Congr. Appl. Mechanics, Delft, 55–63.

HANDIN, J., HAGER, R. V., Jr., FRIEDMAN, M. and FEATHER, J. N. (1963), *Experimental deformation of sedimentary rocks under confining pressure*. Bull. A.A.P.G. *47*, 717–754.

INGLIS, C. E. (1913), *Stresses in a plate due to the presence of cracks and sharp corners*. Tran. Instn. Nav. Archit., London *55*, 219–341.

JAEGER, J. C. and COOK, R. G. W., *Fundamentals of Rock Mechanics*. Halstead Press, New York, 1979.

KING, G. (1983), *The accommodation of large strains in the upper lithosphere of the earth and other solids by self-similar fault systems: The geometrical origin of b-value*. Pure Appl. Geoph. *121*, 761–815.

MANDELBROT, B. B., *The Fractal Geometry of Nature*. W. H. Freeman Co., San Francisco, 1982.

MCCLINTOCK, F. A. and WALSH, J. B. (1962), *Frictional of Griffith cracks under pressure*. Fourth U.S. Nat. Congr. Appl. Mech., Proc. 1015–1021.

MORROW, C. A., SHI, L. Q. and BYERLEE, J. D. (1981), *Permeability and strength of San Andreas fault gouge under high pressure*. Geoph. Res. Lett. *8*, 325–328.

MORROW, C. A., SHI, L. Q. and BYERLEE, J. D. (1982), *Strain hardening and strength of clay-rich fault gouges*. J. Geophys. Res. *87*, 6771–6780.

MORROW, C. A., SHI, L. Q. and BYERLEE, J. D. (1984), *Permeability of fault gouge under confining pressure and shear stress*. J. Geophys. Res. *89*, 3192–3200.

MURRELL, S. A. F. (1963), *A criterion for brittle fracture of rocks and concrete under triaxial stress and the effect of pore pressure on the criterion*. Fifth Rock Mech. Symp., Univ. Minnesota, in *Rock Mechanics*, C. Fairhurst (ed.), Pergamon, Oxford, p. 563–577.

MURRELL, S. A. F. (1965), *The effect of triaxial stress systems on the strength of rocks at atmospheric temperatures*. J. Geophys. *10*, 231–281.

NEMAT-NASSER, S. and HORII, H. (1982), *Compression-induced nonplanar crack extension with application to splitting, exfoliation and rockburst*. J. Geophys. Res. *87*, 6805–6821.

OLGAARD, D. L. and BRACE, W. F. (1983), *The microstructure of gouge from a mining-induced seismic shear zone*. Intl. J. Rock Mech. Min. Sci. Geomech. Abstr. *20*, 11–19.

PENG, S. and JOHNSON, A. M. (1972), *Crack growth and faulting in cylindrical specimens of Chelmsford granite*. Intl. J. Rock Mech. Min. Sci. *9*, 37–86.

PHILLIPS, J. C. (1982), 'Character and origin of cataclasite developed along the low-angle Whipple detachment fault, Whipple Mountains, California,' in *Mesozoic-Cenozoic Tectonic Evolution of the Colorado River Region, California, Arizona, and Nevada*. F. E. Frost and D. L. Marting (eds.), Cordilleran Pubs., San Diego, 1982, pp. 109–116.

ROBERTSON, E. C. (1982), 'Continuous formation of gouge and breccia during fault displacement,' in *Issues in Rock Mechanics*. R. E. Goodman and F. E. Henze (eds.), A.I.M.M.P.E., New York, 1982.

SAMMIS, C. G. and ASHBY, M. F. (1986), *The failure of brittle porous solids under compressive stress states*. Acta Metall. *34*, 511–526.

SAMMIS, C. G. and OSBORNE, R. H. (1982), *Textural and petrographic modal analysis of gouge from a depth of 165 m in the San Andreas fault zone* (abstr.). EOS *63*, 1109.

SCHOLZ, C. H. (1968), *Microfracturing and the inelastic deformation of rock in compression*. J. Geophys. Res. *73*, 1417–1432.

SONDERGELD, C. H. and ESTEY, L. H. (1981), *Acoustic emission study of microfracturing during the cyclic loading of Westerly granite*. J. Geophys. Res. *86*, 2915–2924.

SUMMERS, R. and BYERLEE, J. D. (1977), *A note on the effect of fault gouge composition on the stability of frictional sliding*. Intl. J. Rock Mech. Min. Sci. Geomech. Abstr. *14*, 155–160.

Wang, C., Mao, N. and Wu, F. T. (1980), *Mechanical properties of clays at high pressure.* J. Geophys. Res. *85*, 1462–1468.

Wu, F., Blatter, L. and Roberson, H. (1975), *Clay gouges in the San Andreas fault system and their possible implications.* Pure Appl. Geophys. *113*, 87–95.

(Received 25th September 1985, revised 24th March 1986, accepted 28th March, 1986)

PAGEOPH, Vol. 124, Nos. 1/2 (1986)

0033-4553/86/020079-28$1.50 + 0.20/0

Implications for Mechanical Properties of Brittle Faults from Observations of the Punchbowl Fault Zone, California

F. M. CHESTER[1] and J. M. LOGAN[1]

Abstract—Field observations of the Punchbowl fault zone, an inactive trace of the San Andreas, are integrated with results from experimental deformation of naturally deformed Punchbowl fault rocks for a qualitative description of the mechanical properties of the fault and additional information for conceptual models of crustal faulting. The Punchbowl fault zone consists of a single, continuous gouge layer bounded by zones of extensively damaged host rock. Fault displacements were not only localized to the gouge layer, but also to discrete shear surfaces within the gouge. Deformation in the exposure studied probably occurred at depths of 2 to 4 km and was dominated by cataclastic mechanisms. Textural data also suggest that significant amounts of pore fluids were present during faulting, and that fluid-assisted mechanisms, such as dissolution, diffusion, and precipitation, were operative.

The experimental data on specimens collected from the fault zone suggest that there is a gradual decrease in strength and elastic modulus and an increase in relative ductility and permeability toward the main gouge zone. The gouge layer has fairly uniform mechanical properites, and it has significantly lower strength, elastic modulus, and permeability than both the damaged and the undeformed host rock.

For the Punchbowl fault and possibly other brittle faults, the variations in loading of the gouge zone with time are primarily governed by the morphology of the fault and the mechanical properties of the damaged host rock. In addition, the damaged zone acts as the permeable unit of the fault zone and surrounding rock. It appears that the gouge primarily governs whether displacements are localized, and it therefore may have a significant influence on the mode of slip.

Key words: Deformation, faults, cataclasis, gouge, rock mechanics.

Introduction

Seismogenic crustal faulting occurs in a complex deformational regime where brittle (cataclastic) and intracrystalline plastic mechanisms are operative to varying degrees, both temporally and spatially (e.g., SIBSON, 1979, 1984; WHITE and WHITE, 1983; MAWER and WILLIAMS, 1985). One approach to gaining insight into such deformation is first to understand the brittle and plastic end members. This paper reports on a combined field and laboratory study of the structure and mechanical properties of a fault representative of the brittle end member: the Punchbowl fault

[1] Departments of Geology and Geophysics and Center for Tectonophysics, Texas A and M University, College Station, Texas 77843.

zone in the vicinity of the Devil's Punchbowl, Los Angeles County Park, southern California (Fig. 1). The Punchbowl fault is an inactive, exhumed trace of the San Andreas system. It is studied along 5 km of its length as an analogue of active faults in the southern California area.

The field phase of this study involved characterization of the structure of the fault zone and textures of the fault rocks, to determine the distribution of strain, the deformation mechanisms operative, and the environmental conditions during faulting. These data indicate that the latest stage of faulting took place under brittle conditions in the presence of chemically active pore fluids. The laboratory work involved the experimental deformation of several representative natural fault rocks under brittle conditions. The experiments were designed to investigate the variation in mechanical properties (such as ultimate strength, elastic modulus, relative ductility, and frictional behavior) both along and across the fault zone under wet and dry conditions. The experimental data are integrated with the field observations to further develop conceptual mechanical models of brittle fault zones.

The Punchbowl fault is a segment of the Punchbowl–Nadeau fault system at the juncture of the San Gabriel Mountains and the Mojave Desert, only 4 km from the historically active trace of the San Andreas fault (Fig. 1). The Punchbowl and San Andreas faults have approximately parallel trends and join to the northwest and southeast (DIBBLEE, 1968). The Punchbowl fault was an active trace, possibly the main trace, of the San Andreas system during much of the Quaternary (SHARP and SILVER, 1971; WOODBURNE, 1975). Faulting at that time accounts for approximately 20 km of right-lateral displacement (WOODBURNE, 1975).

The specific portion of the fault investigated is that within and adjacent to the Devil's Punchbowl, Los Angeles County Park. At this locality it is the northeastern trace of the Punchbowl–Nadeau system and juxtaposes arkosic Miocene–Pliocene sedimentary strata of the Punchbowl Formation and a belt of highly deformed crystalline basement (Fig. 2). The deformed belt ranges to approximately 0.5 km in width and marks the extent of the Punchbowl–Nadeau system. Other fault traces of the system occur within the deformed belt but are not as clearly defined or as continuous as the Punchbowl fault (e.g., NOBLE, 1954; COX et al., 1983).

Description of the Fault Zone

The Punchbowl fault is extremely well exposed within the study area. It consists of a single, continuous gouge layer bounded by zones of extensively damaged host rock (Fig. 3). The fault zone and surrounding rock are divided into three units defined by relative deformation intensity: the main gouge zone, the damaged host rock, and the undeformed host-rock. At the macroscopic scale the main gouge zone

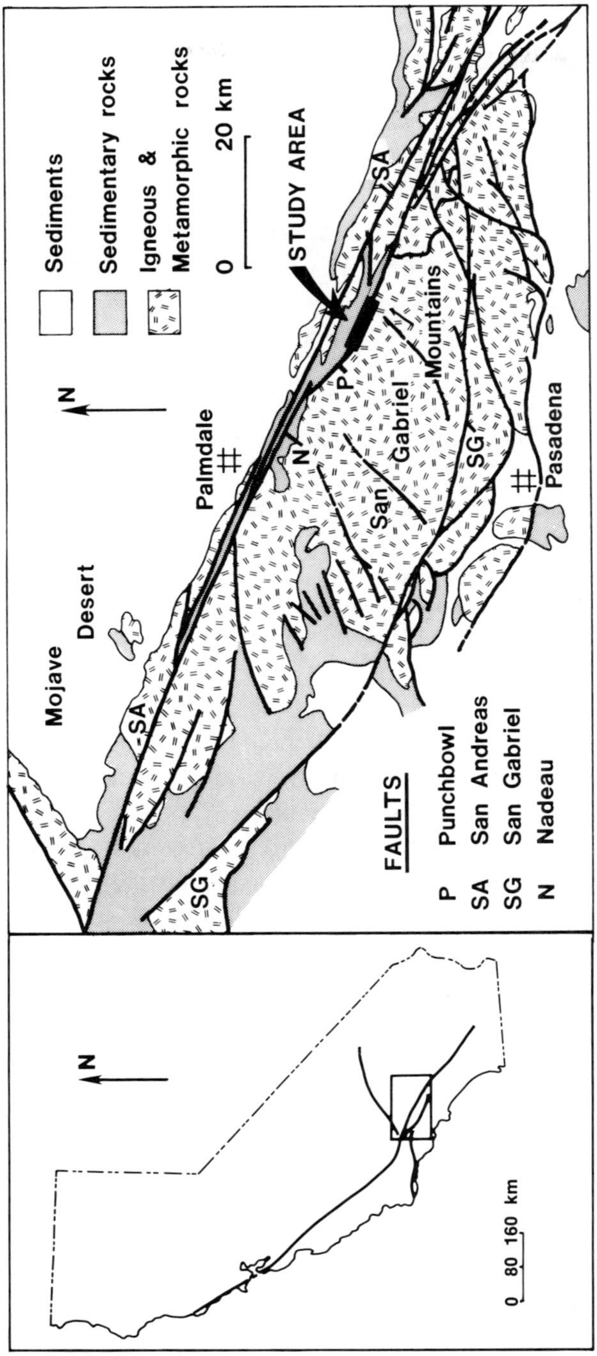

Figure 1

Map showing trace of Punchbowl–Nadeau fault system and location of study area on northeast side of San Gabriel Mountains, California.

Figure 2
Canyon exposure of Punchbowl fault zone within Devil's Punchbowl, Los Angeles County Park. Fault juxtaposes basement rocks and Punchbowl Formation along continuous gouge layer. Location and orientation of gouge layer shown by arrow. Canyon wall approximately 75 m high.

represents the Punchbowl fault surface. Although this surface is continuous, it changes 13° in strike and 28° in dip within the study area. Small faults within the damaged zone are referred to as subsidiary faults.

The structure of the damaged zone was investigated by quantifying subsidiary fault intensity along three traverses across the zone (CHESTER, 1983). The structures of the main gouge zone and gouge–host-rock contacts were studied by excavating

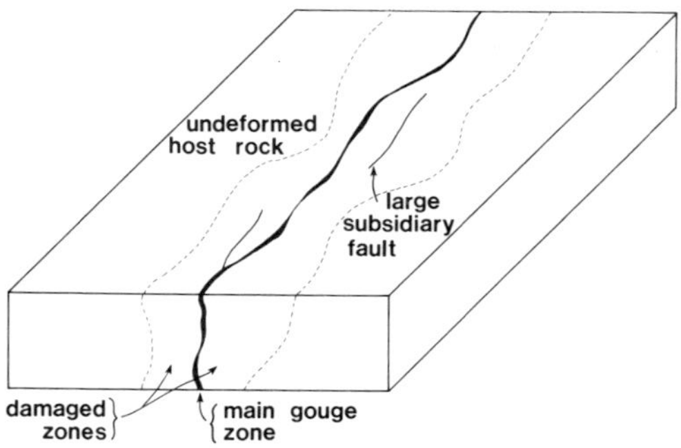

Figure 3
Block diagram of structural units of Punchbowl fault zone.

the gouge zone at two localities (CHESTER and LOGAN, 1986). Representative samples of the fault rocks were collected on the traverses and at the excavations for compositional analysis, microscopic study, and experimental deformation. At the present time only the main gouge zone and damaged zone of the Punchbowl Formation have been studied in detail.

Undeformed host rock

The Punchbowl Formation consists of an interbedded sequence of conglomerate, sandstone, and siltstone; the type section is located in the study area (NOBLE, 1954). Bedding thickness is highly variable; the conglomeratic sandstone portions are generally more thickly bedded (1 to 5 m) than the sandstone and siltstone sections (1 cm to 1 m). Sorting within the formation is poor, and grain shapes are subangular to subrounded. Fine laminations and sedimentary structures are common in the sandstone beds, and the rocks are moderately indurated. The maximum depth of faulting recorded in the exposure studied is estimated to be no more than 4 km, judged by the thickness of the sedimentary section above the Punchbowl Formation (CHESTER, 1983). As such, temperatures and pressures in the fault zone during faulting could have been up to approximately 125 °C and 45 MPa.

The belt of deformed basement consists of a heterogeneous group of Precambrian, generally biotite-rich, gneissic rocks, and Cretaceous quartz diorite, tonalite, granodiorite, and biotite monzogranite plutonic rocks (NOBLE, 1954; COX et al., 1983). The orientation of foliations and gneissic banding is highly variable throughout the belt. Textures indicate that the belt has undergone a complex deformational history with episodes of microscopic ductile deformation, semibrittle deformation with retrograde metamorphism, and brittle deformation. Pervasive fracturing at the grain scale is the feature distinguishing the belt from other basement rocks in the area. Probably much of this deformation occurred prior to deposition of the Punchbowl Formation (NOBLE, 1954) and therefore prior to formation of the fault rocks investigated herein.

Damaged host rock

Mesoscopic deformational features in the damaged zone of the Punchbowl Formation include subsidiary faults, disrupted bedding, and cataclastically reduced conglomerate clast and grain size. These features increase in intensity towards the main gouge zone and define a zone of intense deformation that is laterally continuous but variable in width, averaging 15 m (Fig. 4). The contact with the undeformed host rock is irregular and gradational. Subsidiary faults range from microscopic to some greater than 15 m in length and have gouge layers as thick as approximately

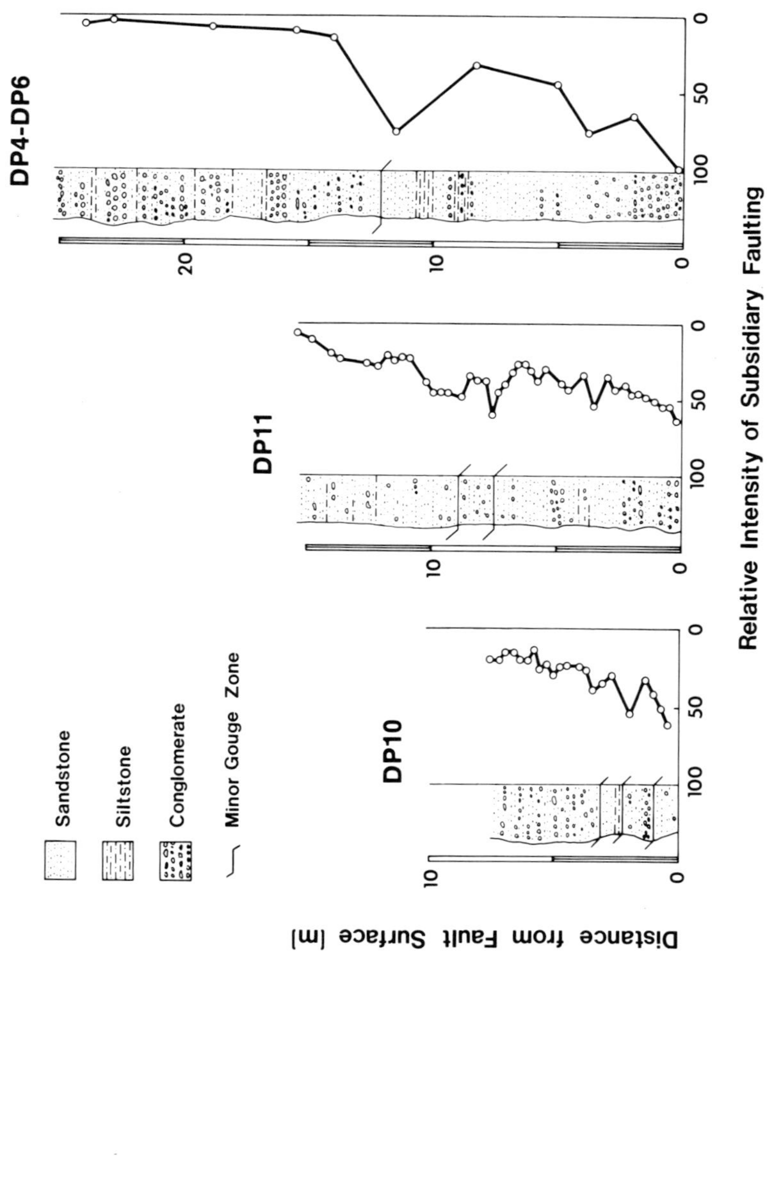

Figure 4

Fault and fracture intensity measurements on three transects across damaged zone of Punchbowl Formation. Distance from main gouge zone, measured normal to plane of fault, is shown on the vertical axes; relative intensity of faults and fractures shown on horizontal axes. Measurements indicated by open circles on plots. Values determined by summing number of fault and fracture intercepts with two mutually perpendicular line segments 30 cm in length. Key to lithologic columns is shown.

10 mm. The most common are 1 to 3 m long, with gouge layers approximately 1 mm thick. Qualitatively, the thickness of the gouge layer increases with fault length and displacement. Mesoscopic fractures without shear displacement that are greater than 1 m in length are not common. The subsidiary faults form a diffuse, but generally conjugate, set. The fault fabric is compatible with right-lateral displacement on the Punchbowl fault and suggests plane strain within the exposures studied (CHESTER and LOGAN, 1986).

Local variations in subsidiary fault intensity are superimposed on the overall progressive increase in deformation intensity towards the main gouge zone. These are manifested as local peaks in intensity occurring near relatively large (greater than 5 mm thick) subsidiary faults (Fig. 4). Minor variations also appear to be related to changes in lithology and layering. The conglomeratic sandstones tend to have a relatively low subsidiary fault intensity and show a more rapid decrease in intensity away from the main gouge zone. The interbedded siltstone, sandstone, and conglomerate units are more intensely faulted and cataclasized, and are highly deformed at greater distances from the main gouge zone.

Like the mesoscopic features, the intensity of microfractures also increases towards the main gouge zone (Fig. 5 a, b, c). In small cataclastic zones directly adjacent to the main gouge zone, however, the grain size is significantly reduced through cataclasis and the number of microfractures per grain is decreased (Fig. 5d). The greatest cataclastic reduction of grain size in the damaged host rock is in the subsidiary fault gouge (Fig. 5e). By comparison, though, the subsidiary fault gouge is significantly coarser-grained than that in the main gouge zone (Fig. 5, e, f).

Open, healed, and sealed microfractures are present in the Punchbowl Formation. Open fractures within single grains are the most common. They often occur in subparallel sets and may or may not transect the entire grain. Some are lined with iron oxide or other minerals. Healed fractures in quartz appear as irregular zones of minute bubbles and other fine inclusions and generally do not transect entire grains.

Sealed microfractures may transect several grains. Quartz-sealed fractures and cement occur throughout the damaged zone and are most evident in and near the subsidiary faults (Fig. 6a). Calcite-sealed fractures and subsidiary faults and calcite cement occur only in the Punchbowl Formation directly adjacent to the main gouge zone. Most of the calcite is extensively twinned. Where quartz and calcite are present, they are mutually crosscutting. Dilation of brecciated grains is suggested by extension, but no rotation, of fragments along sealed fractures (Fig. 6b). Dilational features in the host rock are more prevalent directly adjacent to the main gouge zone.

Textures in the subsidiary faults record at least several episodes of deformation followed by cementation. Many of the subsidiary faults also show concentrations of opaque material (Fig. 6c) and interlocking (sutured) porphyroclasts (Fig. 6d). These features suggest that pressure solution was operative in conjunction with cataclasis.

Representative samples of the Punchbowl Formation sandstones from the

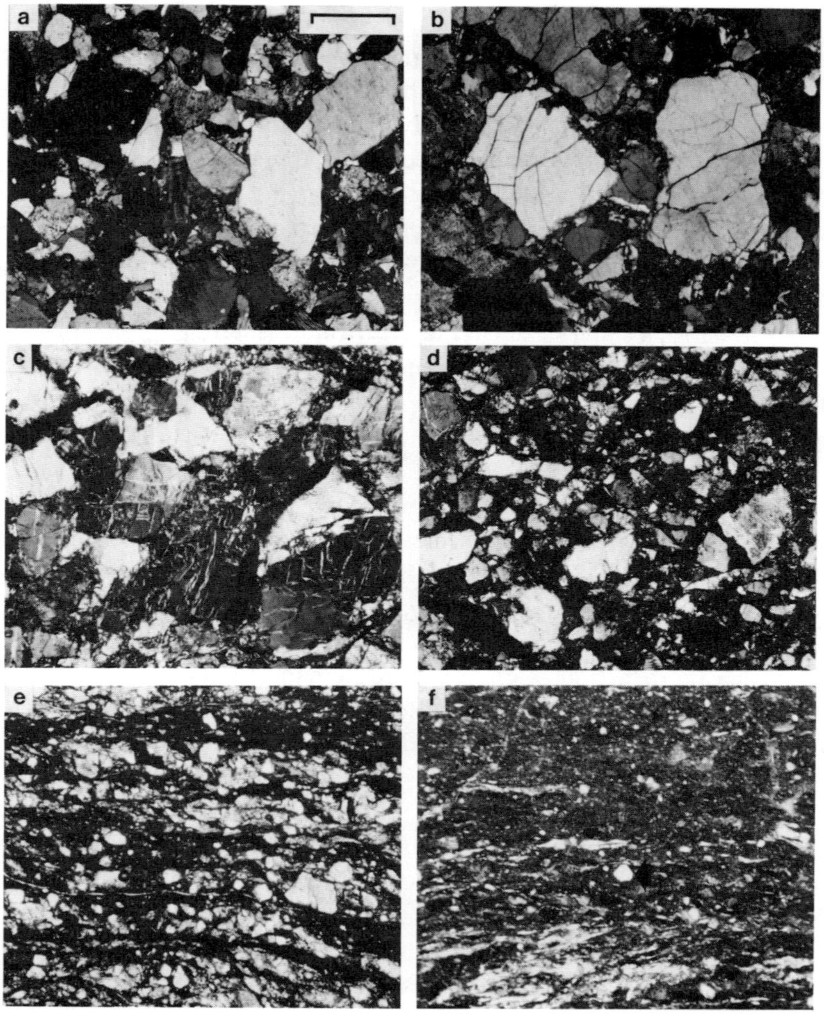

Figure 5

Progressive change in texture across fault zone from undeformed Punchbowl Formation to main gouge zone. Scale bar applies to all photos and is 0.4 mm; views are under crossed polarizers in (a), (b), (c), and (d) and under plane-polarized light in (e) and (f). (a) Undeformed Punchbowl Formation (DP13A) collected approximately 100 m from main gouge zone. (b) Microfractured sandstone 2.7 m from main gouge zone. (c) Extensively fractured sandstone (DP10G) 0.6 m from main gouge zone. (a), (b), and (c) show progressive increase in microfracture intensity. (d) Cataclastic sandstone 0.1 m from the Punchbowl Formation and gouge contact. Microfracture intensity and grain size dramatically reduced from (c). (e) Subsidiary fault gouge (DP10A) located 0.1 m from main gouge zone. Foliation parallel to subsidiary fault. (f) Punchbowl fault gouge (DP10D) from main gouge zone. Note extreme grain-size reduction. Large and small arrows point to single-crystal porphyroclasts and irregularly shaped microcrystalline porphyroclasts, respectively.

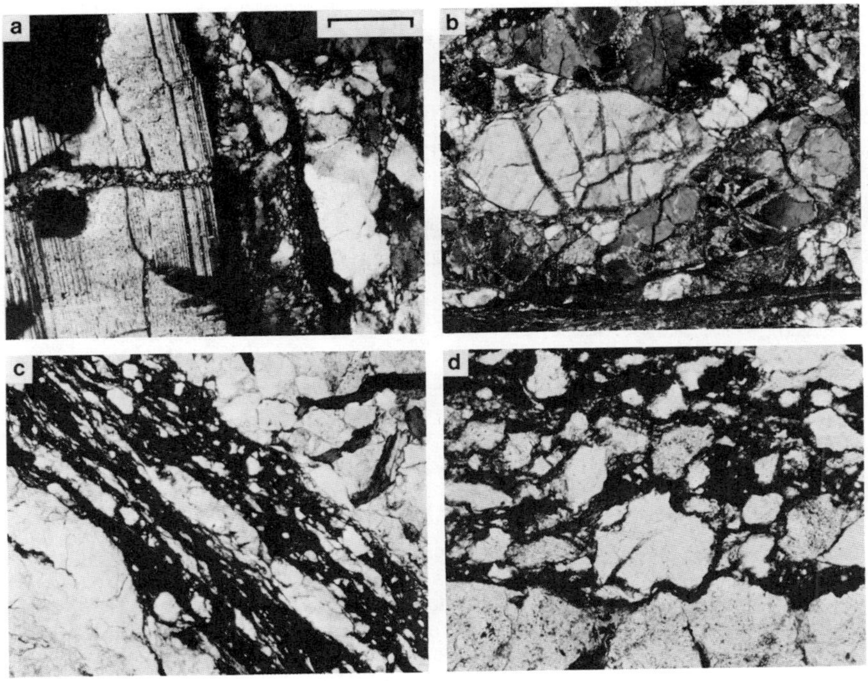

Figure 6

Photomicrographs of subsidiary faults and microfractures in damaged Punchbowl Formation, showing textural evidence of fluid-assisted processes. Scale bar applies to all photos and is 0.4 mm in (b) and (c) and 0.13 mm in (a) and (d); views are under crossed polarizers in (a) and (b) and under plane-polarized light in (c) and (d). (a) Quartz-sealed fracture adjacent to quartz-cemented subsidiary fault. (b) Dilation of quartz grain by extension on network of fractures; uniform extinction of quartz fragments shows fragments were not significantly rotated, i.e., pure dilation. Cement is calcite. (c) Concentration of opaques and elongate shape of quartz grains in subsidiary fault gouge (sample DP10F). (d) Close-up of subsidiary fault in (c) showing embayed and interlocking grain boundaries. Features in (c) and (d) suggest dissolution of silica during faulting.

damaged and undeformed host rock and of gouge from a subsidiary fault within the damaged zone are all relatively similar in composition (Table 1).

The damaged zone of the basement is also defined by an obvious increase in subsidiary fault and fracture intensity toward the main gouge zone, and ranges between less than 1 m to greater than 10 m in width. Associated with the increase in fault and fracture intensity is a reduction in grain size and a disruption of igneous and metamorphic textures. The melanocratic rocks tend to show a greater reduction in grain size and lower apparent cohesion than the leucocratic rocks. Compositional layering that is oriented parallel to the main gouge zone is produced in the damaged zone from offset on the subsidiary faults and mesoscopic cataclastic flow (Fig. 2).

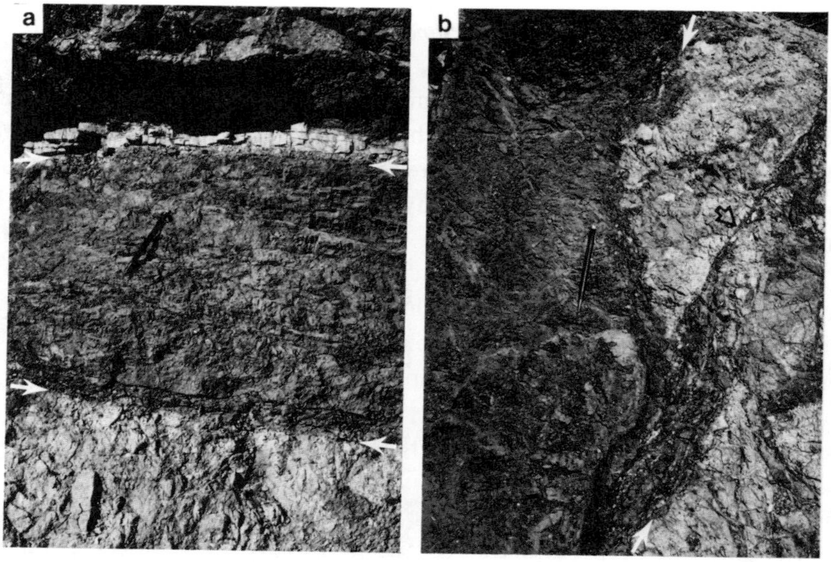

Figure 7
Excavations of main gouge zone at DP4. Pencil is 14 cm long. (a) Nearly horizontal exposure of gouge zone with basement at top and Punchbowl Formation at bottom. Contacts with host are distinct (white arrows). Punchbowl Formation directly adjacent to gouge is intensely fractured. Both ends of pencil point to prominent surfaces in gouge layer, interpreted as sites of shear. (b) View of gouge and Punchbowl Formation contact, showing characteristic irregular geometry. Host embayed at point of intersection with subsidiary fault (black arrow).

Table 1

Mineralogy[1] of Undeformed and Deformed Punchbowl Formation Sandstones[2]

Sample no.	Quartz	Feldspar	Calcite	Chlorite	Illite	Smectite	Laumontite	Clinoptilolite
DP4B[3]	54	31	1	1	—	5	8	—
DP4C[3]	62	29	—	2	2	3	2	—
DP10A[4]	55	27	—	2	10	6	—	trace
DP10G	55	36	2	trace	—	2	—	5
DP10J	52	32	—	2	2	5	7	—
DP13A[5]	54	32	1	—	5	3	3	2
DP13B	53	28	1	—	3	6	7	2

[1] Mineralogy determined by x-ray diffraction analysis and given in percent by volume.
[2] All samples except DP13 are from the damaged zone of the Punchbowl Formation.
[3] DP4B and DP4C are 0.1 and 1 m, respectively, from the contact between gouge and Punchbowl Formation.
[4] Subsidiary fault gouge used in friction tests.
[5] Sample used in triaxial compression tests.

Main gouge zone

The main gouge zone is typically 30 cm wide, but it varies from approximately 5 cm to 1 m. The gouge is generally dark brown and very fine-grained and appears similar in texture and color along the entire segment of the fault studied. The bulk of the gouge is free of any mesoscopic porphyroclasts.

The contact between the main gouge zone and the Punchbowl Formation is sharp and distinct (Fig. 7). The contact surface may be somewhat planar or curviplanar for several meters but is generally more irregular (Fig. 7b). The irregularities appear to be related, in part, to the lithology of the Punchbowl Formation. Conglomeratic sandstone units tend to form positive irregularities into the gouge, and siltstone or thinly bedded sandstone units tend to be locally embayed. Many subsidiary faults in the host rock merge with the main gouge zone but do not offset the zone (Fig. 7b).

The contact between the gouge and the basement varies in morphology, but it is easily located in the field by color and composition. Very distinct, sharp, and planar contacts occur where aplite or other rocks rich in felsic minerals are present (Fig. 7a). The most gradational contacts noted occur where rocks rich in mafic minerals are present. Texture and fabric across such contacts appear similar at the mesoscopic scale. Where the basement is composed primarily of felsic minerals but contains thin layers that are rich in mafic minerals, a different type of gradational contact is formed. In such cases thin gouge zones splay off the main gouge zone and isolate pods of basement rock. These are formed preferentially within thin mafic-rich layers that are subparallel to the main gouge zone, and they decrease in number away from the main gouge zone.

The gouge is highly fractured at orientations normal and parallel to the layer and is very easily eroded. In excavations it is apparent that the intensity of fracturing is dramatically reduced 0.5 to 1 m below the air–rock interface, suggesting that most fractures are a result of stress relaxation upon uncovering; some surfaces within the gouge, however, continue to be present at depth (Fig. 7a). These surfaces are probably related to differences in texture and induration. They occur throughout the gouge and at the gouge–host-rock contact and have a preferred orientation parallel, or slightly oblique, to the Punchbowl fault surface. The surfaces are generally planar but may be curved, and they can be greater than 10 m long. One such surface noted is extremely planar and juxtaposes gouge of slightly different color and induration. It is parallel to the main gouge zone and spans the irregularities of the contact between the gouge and the Punchbowl Formation. Although mesoscopic sealed fractures, veins, or other markers that might show offsets within the main gouge zone are not evident, the relations above suggest that the surfaces are sites of displacement, i.e., mesoscopic shear surfaces.

Microscopically, the main gouge zone is texturally similar throughout the length of the fault studied (CHESTER and LOGAN, 1986). Approximately 70% of the

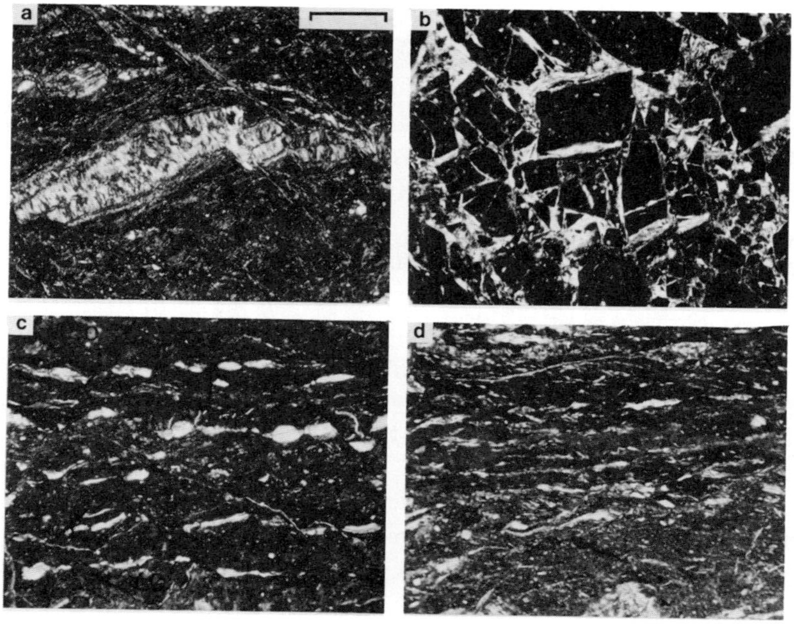

Figure 8
Texture of Punchbowl fault gouge. Scale bar applies to all photos and is 0.4 mm; views are under crossed polarizers in (a) and under plane-polarized light in (b), (c), and (d). (a) Symmetric bands of fibrous quartz cement in sealed fracture, which is offset by microscopic shear bands. (b) Quartz-cemented microbreccia. (c) Layering defined by distribution of microcrystalline quartz porphyroclasts. Each layer composed of porphyroclasts that are sheared and offset remnants of ancient, once continuous sealed fractures. (d) Example of typical foliation present in gouge.

gouge consists of a brown matrix composed of grains less than 10 μm in size (Fig. 5f). Matrix grains that can be optically distinguished are typically equant, angular porphyroclasts of quartz or feldspar. Porphyroclasts larger than 10 μm and composed of a single crystal constitute approximately 5% of the gouge. The remainder of the gouge is composed of microcrystalline quartz or calcite, either as sealed fractures or as irregularly shaped porphyroclasts. The sealed fractures typically constitute 5 to 10% of the gouge. Open fractures also are present and appear to be related to weathering.

Sealed fractures occur either as irregular veins having segments with different thicknesses and orientations or as very planar, smoothly tapered cracks that have extremely high-aspect ratios. The grain size of the calcite cement increases towards the center of the fractures. In some cases this filling is vuggy. The quartz cement is both massive and fibrous. The fibrous quartz occurs in bands symmetric about the center of the fracture (Fig. 8a) and probably records a crack-seal type of formation (RAMSAY, 1980). In some areas the sealed fractures are so extensive that the gouge is a cemented microbreccia in which the breccia clasts are composed of gouge matrix (Fig. 8b).

Table 2

Mineralogy[1] of the Punchbowl Fault Gouge[2]

Sample no.	Quartz	Feldspar	Calcite	Chlorite	Illite	Smectite	Laumontite	Clinoptilolite
DP4D[3]	47	34	—	—	—	14	—	5
DP4F[3]	50	29	1	—	—	15	—	5
DP4H[4]	55	32	—	—	—	10	2	1
DP10D[3]	49	31	—	trace	—	15	—	5
DP10K[5]	42	35	2	4	—	16	1	—

[1] Mineralogy determined by x-ray diffraction analysis and given in percent by volume.
[2] All samples from the main gouge zone.
[3] Gouge samples used in friction tests.
[4] Intact gouge used in triaxial compression tests.
[5] From gouge–basement contact.

There is a complete record of cataclastic degradation of sealed fractures present in the gouge. Pristine sealed fractures, offset and disrupted sealed fractures, and completely separated irregularly shaped porphyroclasts are present. The porphyroclasts also have the same internal structures as the pristine sealed fractures. These features suggest that all the microcrystalline porphyroclasts are the sheared and offset remnants of ancient, once continuous sealed fractures (Fig. 8 a, c). Moreover, such features in conjunction with the crosscutting relations between sealed fractures indicate many episodes of fracturing, each with its own quartz and calcite cementation (CHESTER and LOGAN, 1986).

Foliations in the gouge are defined by textural and compositional layering, preferred orientation of clay within the matrix, and microscopic shear bands (Fig. 8d; CHESTER et al., 1985). The shear bands produce an anastomosing network of discrete shear surfaces that pervade the gouge. Their preferred orientation is similar to that of the R_1 Riedel shear (CHESTER and LOGAN, 1986).

The composition of the main gouge zone within the study area is similar along strike but is significantly different from that of the subsidiary fault gouge and of the Punchbowl Formation sandstone (Tables 1 and 2). The latter contains minor amounts of laumontite and illite, whereas the main gouge zone is depleted of these minerals. Moreover, the main gouge zone contains three times as much smectite (Tables 1 and 2). The mineralogy of the basement is not known in detail, although microscopic observations indicate amounts of quartz and feldspar similar to those in the Punchbowl Formation. Minor amounts of chlorite, calcite, and illite are also present. This mineralogy also is different from that in the main gouge zone.

Deformation mechanisms and environmental conditions of faulting

Microstructures in the fault zone indicate that the brittle processes of microfracture, frictional sliding along fractures and grain boundaries, and rigid rotation of grain fragments were the dominant strain-accommodating mechanisms. In regions of high shear strain these mechanisms collectively lead to cataclastic flow. Features suggesting that dilatancy occurred in the fault zone were probably formed in conjunction with cataclasis (PATERSON, 1978).

Fluid-assisted processes also were operative during faulting. Some fluid-assisted mechanisms aid deformation, whereas others (such as cementation) lead to healing and strengthening of the fault rocks. Solution features and healed microfractures in quartz indicate that dissolution and diffusion of silica occurred at least locally (e.g. RUTTER, 1983; SMITH and EVANS, 1984). Mineral alteration and precipitation also attest to the mobility of other elements and the chemical reactivity of the rock with pore fluids. Thus it is possible, if not probable, that the cataclastic processes of crack growth and frictional grain-boundary sliding were also fluid-assisted by such mechanisms as stress corrosion, diffusion, and dissolution (ATKINSON, 1984).

The difference in composition between the main gouge zone and the Punchbowl Formation host rock coincides with an abrupt change in rock texture at the gouge–host-rock contact. This suggests that the mineral alteration may have been enhanced by the smaller grain size from comminution. The phases of smectite and clinoptilolite in the main gouge zone would be stable with the silica-rich fluids that were present during faulting. Thus the current mineralogy of the gouge probably reflects that which existed during faulting. The well-developed clay foliation in the gouge also suggests that the smectite was present during faulting, although post-faulting alteration cannot be totally ruled out. If the present mineralogy of the gouge represents the stable phases that existed during faulting, then the temperatures probably did not exceed 150 to 200 °C (MULLER, 1967; MIYASHIRO, 1973). The cataclastic and fluid-assisted processes that were operative, the apparent lack of mechanisms of dislocation motion, and the gouge mineralogy, all are consistent with the depth of faulting estimated by the thickness of the stratigraphic cover (2 to 4 km) with a normal to high geothermal gradient.

Mechanical properties of the fault rocks

Experimental design

The variation in mechanical properties across the fault zone was investigated by deforming samples that represent the three structural units of the fault zone and surrounding rock under triaxial compression (Table 3). An undeformed sandstone was chosen to represent the undeformed host-rock unit. Sandstone samples con-

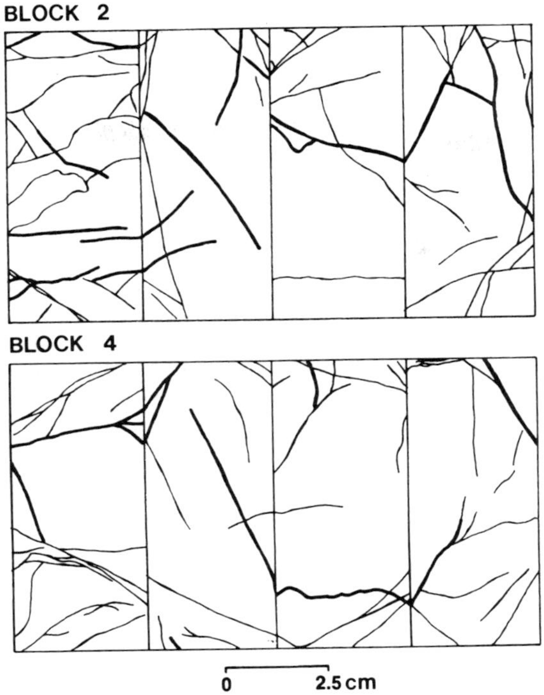

Figure 9
Map of subsidiary faults and fractures in multiple-fault specimens (DP10C) used in triaxial compression tests. Sides of block are 'unfolded' onto page; ends of block not shown. Heavy lines indicate faults open by separation; thin lines indicate closed faults. Block 2 used in dry test and block 4 in wet test.

taining single or multiple subsidiary faults were chosen to represent the damaged host-rock unit. A single intact block of gouge was chosen to represent the main gouge-zone unit.

Specimens of the undeformed sandstone and sandstone with a single natural fault were cylindrical, approximately 2.5 cm in diameter, and 5 cm in length. Those containing a natural fault were oriented with the fault approximately 30° to the cylinder axis. The multiple-fault and intact gouge samples were prepared by cutting prisms, to ensure they remained intact. To give two multiple-fault specimens that were as similar as possible, the specimen size was increased slightly for a more random fault distribution. The fault geometry was mapped for several specimens, and the two most similar were chosen (Fig. 9).

The frictional behavior of the Punchbowl fault was investigated by testing natural gouge samples in a standard configuration for defining rock friction with a triaxial apparatus (CHESTER, 1983). The gouge samples were chosen to represent the greatest contrasts in gouge texture and color noted in the field (Table 4). The specimen

Table 3

Fault Rock Samples Used in Triaxial Compression Tests

Structural unit	Sample, no.	Description	Remarks
Undeformed host rock	Undeformed sandstone (DP13A)	Medium-grained Punchbowl Formation sandstone collected approximately 100 m from main gouge zone at locality DP10; porosity of sample approximately 11%.	Composition: see Table 1. Photomicrograph: Fig. 5a.
Damaged host rock	Single natural fault (DP10F)	Medium- to coarse-grained Punchbowl Formation sandstone containing single subsidiary fault; collected approximately 12 m from main gouge zone.	Photomicrograph of subsidiary fault: Fig. 6, c, d.
	Multiple natural faults (DP10C)	Medium- to coarse-grained Punchbowl Formation sandstone containing many subsidiary faults and fractures; collected 1 m from main gouge zone.	Fault and fracture map of experimental specimens in Fig. 9.
Main gouge zone	Intact gouge (DP4H)	Large block of intact, dark-brown, slightly indurated gouge from main gouge zone.	Composition: see Table 2. Sample locality shown in Fig. 7a.

Table 4

Gouge Samples Used in Triaxial Friction Experiments

Gouge type	Sample no.	Description	Remarks
Main gouge zone	DP4D	Moderately indurated brown gouge from the contact between gouge and Punchbowl Formation.	Composition: Table 2. Sample locality shown in Fig. 7a.
	DP4F	Highly fractured light-brown gouge located directly adjacent to block of intact gouge (DP4H) used in compression tests.	Composition: Table 2. Sample locality shown in Fig. 7a.
	DP10D	Well-indurated dark-brown gouge.	Composition: Table 2. Photomicrograph: Fig. 5f.
Subsidiary fault gouge	DP10A	Poorly indurated flakey brown gouge from subsidiary fault 1 m from main gouge zone and sample DP10D.	Composition: Table 2. Photomicrograph: Fig. 5e.

assembly consists of gouge between a 35° sawcut in a cylindrical core, 5 cm in diameter, of thermally cracked Tennessee sandstone. The gouge samples were disaggregated to a maximum particle size of 63 μm with a mortar and pestle; 0.8 gm portions of gouge were used in each experiment.

All experiments were conducted at a constant displacement rate of $2.5(10)^{-4}$ or

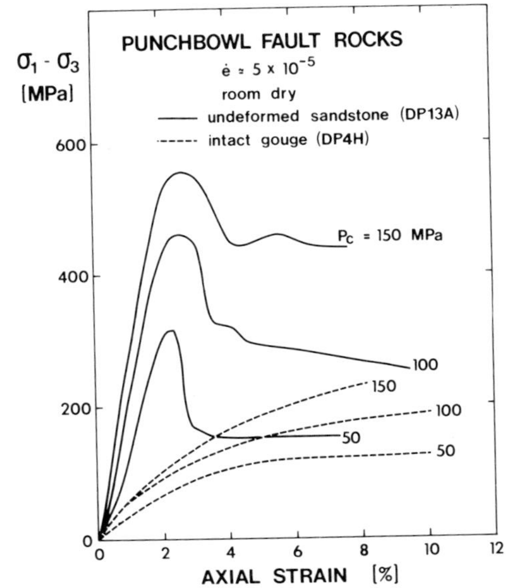

Figure 10
Differential stress versus axial strain for triaxial compression of undeformed Punchbowl Formation sandstone (DP13A) and intact gouge (DP4H).

$2.5(10)^{-6}$ cm/sec in a large-specimen triaxial apparatus described by HANDIN et al. (1972). To simulate grossly the environmental conditions in the fault zone during natural deformation, the experiments were conducted at an effective confining pressure of 50 MPa and at room temperature. Several additional experiments were conducted at 100 and 150 MPa. All the fault rocks were deformed under nominally dry and water-saturated conditions.

Axial displacement was measured with a displacement transducer (accurate to within 10 μm) located across the specimen inside the pressure vessel; force was measured with an external load cell (accurate to within 2%). For the wet tests the specimens were saturated with distilled water under a vacuum for 24 hr. The axial displacement rates employed were slow enough to ensure that the effective pressure remained constant within the specimens throughout the wet tests. The volume of pore fluid added or removed from the specimen to maintain constant pore pressure was recorded throughout the experiments as a qualitative measure of dilatancy, i.e. change in porosity and relative permeability.

Experimental results

The undeformed sandstone and intact gouge show a great contrast in behavior under triaxial compression at confining pressures of 50, 100, and 150 MPa (Fig. 10).

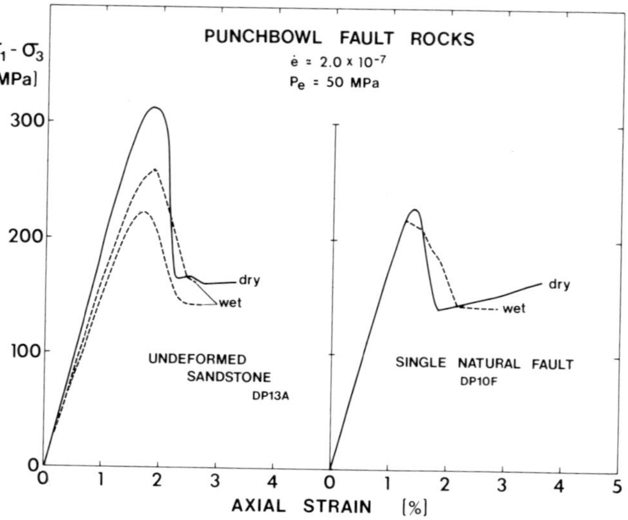

Figure 11
Differential stress versus axial strain for triaxial compression of undeformed-sandstone and single-fault samples under wet and dry conditions.

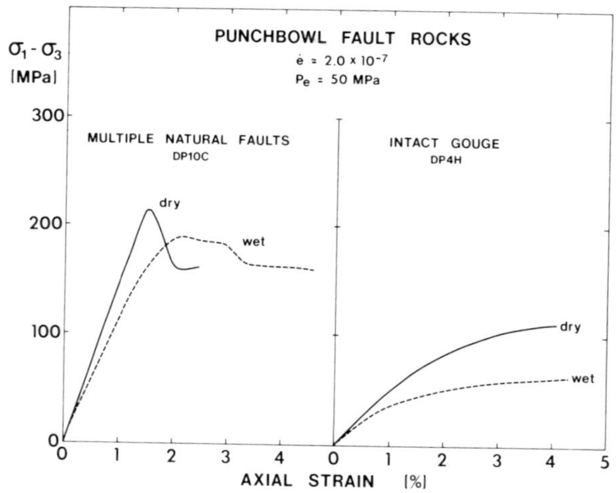

Figure 12
Differential stress versus axial strain for triaxial compression of multiple-fault and intact-gouge samples under wet and dry conditions.

In all cases the sandstone is brittle and fails by macroscopic faulting. The gouge is mechanically ductile, as determined from the curves of differential stress versus axial strain, but it fails by macroscopic faulting at 50 MPa and by uniform flow at 100

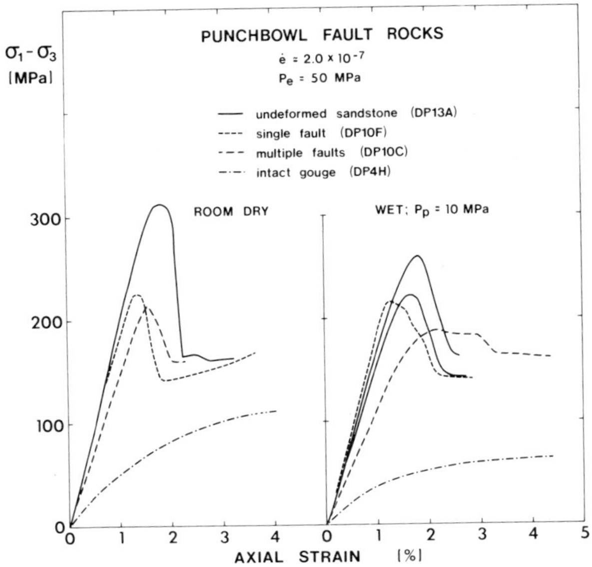

Figure 13
Summary plot of Figures 11 and 12, showing differential stress versus axial strain for triaxial compression of Punchbowl fault rocks under wet and dry conditions.

and 150 MPa. The difference in post-failure strength of the sandstone and of the gouge increases with increasing confining pressure (Fig. 10).

All samples deformed at an effective pressure of 50 MPa show a consistent difference in behavior between wet and dry conditions (Figs. 11 and 12). Under wet conditions all four rock types show a reduction in elastic modulus (except for the single fault sample), a reduction in ultimate strength and, for experiments with stress drops, a lower negative slope for the post-failure portion of the curve of differential stress versus axial strain. The strain at failure also varies slightly between samples (Figs. 11, 12, and 13). The stress drops are sudden under dry conditions, but they take several hours when the rocks are saturated. The preëxisting natural faults in the samples were reactivated during the experimental deformation.

The post-failure strengths of the three sandstone samples are similar under dry conditions and are not appreciably changed under saturated conditions (Fig. 13). The intact gouge experiences a dramatic reduction in strength when wet, such that there is a greater variation in properties between the gouge and sandstone under wet conditions (Fig. 13). The changes in pore volume measured in the wet tests indicate that all three sandstone samples experience dilatancy prior to reaching ultimate strength, whereas the porosity of the intact gouge is continuously reduced throughout the entire test. Relative permeabilities, deduced from equilibration times after abrupt changes in pore pressures, indicate that the multiple fault samples have the highest

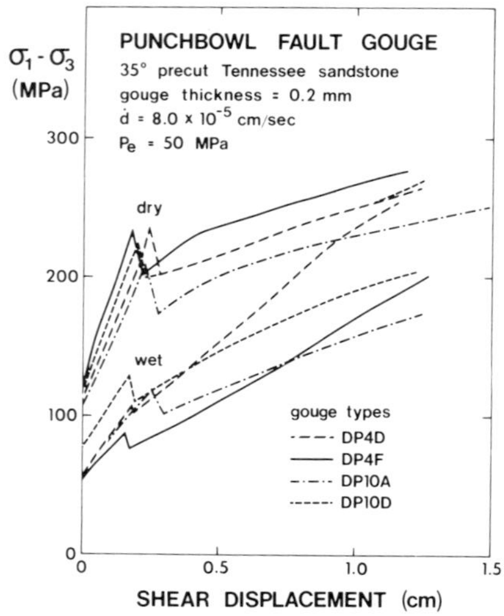

Figure 14
Differential stress versus shear displacement for Punchbowl fault gouges between a 35° sawcut in cylindrical cores of Tennessee sandstone under wet and dry conditions.

permeability and that the intact gouge has significantly lower permeability than the sandstones. The method employed, however, did not allow absolute permeability to be determined.

The different fault gouges in shear have similar sliding behavior under wet and dry conditions, and the dry gouges have similar strengths (Fig. 14). Under wet conditions the initial sliding strengths of the gouges are much lower than those of the corresponding dry tests. The four wet gouges strain-harden at slightly different rates, producing a greater variation in sliding strengths at high shear strains under wet conditions than under dry conditions. No correlation between behavior and clay composition of the gouge is demonstrated (Table 2 and Fig. 14).

Effect of natural deformation and water on rock properties

The triaxial tests on the Punchbowl fault rocks demonstrate a variation in mechanical behavior between samples for different degrees of prior natural deformation. In general, with an increasing degree of natural deformation there is a reduction in ultimate strength and elastic modulus and an increase in relative ductility. The experiments show a gradual change in behavior for the suite of sandstone samples (undeformed, single fault, and multiple faults) and a marked change in behavior between the sandstones and the intact gouge.

The differences in behavior in both types of test may be attributed to the changes in intrinsic parameters (composition and texture) among the samples, because extrinsic parameters (such as temperature, pressure, and pore fluids) are held constant, and the sample preparation techniques and sample sizes are essentially the same. The composition of all the sandstone samples are similar, so differences in mechanical behavior may be correlated with changes in texture associated with the deformational features present. The difference in texture and composition (i.e., amount of clay) between the gouge and sandstone produces the marked difference in mechanical response.

Previous studies have shown that pore water may influence the behavior of clay minerals in two ways: (a) pore-water pressure reduces the magnitude of the effective confining pressure transmitted through the load-bearing framework (HANDIN et al., 1963) and (b) the pore water changes the state of hydration of the expanding layer clays and the ions on the surface of the clay particles such that the intrinsic properties of the clays are modified (e.g., WU, 1978). These effects have been demonstrated under high pressures in clay-rich simulated-gouge and natural-gouge rock-friction experiments (WANG and MAO, 1979; MORROW et al., 1982).

Under wet conditions the behavior of the Punchbowl fault rocks is dramatically influenced by the expanding-layer clay fraction of the rocks, which generally amounts to less than 15%. The water weakening demonstrated is interpreted as resulting from a change in state of hydration, because all the experiments were conducted at very slow displacement rates which allow the effective pressure within the specimens to be maintained at a constant value.

Discussion

Strain localization and formation of the damaged zones

Many studies have shown that displacements in brittle fault zones are primarily accommodated within well-defined, narrow layers of cataclasite or gouge, which are bounded by zones of damaged host rock pervaded by fractures and breccia (e.g., REED, 1964; FLINN, 1977; SIBSON, 1979). The cataclasite or gouge layers represent the fault surface, the site where the majority of displacements are accommodated. The damaged zones record long-continued deformation from movement on those faults; that is, the damaged zones are a consequence of progressive fracturing from stress-cycling caused by continuous or repeated movements on non-parallel or nonplanar faults (FLINN, 1977; SIBSON, 1979; WHITE and WHITE, 1983). The Punchbowl fault structure fits this generality.

Across the study area the Punchbowl fault gouge is similar in texture and composition. The change in texture from the damaged host rock to the Punchbowl fault gouge records a significant and abrupt increase in strain magnitude across the contact. High shear strain within the gouge is also recorded by the presence of well-developed cataclastic foliations (CHESTER and LOGAN, 1986). The similarity in gouge texture and composition is probably a result of the very high shear strain which has effectively comminuted and mixed the gouge components.

These features exemplify the localization of strain in the fault zone at the mesoscopic scale. The localization indicates that the gouge is significantly weaker than the surrounding undeformed and damaged host rock and that strain softening occurred in the fault zone. These observations agree with the relative behavior of the host rock and gouge demonstrated experimentally and support the validity of using relative mechanical properties derived experimentally to describe natural systems (e.g., STEARNS, 1968; HANDIN et al., 1963).

The pervasive occurrence of mesoscopic and microscopic shears within the main gouge zone attests to the fact that fault displacements were localized not only to the gouge layer but also along discrete surfaces within the gouge. Probably some fault displacements also were localized along the gouge–host-rock contacts that are locally planar or curviplanar. At most localities observed, however, the contacts are irregular and the mesoscopic shear surfaces span the host-rock protrusions into the gouge. This appears to provide a more planar sliding surface, of which only a small percentage is not contained totally within the gouge.

Structures similar to those showing localization of displacements to discrete shears in the gouge of the Punchbowl fault are also seen in many simulated-gouge rock-friction experiments for investigating the frictional properties of fault zones (e.g., LOGAN et al., 1979, 1981). Experimental and theoretical work has led to the suggestion that sliding mode is related to such localization (KRAGELSKII, 1965; RUINA, 1980; LOGAN et al., 1981; CHESTER, 1985). Strain localization probably reflects displacement-, time-, or velocity-dependent weakening, any of which is a necessary, but not sufficient, condition for seismic slip (e.g., RICE, 1983). A negative dependence of frictional resistance on sliding rate has been interpreted as developing out of competing mechanisms of decreasing intrinsic shear strength and increasing true area of contact with time (DIETERICH, 1978; DIETERICH and CONRAD, 1984). In the Punchbowl fault such time-dependent frictional phenomena could arise from mechanisms of fluid-assisted crack growth, healing, and sealing of microfractures and subsidiary faults, and alteration of gouge minerals.

The amount of quartz and calcite cement in intact and relict sealed fractures suggests that at least 30% of the gouge was precipitated from a fluid phase. As mentioned previously, the presence of a significant amount of cement and of cement showing various degrees of cataclastic degradation suggests that the gouge was repeatedly cemented and fractured during faulting. Repeated deformation of the host rock throughout the Punchbowl faulting episode is also indicated by the textural features showing cementation and reactivation of the subsidiary faults. Cementation of fractures presumably strengthens the gouge as well as the host rock, although the latter must have always been relatively stronger. If cementation occurred between slip events on the fault, then cementation could be an important time-dependent strengthening process that allows the apparent repeated strain softening and localization of displacements within the gouge layer.

It appears that almost all of the fault displacements were accommodated in the gouge layer, and that only a very small volume of the host rock is deformed to the point of forming gouge. This is consistent with other large faults which have kilometers of displacement along very narrow gouge or cataclasite layers (ENGELDER, 1974), but it is inconsistent with the linear relation between the thickness of the gouge layer and fault displacement noted for small, natural faults (ROBERTSON, 1982) and experimental faults (ENGELDER, 1974). Apparently, in small faults cataclastic degradation of the host is necessary for displacement because geometric irregularities in the host rock continually impinge (ENGELDER, 1974). For larger faults in which a well-developed gouge layer separates and blankets the host-rock asperities, host degradation probably occurs at slower rates or possibly not at all (ENGELDER, 1974; JACKSON and DUNN, 1974). In the Punchbowl fault there is good evidence that displacements were primarily accommodated within the gouge and the gouge was continually reworked such that continued degradation of the host was minimized.

Recognition of mechanical units

From the experimental data showing the relationship between degree of deformation and mechanical behavior, the structurally defined units of the Punchbowl fault zone and surrounding rock may be characterized as discrete mechanical units. The irregular and gradational contact between the undeformed host rock and the damaged host rock, and the overall progressive increase in deformation intensity towards the Punchbowl fault surface, may be viewed mechanically as an overall continuous reduction in ultimate strength and elastic modulus, from the undeformed host rock to the gouge–host-rock contact (Fig. 15). The irregular geometry of the damaged zone, combined with significant local changes in lithology and intensity of deformation, represents a damaged zone having regions of different stiffness and strength. In general, it is suggested that there is a variation in properties within the damaged host rock along strike and a significant variation in mechanical properties across the fault zone transverse to strike (Fig. 15). The very sharp contact between the main gouge zone and the damaged host rock is a distinct mechanical boundary, across which the most significant difference in properties exists.

The results of the friction tests show similar strengths and sliding behavior for different gouge samples. Moreover, the similarity in texture and composition of the gouge throughout the study area also suggests that the Punchbowl fault gouge has relatively little variation in mechanical properties. The uniform mechanical properties of the gouge greatly contrasts with the variation in properties inferred for the host rock. During slip on the fault, soft and stiff portions of the host will be juxtaposed

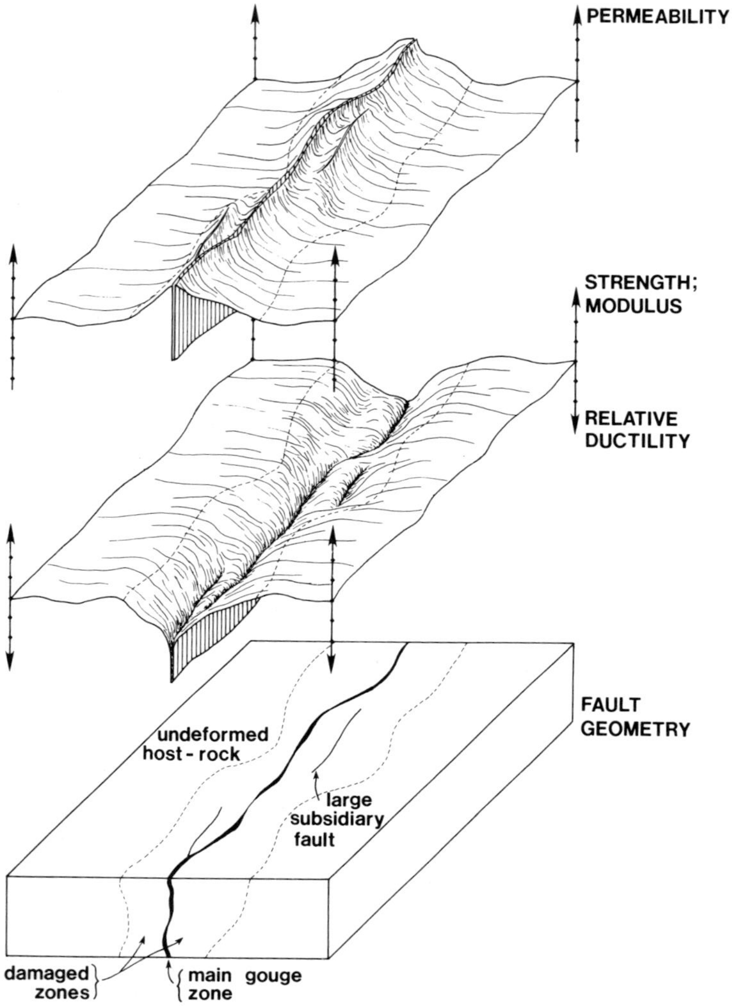

Figure 15
Schematic diagram of structure and relative properties of idealized brittle fault zone based on observations of Punchbowl fault. Block diagram indicates structural-mechanical units of fault. Surfaces indicate relative variation in properties along and across fault. Note that increasing direction of particular properties may be up or down, as shown on axes. Mesoscopic or macroscopic scale intended.

and lead to variations in loading of the gouge along strike. This may also enhance geometric irregularities in the fault surface and promote impingements of the host across the gouge layer. The fabric of the subsidiary faults in the damaged Punchbowl Formation suggests that subsidiary faults were formed and reactivated by stress concentrations generated at constraining and releasing bends as displacement progressed (CHESTER and LOGAN, 1986). Activation and formation of subsidiary faults in the damaged zones is probably the dominant process by which the host accommodated sliding past bends on the fault surface, consistent with suggestions made by SIBSON (1979) and FLINN (1977). If gouge behavior is dependent on parameters such as mean stress, then the uneven loading would produce variations in response of the gouge layer along strike, even though it is considered relatively uniform mechanically.

Based on the relative permeabilities of the Punchbowl fault rocks inferred from the experiments, the permeability of the host would be expected to increase up to the main gouge zone but to abruptly decrease within the gouge (Fig. 15). This might also occur to a lesser extent near the larger subsidiary faults that contain gouge layers. The preferred orientation of smaller subsidiary faults and fractures, which presumably enhance permeability, would produce a permeability anisotropy in the damaged host. Although on the average the main gouge zone is expected to have significantly lower permeability than the damaged and undeformed host rock, the extensive and repeated fracturing and cementation in the gouge suggests that permeability in this unit probably changes with time (e.g., STIERMAN, 1984).

Although these inferences about the properties and processes of brittle faulting in the Punchbowl fault have been drawn primarily from observations at the mesoscopic scale, many may apply to brittle faults at a variety of scales. If this is so, then the idealized model of the Punchbowl fault (Fig. 15) may be incorporated into general conceptual models of crustal fault zones. Such an extension may be justified since many aspects of brittle deformation appear to be scale-invariant (e.g., TCHALENKO, 1970; ALLEGRE et al., 1982).

Summary and conclusions

Cataclastic deformational processes must be dominant within upper crustal, seismogenic faults. The mode of fault slip, however, may be dependent upon how cataclasis is modified by time-dependent processes, such as fluid-assisted diffusional mechanisms and intracrystalline plasticity. Data from the Punchbowl fault suggest that the former may be extremely important for some shallow faulting, which is in agreement with the suggestion of others (e.g., SIBSON, 1979; STIERMAN, 1984), but studies of other shallow fault zones in southern California indicate that the influence of pore fluids varies greatly (e.g., ANDERSON et al., 1983; WANG, 1984).

It is generally recognized that fault displacements in brittle zones are localized to a gouge or cataclasite layer (e.g., SIBSON, 1979), but structures in the Punchbowl fault also suggest that displacements may be further localized to discrete shear surfaces within the gouge. The structures recording localization of displacements in the Punchbowl fault gouge are geometrically similar to those produced in rock-friction experiments on simulated gouge (e.g., LOGAN et al., 1979, 1981). For the Punchbowl fault, the localization possibly occurred as a result of strain softening of gouge that was repeatedly strengthened by cementation.

Conceptual models of brittle fault zones and fault processes should incorporate a significant variation in mechanical properties along and transverse to the fault zone. Models could be based on three mechanical units: an undeformed host rock, a damaged zone, and a gouge layer. The damaged host rock is probably the most important to generating variations in loadings of the gouge and producing geometric impingements along the fault. Features in the Punchbowl fault support the suggestion that the pressure, temperature, and fluid regime of faulting can modify the style of deformation in the fault zone, which may affect such mechanical and physical properties of the fault rocks as strength and permeability (e.g., WU, 1978; SIBSON, 1979; WANG, 1984; STIERMAN, 1984). For example, in the Punchbowl fault the style of deformation of the host rock causes an increase in permeability towards the gouge layer. As suggested by others, however (e.g., WU, 1978; STIERMAN, 1984), such processes in the gouge as comminution, alteration of the gouge to clays, and sealing of fractures abruptly reverse this trend, so that the permeability of the gouge is significantly lower than that of the surrounding rock.

Acknowledgments

We would like to thank J. S. Chester, M. Friedman, and two anonymous reviewers for their critical reading of the manuscript. F. M. Chester gratefully acknowledges J. S. Chester, V. Chester, and A. Delbanco for assistance in the field and J. N. Magouirk for technical assistance in the laboratory. Field study of the Punchbowl fault zone was supported by U.S. Geological Survey Contracts 14-08-0001-21181 and 14-08-0001-G-820.

REFERENCES

ALLEGRE, C. J., LE MOUEL, J. L. and PROVOST, A. (1982), *Scaling rules in rock fracture and possible implications for earthquake prediction.* Nature *297*, 47–49.

ANDERSON, J. L., OSBORNE, R. H. and PALMER, D. F. (1983), *Cataclastic rocks of the San Gabriel fault: An expression of deformation at deeper levels in the San Andreas fault zone.* Tectonophys. *98*, 209–251.

ATKINSON, B. K. (1984), *Subcritical crack growth in geological materials.* J. Geophys. Res. *89*, 4077–4114.

CHESTER, F. M., *Mechanical Properties and Fabric of the Punchbowl Fault Zone, California.* M. S. Thesis, Texas A&M Univ., Texas, 1983.

CHESTER, F. M. (1985), *Correlation of halite gouge texture with sliding mode and velocity dependence in experimental faults* (Abstr.). EOS, Trans. Am. Geophys. Union. *66*, 1100–1101.

CHESTER, F. M., FRIEDMAN, M. and LOGAN, J. M. (1985). *Foliated cataclasites.* Tectonophys. *111*, 139–146.

CHESTER, F. M. and LOGAN, J. M. (1986), *Cataclastic shear fabric of the Punchbowl fault zone, California.* Submitted to P. R. Cobbold, D. Gapais and W. D. Means (eds.), Spec. issue on Shear Criteria in Rocks. J. Struct. Geol.

COX, B. F., POWELL, R. E., HINKLE, M. E. and LIPTON, D. A. (1983), *Mineral resource potential map of the Pleasant View roadless area, Los Angeles County, California.* U.S. Geol. Surv. Map MF-1649-A, scale 1:62,500.

DIBBLEE, T. W., Jr. (1968), 'Displacements on the San Andreas fault system in the San Gabriel, San Bernardino, and San Jacinto Mountains, southern California,' in *Proc. Conf. Geol. Problems San Andreas Fault System*. Stanford Univ. Pubs. Geol. Sci. *11*, 260–276.

DIETERICH, J. H. (1978), *Time-dependent friction and the mechanics of stick-slip*. Pure Appl. Geophys. *116*, 790–806.

DIETERICH, J. H. and CONRAD, G. (1984), 'Effect of humidity on time- and velocity-dependent friction in rocks,' in S. H. Kirby and C. H. Scholz, (eds.), *Chemical Effects of Water on the Strength and Deformation of Crustal Rocks*. J. Geophys. Res. *89*, 4196–4202.

ENGELDER, J. T. (1974), *Cataclasis and the generation of fault gouge*. Geol. Soc. Am. Bull. *85*, 1515–1522.

FLINN, D. (1977), *Transcurrent faults and associated cataclasis in Shetland*. J. Geol. Soc. London *133*, 231–248.

HANDIN, J., FRIEDMAN, M., LOGAN, J. M., PATTISON, L. J. and SWOLFS, H. S. (1972), 'Experimental folding of rocks under pressure—Buckling of single layer beams,' in *Flow and Fracture of Rocks*. Geophys. Monogr. Am. Geophys. Union *16*, 1–28.

HANDIN, J., HAGER, Jr., R. V., FRIEDMAN, M. and FEATHER, J. N. (1963), *Experimental deformation of sedimentary rocks under confining pressure pore pressure tests*. Am. Assoc. Pet. Geol. *47*, 717–755.

JACKSON, R. E., and DUNN, D. E. (1974), *Experimental sliding friction and cataclasis of foliated rocks*. Int. J. Rock Mech. Min. Sci. Geomech. Abstr. *11*, 235–249.

KRAGELSKII, I. V., *Friction and Wear*. Butterworths, Washington, D.C. 1965.

LOGAN, J. M., FRIEDMAN, M., HIGGS, N. G., DENGO, C. and SHIMAMOTO, T. (1979), 'Experimental studies of simulated gouge and their application to studies of natural fault gouge,' in *Proc. Conf. VIII Anal. Actual Fault Zones Bedrock*. U.S. Geol. Surv. Open-file Rept. 79-1239, 305–343.

LOGAN, J. M., HIGGS, N. G., and FRIEDMAN, M. (1981), 'Laboratory studies on natural fault gouge from the U.S. Geological Survey Dry Lake Valley No. 1 Well, San Andreas fault zone,' in *Mechanical Behavior of Crustal Rocks: The Handin Volume*. Geophys. Monogr. Am. Geophys. Union *24*, 121–134.

MAWER, C. K. and WILLIAMS, P. F. (1985), *Crystalline rocks as possible paleoseismicity indicators*. Geology *13*, 100–102.

MIYASHIRO, A., *Metamorphism and Metamorphic Belts*. Allen and Unwin, London, 1973.

MORROW, C. A., SHI, L. Q., and BYERLEE, J. D. (1982), *Strain hardening and strength of clay-rich fault gouges*. J. Geophys. Res. *87*, 6771–6780.

MULLER, G., 'Diagenesis of argillaceous sediments,' in *Diagenesis in Sediments, Developments in Sedimentology 8*, Elsevier, Amsterdam, 1967, Chap. 4.

NOBLE, L. F. (1954), *Geology of the Valyermo quadrangle and vicinity, California*. U.S. Geol. Surv. Quad. Map GQ-50, scale 1:24,000.

PATERSON, M. S., *Experimental Rock Deformation: The Brittle Field*. Springer-Verlag, Berlin, 1978.

RAMSAY, J. G. (1980), *The crack-seal mechanisms of rock deformation*. Nature *284*, 135–139.

REED, J. J. (1964), *Mylonites, cataclasites, and associated rocks along the Alpine fault, South Island, New Zealand, N.Z. J. Geol. Geophys. 7*, 645–684.

RICE, J. R. (1983), *Constitutive relation for fault slip and earthquake instabilities*. Pure Appl. Geophys. *121*, 443–475.

ROBERTSON, E. C. (1982), 'Continuous formation of gouge and breccia during fault displacement,' in *Issues in Rock Mechanics*. Proc. Symp. Rock Mech., Am. Inst. Min. Eng. *23*, 397–404.

RUINA, A. L., *Friction laws and instabilities: A quasistatic analysis of some dry frictional data*. Ph.D. Dissertation, Brown Univ., Rhode Island, 1980.

RUTTER, E. H. (1983), *Pressure solution in nature, theory and experiment*. J. Geol. Soc. London *140*, 725–740.

SHARP, R. V., and SILVER, L. T. (1971), *Quaternary displacement on the San Andreas and Punchbowl faults at the San Gabriel Mountains, southern California*. Geol. Soc. Am. Abstr. Prog. 3, p. 191.

SIBSON, R. H. (1979), 'Fault rocks and structure as indicators of shallow earthquake source processes,' in *Proc. Conf. VIII, Anal. Actual Fault Zones in Bedrock*. U.S. Geol. Surv. Open-file Rept. 79–1239, 276–304.

SIBSON, R. H. (1984), *Roughness at the base of the seismogenic zone: Contributing factors*. J. Geophys. Res. *89*, 5791–5799.

SMITH, D. L., and EVANS, B. (1984), *Diffusional crack healing in quartz*. J. Geophys. Res. *89*, 4125–4135.

STEARNS, D. W., 'Certain aspects of fracture in naturally deformed rocks,' in *Rock Mechanics Seminar: Special Report*. Air Force, Cambridge Res. Lab., 1968, p. 97–188.

STIERMAN, D. J. (1984), *Geophysical and geological evidence for fracturing, water circulation and chemical alteration in granite rocks adjacent to major strike-slip faults*, J. Geophys. Res. *89*, 5849–5857.

TCHALENKO, J. S. (1970), *Similarities between shear zones of different magnitudes*. Geol. Soc. Am. Bull. *81*, 1625–1640.

WANG, C. Y. (1984), *On the constitution of the San Andreas fault zone in central California*. J. Geophys. Res. *89*, 5858–5866.

WANG, C. Y., and MAO, N. (1979), *Shearing of saturated clays in rock joints at high confining pressures*. Geophys. Res. Lett. *6*, 825–828.

WHITE, J. C., and WHITE, S. H. (1983), *Semi-brittle deformation within the Alpine fault zone, New Zealand*. J. Struct. Geol. *5*, 579–589.

WOODBURNE, M. O. (1975), *Cenozoic stratigraphy of the Transverse Ranges and adjacent areas, southern California*. Geol. Soc. Am. Spec. Paper 162.

WU, F. T. (1978), *Mineralogy and physical nature of clay gouge*. Pure Appl. Geophys. *116*, 655–689.

(Received 26th September 1985, revised 24th March 1986, accepted 26th March 1986)

PAGEOPH, Vol. 124, Nos. 1/2 (1986)
0033-4553/86/020107-19$1.50 + 0.20/0

Characteristics of Faults and Shear Zones in Deep Mines

ROBERT E. WALLACE[1] and HAL T. MORRIS[1]

Abstract—The characteristics of fault and shear zones to depths of 2.5 km are well documented in deep mines in North America. The characteristics may be summarized as follows. (a) Fault zones usually are irregular, branched, anastomosed, and curved rather than simple and planar. (b) Faults are generally composed of one or more clay or clay-like gouge zones in a matrix of sheared and foliated rock bordered by highly fractured rock. (c) The widths of fault zones appear to be greater when faults have greater displacement, probably as a result of a long history of repeated minor movements. Fault zones with kilometers of displacement tend to be 100 m or more wide, whereas those with only a few hundred meters of displacement commonly are only 1 m or less wide. (d) Some zones represent shear distributed across hundreds of meters without local concentration in a narrow gouge zone. (e) Many fault zones are wet even above the water table, and water moves along them at various rates, but some also serve as subsurface dams, ponding ground water as much as several hundred meters higher on one side than on the other. No striking differences in the characteristics of faults over the vertical range of 2.5 km are documented.

Key words: Faults, seismogenic faults, deep faults.

Introduction

Investigations of the processes of faulting that generate earthquakes require information about the nature of fault zones at hypocentral depths. Deep mines provide the best access to faults at such depths. Mine workings, tunnels, and shafts permit an investigator to see, feel, sample, and test fault-zone materials and their geometric relations in situ. This paper is intended to provide some insight into the extreme complexity of faults and shear zones in nature for those students of earthquakes who have not been able to examine many faults in mines.

We present descriptions of a few faults in selected mines extending to depths of as much as 2.5 km below the present ground surface and which were probably as deep as 5 km below the ground surface at the time of faulting. The Coeur d'Alene mining district in northern Idaho provides some excellent examples, but additional data about faults in the Tintic mining district in central Utah, the Kirkland Lake district of northeastern Ontario, and the O'Brien mine in Quebec are also informative.

[1] U.S. Geological Survey, 345 Middlefield Road, Menlo Park, California 94025.

The field investigations that provided the data for this paper were not conducted with earthquake-source problems in mind. Most of these data and interpretations are an outgrowth of studies of mineral deposits which were conducted over many years by the authors and their colleagues with the U.S. Geological Survey. Most of the information is exerpted from existing published literature and is provided in a form that we hope will be useful to investigators concerned with earthquake sources.

Material in this review was first presented in 1979 at a conference, 'Analysis of actual fault zones in bedrock' (WALLACE, and MORRIS, 1979).

Fault and shear zones in the Coeur d'Alene mining district, Idaho

The Coeur d'Alene mining district in northern Idaho lies in an intensely faulted and sheared structural knot (Fig. 1), an important structural element of which is the Osburn fault. The Osburn fault has 26 km of right-lateral strike slip (WALLACE et al., 1960; HOBBS et al., 1965). In the process of recovering billions of dollars' worth of ore from the district, the region has been cut by mine galleries, adits, and shafts totalling several hundreds of kilometers in length. Furthermore, ore has been mined over a vertical range of more than 2.2 km, so faults and shear zones have been exposed over that vertical extent. At few places in the world is there a set of faults as well exposed both in lateral and vertical extent as in the Coeur d'Alene district. As a fault hundreds of kilometers long and having many kilometers of strike slip, the Osburn fault provides a striking analogue of the San Andreas fault in California, one of the great seismogenic faults of the world.

At present faults are exposed at the surface at elevations above 2,000 m and reach to depths more than 300 m below sea level. How deep the faults were buried at their time of formation is unknown, but FRYKLUND (1964), from a study of the environment of ore formation, suggests that 5 km may be a likely depth for the deepest ore bodies.

Country rocks involved in the faulting of the Coeur d'Alene district described here are of Precambrian age and range from fine-grained argillities and siltites (quartzites having silt-sized grains) to fairly coarse-grained quartzites. Low-grade metamorphism of these strata is common. Beds are from a few centimeters to several meters thick, formations are hundreds to thousands of meters thick, and the total sedimentary pile exposed is more than 15 km thick. The beds have been compressed into folds having amplitudes of several kilometers so that from the highest unit on the crest of the largest anticline to the lowest unit in the trough of the largest syncline could represent several times 15 km, and perhaps the faults described typify the characteristics of faults at depths of 30 km or more.

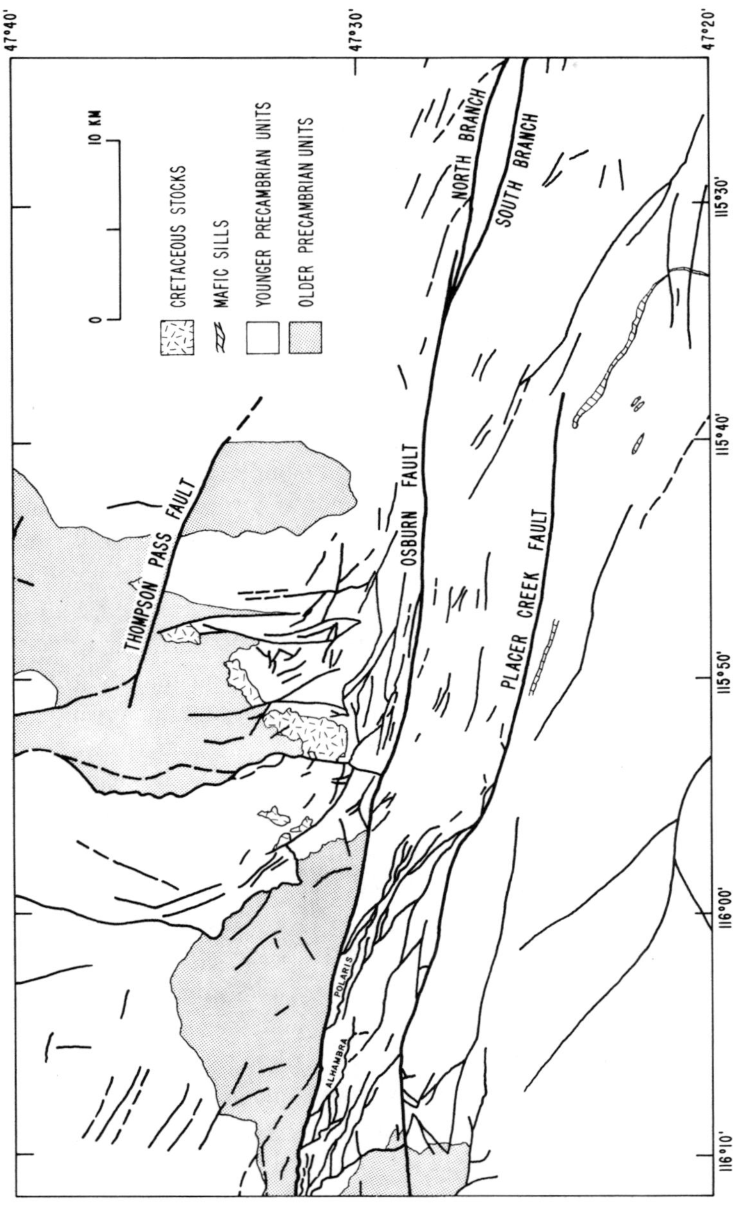

Figure 1

The Osburn fault in northern Idaho displays a 26 km right-lateral strike slip and the Thompson Pass fault an approximately 5 km right-lateral. Compare density of faulting mapped at this scale with that shown in Figures 2 and 3. At both large and small scales the fault pattern is complex.

Continuity and irregularities of faults in depth and laterally

Faults in the Coeur d'Alene district characteristically are complex both in their lateral continuity and with depth. A pattern of branching and anastomosing shears describes faults both in vertical section and in map plan, and at scales ranging from faults several kilometers long to fractures and shears a few meters or tens of meters long. At all scales individual faults are complex and undulating, not planar, and at all scales a network of fractures rather than a single rupture surface is characteristic.

The complexity of faults at various scales may be demonstrated by a comparison of maps and cross-sections prepared at different scales. At a scale of a few kilometers to a centimeter a map of the larger faults, such as the Osburn and Placer Creek faults, and their related branching faults, including the Alhambra and Polaris, shows a complex zone of shearing from 4 to 6 km wide (Fig. 1). At a scale of a few hundred meters to a centimeter a cross section through the Coeur d'Alene mine shows a complex network of faults as seen in a vertical plane (Fig. 2). Significantly, the density of fault traces shown is greatest where the information is densest, that is, where the fault system has been intersected by workings at vertical intervals of 61 m (200 feet). Such a density of faulting and degree of complexity cannot be confirmed where faults are intersected only at the surface and at one or two levels below; for example, note the simple, planar interpretations of the Polaris and St. Elmo faults.

The complexities of fault zones in detail are represented on mine maps at original scales of 1 cm to 6 m (1 inch to 50 feet) and larger. A detail from the Coeur d'Alene mines is shown in Figure 3. Note especially the differences in the fault at points A, B, C, and D. The fault changes in strike and width over short distances, in number of strands, and in the dip of individual strands of the fault in the same exposure of the fault zone. Projection of such a fault to more than a few hundred meters thus is difficult.

Because of large vertical and horizontal dimensions and complexities in both fault and vein zones, exploration in the Coeur d'Alene district has depended in large measure on extensive shaft-sinking, crosscutting, and drifting. Drilling and coring are helpful in exploring to distances of a relatively few hundred meters, but incomplete recovery of core and ambiguities in resolving the complexities of the fault and vein zones where cut, have led to the practice of obtaining the larger samples and more complete record afforded by mine drifts and crosscuts. Such a practice may also be desirable in studying source zones of earthquakes.

Continuity at depth cannot be assumed even for some of the largest faults. Many large strike-slip faults, for example, are in reality tear faults in the upper plates of low-dipping thrust sheets. It is commonly assumed that such faults terminate at the thrust plane at a steep angle, but in many areas this is not true. In the Burgin mine of the East Tintic mining district of central Utah (Morris and Lovering, 1979) the ore-bearing Eureka Standard tear fault strikes N45°E and dips about 55°NW from the surface to the 1,400-foot level of the Eureka Standard mine, where it has about

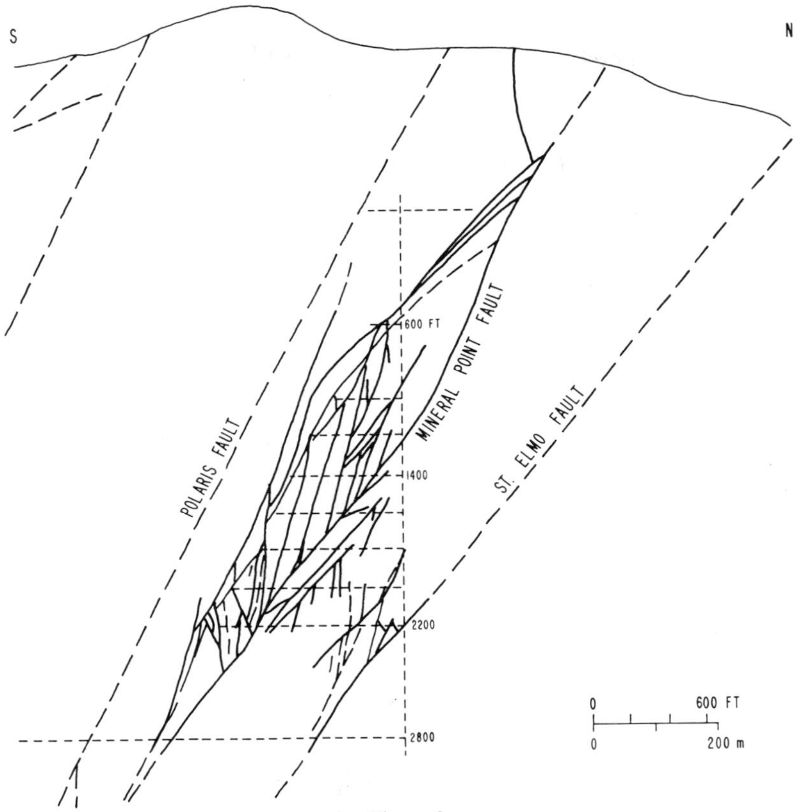

Figure 2
Cross section through the Coeur d'Alene mine showing complex anastomosing fault pattern exposed at 200-foot intervals in workings (short dashes). By contrast, note simplistic, planar interpretation of the St. Elmo and Polaris faults. Where adequately exposed, the faulting can be shown to be relatively closely spaced and complex. Compare with Figures 1 and 3.

762 m (2,500 feet) of right-lateral horizontal displacement. As this fault is followed in mine galleries eastward toward the concealed trace of the northwest-striking, southwest-dipping East Tintic thrust, the dip gradually decreases and the strike gradually turns northward, until the fault merges imperceptibly with the East Tintic thrust (Fig. 4). Other tear faults in the Burgin mine merge into the thrust in the same manner, accounting for different amount of tectonic transport on different parts of the thrust plate.

Few faults are truly planar features. One of the best documented deep-breaking undulating strike-slip faults is the Lakeshore fault of the Kirkland Lake gold-mining district in Ontario, Canada (Fig. 5). This district contains the deepest mines in North America—the Lake Shore and Wright-Hargreaves mines—reaching depths of 2,461

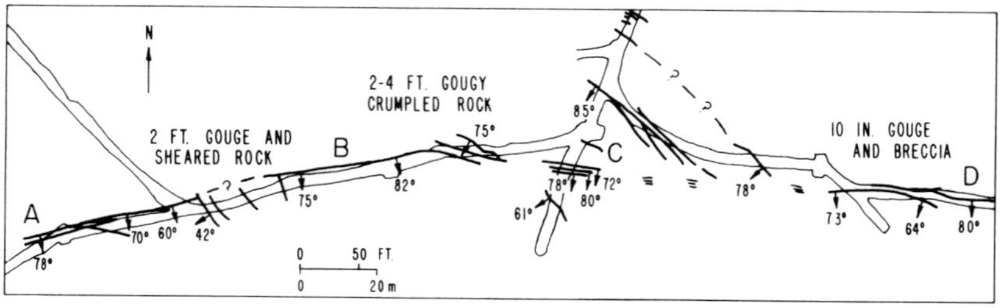

Figure 3
Characteristic irregular pattern of faults exposed in mine workings of Coeur d'Alene mine. Note change in strike and dip over short distances and branching habit. Near A, dips range from 60° to 78° over a strike length of 30 m (100 feet); near C, strands within the fault zone range in dip from 72° to 80°; between B and D the width of gouge ranges from 4 cm to 1.2 m (10 inches to 4 feet). Note density and complexity of faulting, as mapped at this large scale, compared with smaller scales shown in Figures 1 and 2.

m (8,075 feet) and 2,491 m (8,172 feet) respectively in 1963 (CHARLEWOOD, 1964). In these mines the Lakeshore fault trending N12°–25°E has been followed to a depth of more than 2,332 m (7,650 feet) and, according to HOPKINS (1950), shows no evidence of termination or even of diminished throw in the deepest mine workings. From the surface to a depth of about 500 m the fault dips east-southeast at about 70°; at this

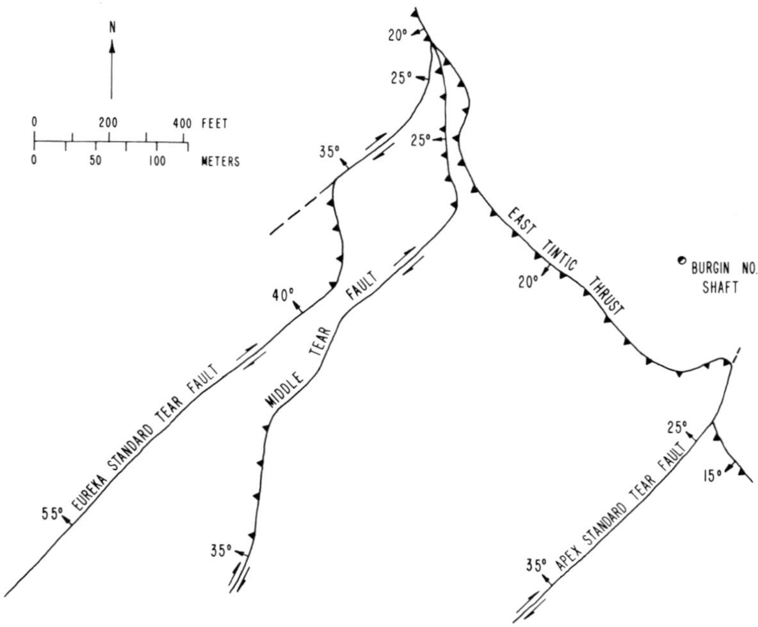

Figure 4
Structural diagram of part of 1,050-foot level of the Burgin mine, East Tintic district, Utah, showing relationship of the Eureka Standard and other tear faults to the East Tintic thrust.

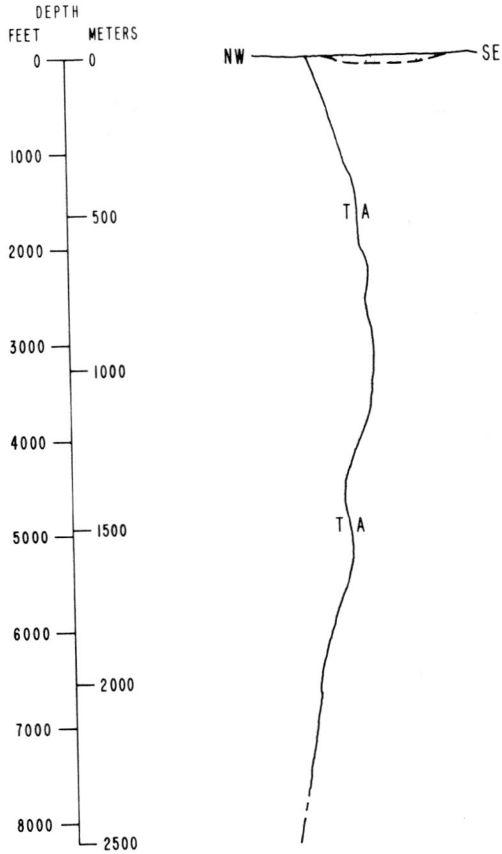

Figure 5
Cross section of Lakeshore fault in syenite porphyry, Wright-Hargreaves mine, Kirkland Lake district, Ontario, showing curving fault plane: T, block that moved, toward viewer; A, movement away from viewer. Redrawn from CHARLEWOOD, 1964, cross section 18.

point the dip steepens, and at a depth of about 900 m the fault zone becomes somewhat sinuous but rolls over and assumes a dip of about 80° west-northwest. Locally in both of the mines the fault plane is vertically under its surface position at a depth of about 1,800 m. Horizontal displacement on the Lakeshore fault is about 198 m (650 feet) and vertical displacement is about 99 m (325 feet). The east side has moved generally north, and north-plunging striae occur on the fault plane. The fault, which mainly cuts syenite porphyry, is sharply defined and generally contains zones of clayey gouge 1 to 5 cm thick and locally some pods of chloritic gouge about 30 cm thick.

As shown in Figures 6 and 7, even greater vertical irregularities are displayed in the East Beck and Centennial faults of the main Tintic mining district, Utah (MORRIS,

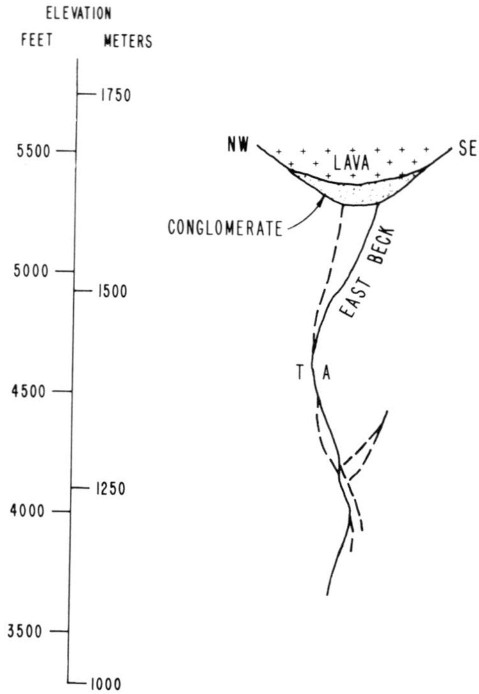

Figure 6
Cross section of East Beck fault between 1,200- and 2,700-foot levels of Chief No. 1 mine, Tintic mining district, Utah, showing sinuosity of fault plane. Fault chiefly cuts limestone and dolomite. Dashed lines, minor related fractures; T, block that moved, toward viewer; A, movement away from viewer.

1968). Both of these faults are exposed in mine workings to depths of 825 and 550 m, where they cut the vertical west limb and nearly flat trough of the Tintic syncline. These faults both strike northeasterly; the dip of the Beck fault ranges from 60°NW to 70°SE, and that of the Centennial fault, from 37°SE to 70°NW. The displacement on both faults is left lateral and is about 550 m on the Beck and 335 m on the Centennial. Horizontal mullions 1 to 3 m deep on the Beck fault indicate virtually no vertical component of displacement.

Irregularities on the mineralized fracture (vein) of the O'Brien gold mine, Quebec, are depicted by structural contours (Fig. 8). Ore has tended to concentrate in elongate corrugations on the original fracture, presumably in zones of relative extension. The irregularities on this fracture surface might exemplify one type of asperity, which on active faults accounts for stress concentrations, locations of microearthquakes, or barriers separating slip segments.

Some fault and shear zones have many forms other than simple planar or curved surfaces. The term *shear zone* usually distinguishes these relatively broad brecciated or sheeted zones from those on which the tectonic movement is concentrated in a

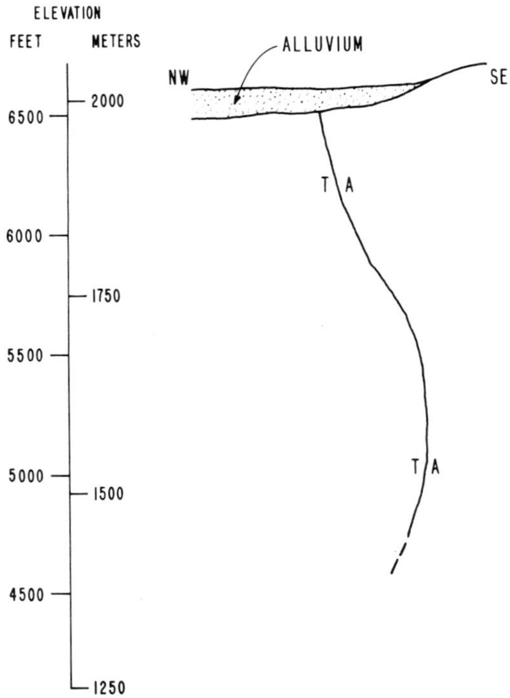

Figure 7
Cross section of Centennial fault in southern part of Chief No. 1 mine, Tintic mining district, Utah, showing curving nature of fault plane. Fault chiefly cuts limestone and dolomite. T, block that moved, toward viewer; A, movement away from viewer.

narrower zone. Displacement on shear zones is not uncommonly distributed through a volume of rock that is wide compared to its lateral dimensions or its total movement.

Shear zones that display strong schistosity may develop in any of a variety of rock types ranging from fine-grained quartzites and siltites to argillites, but generally shear zones are not as pronounced in hard, coarse-grained, relatively brittle quartzites or massive carbonate rocks. The platey, micaceous minerals of the argillites favor the development of foliated shear zones, and where smaller bodies of quartzite are surrounded by argillite, very thin tablets and lenses of quartzite bounded by shear planes (Fig. 9) may develop in response to the pervasive shearing through the argillite, which becomes phyllitic under shearing.

Some ore bodies display a foliated structure indicating shearing, even in massive galena (FRYKLUND, 1964). Such shearing has developed across bodies of ore many meters wide, even though displacement of rock units across the zone demonstrate relatively little total offset. Thus the zone at first may have been one of relative

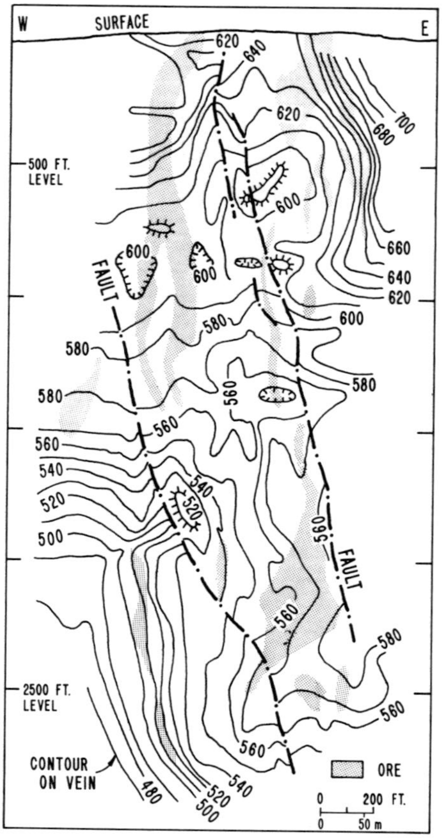

Figure 8
Structural contours (relative to an assumed, nearly vertical plane) on mineralized fracture (vein) at the O'Brien gold mine, Quebec, showing that fracture is very irregular rather than planar. The ore bodies are localized along elongate corrugations on the fracture which plunge nearly vertically. After MILLS, 1950.

extension, which served as a host for mineralization, but which presented a zone of weakness to later stresses of slightly different orientation having a component of shear parallel to the fracture.

Zones of fault gouge, sheared rock, and Breccia

Observations of fault gouge zones in mines demonstrate considerable variation in geometry and width. Of more than 60 faults in the Coeur d'Alene mining district described by HOBBS et al. (1965), essentially all are composed of three elements: a discrete clay-like gouge zone, a zone of sheared rock, and a zone of fractured rock.

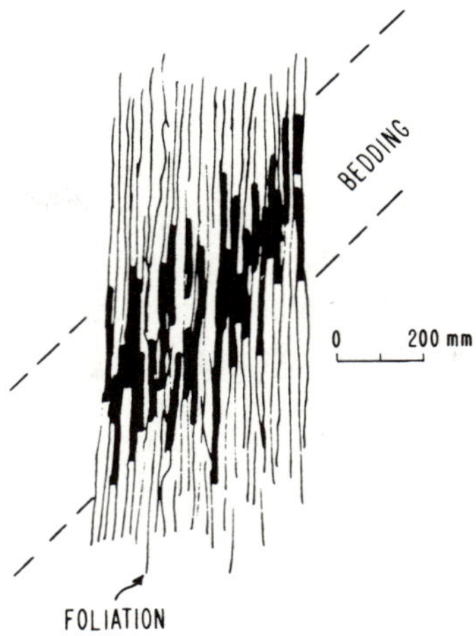

Figure 9
Foliated rock and distributed shear. Cross section of interbedded quartzite (black) and argillitic rocks (white) disassembled by displacement of a few centimeters to a few meters along closely spaced shear planes or foliation, in the steep flank of a major anticline.

The gouge zone generally is a fraction of the total width of the fault zone. Representative values are given in Table 1. In general, the larger the fault displacement,

Table 1

*Relation of displacement and width of fault zone in the Coeur d'Alene mining district, Idaho**

Fault	Displacement (km)	Width of gouge (m)	Width of shear (m)	Width of fractured rock (m)
Osburn	26	20–30	180	?
Dobson Pass	5.6	25	240	
O'Neill Gulch	1±	1±	1±	
Commander	0.6±		4–6	
Carlilse	0.6±	0.1	1	
White Ledge	0.3±	1	6	
French Gulch	0.3±	0.1±	0.3	12–16
Polaris	0.15–1.8	0.6–1.5	0.6–1.5	
Mineral Point	0.15±	1–3	1–3	
Sonora (Sonora Tunnel)	?	0.3		0.2–0.3
Sonora (Copper King), near crosscut	?	0.6		3–7
National	?	0.3		1–10
Blue Star	?	0.6		0.6–6
Standard	?	0.3	2.5	2.5
Frisco	?	0.15–0.45	1±	

* Data summarized from Hobbs et al., 1965.

the wider the zone of disruption. Given several kilometers of displacement, the width of the zone of faulting ranges between 100 and 300 m, of which a tenth or so is gouge. Given several tenths of a kilometer of displacement, the width of the zone of faulting is several meters and the gouge zone is between a tenth and one meter. The ratio of the width of the total fault zone to the width of the fault gouge, thus, is on the order of 10:1, and the ratio of displacement to width of gouge is of the order of 1000:1, through a range of fault displacements and types of faulting. These ratios perhaps are most applicable to faults that are in argillaceous rocks, but in quartzites gouge zones tend to be narrower, perhaps a tenth of the width in argillites, although the width of brecciated and fractured rock may be nearly the same in each lithology. Hobbs et al. (1965, p. 74) report the following about the Osburn fault zone.

"The width of the zone of fractured and sheared rock along the fault varies considerably from place to place, but in most places where the zone has been cut underground its width is more than a hundred feet. This zone contains numerous gouge seams, and quartzitic rocks within it are intensely brecciated and the more argillitic rocks are foliated and crumpled. The width of the zone, although dependent to a large extent upon the competency of the rocks bordering the fault at a particular point, is undoubtedly also greatly affected by the geometry of the fault zone with respect to stress distribution. For example, at a bend in the fault plane stresses may be relatively greater and so extreme brecciation may result. In the main crosscut of the Morning mine, near Mullan, the Osburn fault zone is 800 feed wide, but within this general zone of disturbance a more intensely sheared and brecciated part, 400 feet wide, undoubtedly accommodated most of the displacement. In still finer detail within the 400-foot-wide zone, four separate zones of clay gouge, 10–60 feet thick, are the major loci of movement along the fault. Contrasted to this, only 1.5 miles to the west where the Osburn fault is cut by workings of the Alice mine, Ransome (in Ransome and Calkins, 1908, p. 170) reported: 'About 350 feet from the portal the tunnel cuts through the Osburn fault here marked by 1–2 feet of soft dark clay gouge accompanied by much shattered quartzite. In fact, the Burke beds are broken or shattered for at least 150 feet from the gouge.'

"The reason for this difference in the width of the fault zone and intensity of brecciation and crushing at points so near to one another on the same linear segment of the fault is not obvious.

"Argillitic rocks of the Wallace Formation border the south side of the fault and quartzitic rocks border the north side at both localities, although at the Morning mine the quartzite beds are impure and are part of the St. Regis Formation, whereas at the Alice mine the quartzite beds are relatively pure and are considered to be part of the Revett Quartzite. The difference in competency of the quartzite beds may be partly responsible for the difference in the effect of faulting, but it seems unlikely that this difference accounts for such a wide variance. In the main crosscuts of the Golconda and Granada mines about 1.5 and 2 miles, respectively, west of the Alice mine, the intensity of deformation along the fault compares more nearly to that in the Morning crosscut. At the Golconda mine the fault zone is made up of several strands of gouge plus breccia and foliated rocks in a zone more than 300 feet wide. At the Granada, the zone south of a cave-in which blocked the crosscut, the fault zone is more than 150 feet wide and the rocks are complexly foliated, fractured, and pulverized as in the Golconda crosscut. Argillite of the Wallace Formation borders the south side of the fault at both the Golconda and Granada mines. Impure quartzite of the Burke Formation lies north of the fault at the Golconda mine, and quartzose argillite of the Prichard Formation lies north of the fault at the Granada mine.

"In the main adit of the Silver Dollar mine near the west edge of the Wallace map area, the Osburn fault consists of a zone nearly 200 feet wide that is composed of numerous fault stands, including about 80 feet of gouge and breccia, where the adit is tightly lagged. In the main crosscut at the Bunker Hill mine in the Kellogg map area, the fault zone is reported to be composed of only a few feet of gouge plus some phyllitic rock and breccia, but the details of the fault zone are now hidden by a lining of concrete."

These descriptions indicate the difficulty of correlating variances in width of gouge only to lithology. Local geometry of the fault plane probably has had an important influence.

The gouge zone may lie near the center of the zone of shearing and brecciation, or at one margin of it, or it may be in several discrete bands woven throughout the zone (Fig. 10).

Garth Crosby (personal communication, February 1979) reports that in the Coeur d'Alene mining district many gouge zones have developed at the margins of fault zones and appear to have intruded the unbroken wall rock. According to Crosby, a gradation can be seen, from unbroken wall rock, to wall rock containing thin, incipient shear planes with gouge, to highly sheared and broken rock. Presumably the gouge, behaving plastically, flows into small cracks and breaks off blocks of country rock during displacement. For a while, individual blocks broken off the wall-rock block may retain their integrity even though surrounded by comminuted gouge-like material, but eventually they become completely comminuted, according to Crosby.

In the Osburn Fault zone, as shown in Figure 11, the principal gouge zone, several tens of meters wide, tends to be more or less centered within the fault zone, which is a few hundred meters wide, although small shear and gouge zones splay into the wall rock on either side.

Very commonly gouge zones form an anastomosing pattern throughout the fault zone, according to Hobbs et al. (1965). For example, on the Blue Jay fault zone, they say, "the fault trace .... is along the north side of a broad zone so severely sheared and altered that it is difficult to delineate individual faults. Innumerable shear planes

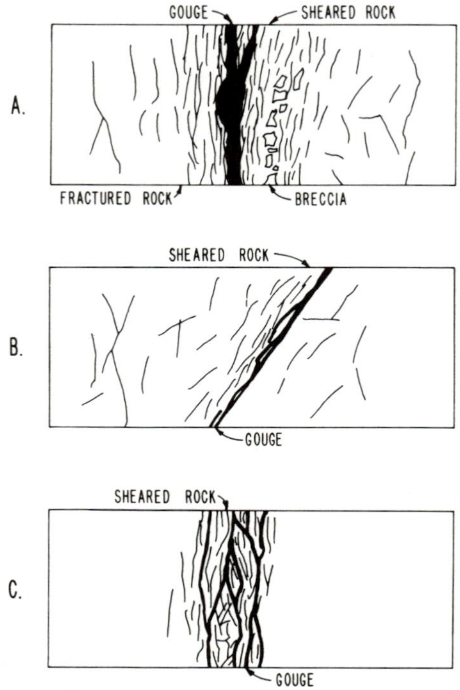

Figure 10
Diagrams showing different positions of gouge within fault zone: A, gouge zone near center of fault zone; B, gouge localized at one boundary of fault zone; C, thin gouge seams forming anastomosing pattern through fault zone.

and gouge-filled seams suggest movement, but there is no way of identifying those in which significantly more movement occurred."

Permeability of fault and shear zones

Faults and shear zones commonly are conduits for ground waters, but range from structures that are tight and dry to wide permeable zones that may contain large volumes of water. In the Coeur d'Alene district the faults and shear zones are almost invariably wet where exposed in the mine workings. The gouge material commonly is saturated with water and is plastic and readily shaped by hand, like modelling clay. Mine openings require timbering to support the moist, highly fractured, and sheared rock. Most such zones 'make' water; that is, water percolates along the fault, in some places slowly, so that only a few liters an hour drain out of the rock, but in other zones faster, at many liters a minute. Some zones are tight and relatively impermeable, ponding ground water to a greater extent on one side than

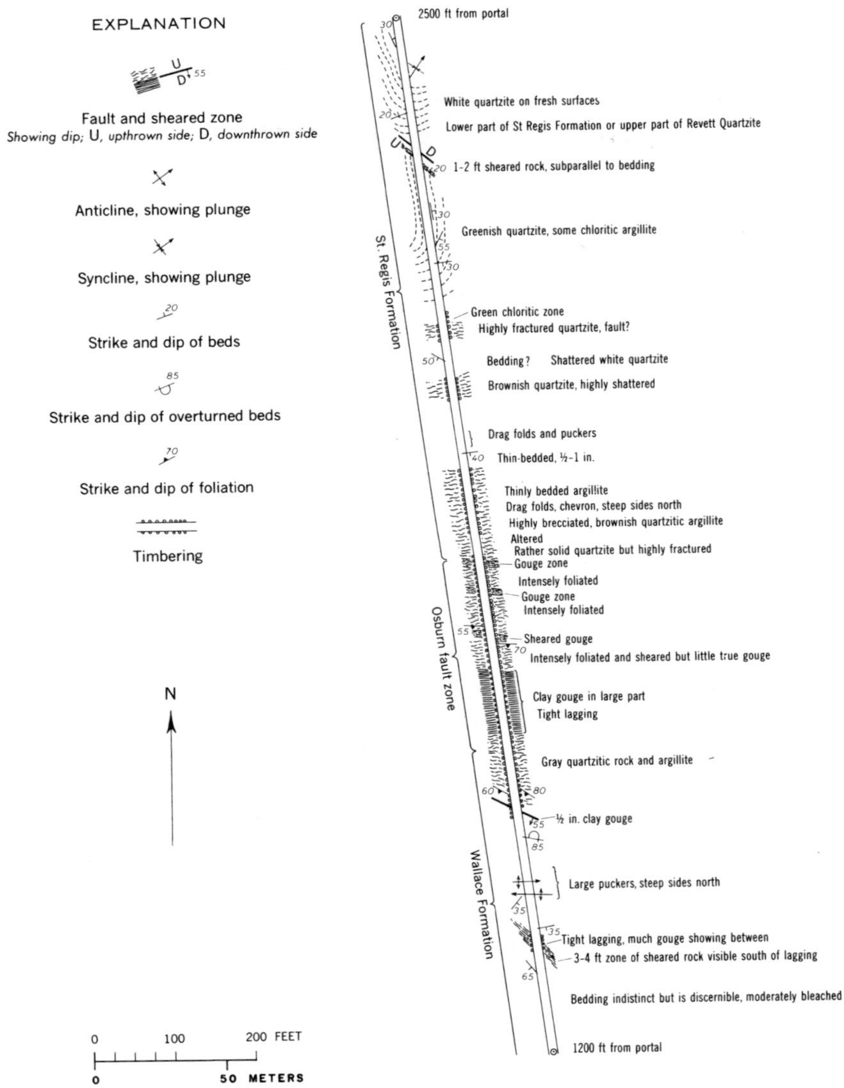

Figure 11
The Osburn fault as exposed in the main crosscut of the Morning mine is between 70 and 125 m wide, 24 m of which are predominantly clay-like gouge. The fault has 26 km right-lateral strike slip. From HOBBS and et al., 1964, p. 75.

on the other. Not uncommonly mine galleries intersect water channels that are in what seems to be only moderately fractured rock, and yet the fractures are obviously highly permeable, because large amounts of water flow from the rocks. By contrast, the nearest large shear zones, perhaps hundreds of meters away, produce water at only a small fraction of the rate produced by the moderately fractured rock. The

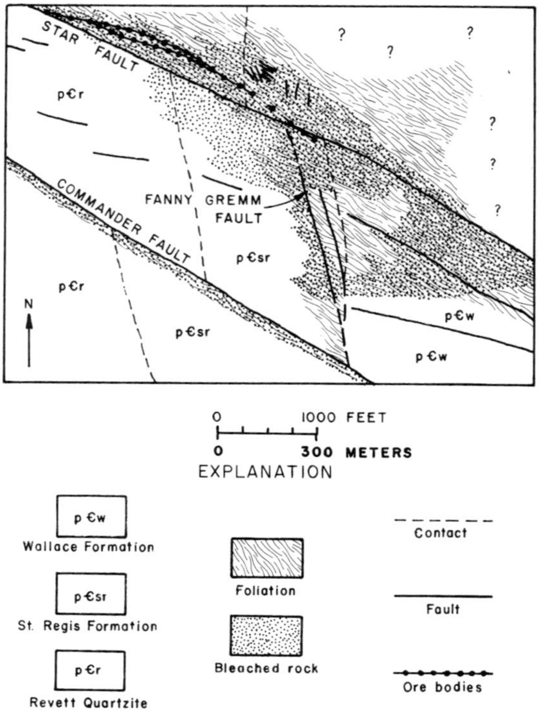

Figure 12
Hydrothermal solutions (see bleached rock) in some places are channelled along faults, and elsewhere faults present a barrier to the fluids. From HOBBS et al., 1965, p. 101.

coarser-grained quartzites of the Coeur d'Alene district appear to be able to support open fractures that provide channels of high permeability.

Similar conditions prevail in the main Tintic district, where mining recently was terminated because of the high pumping costs required to dewater the ore-bearing areas. In the Gemini and Chief No. 1 mines it was discovered that caverns, developed in carbonate rocks along faults and shear zones near the original water tables at the 1,500-foot level, could be used for the disposal of large volumes of water pumped from the 3,100-foot level and workings above it. Obviously the caverns and fault zones provided sealed conduits which extended to water-free areas hydrologically lower than the apparently perched water table in the mining district. In the Tintic district, as in the Coeur d'Alene district, some fault zones are effective dams for ground water. An example is the Eureka Lilly fault in the North Lily mine, northwest of which the ground water stands almost 50 m higher than on the southwest side, although the fault cuts normally permeable quartzite in this area.

In both the Tintic and Coeur d'Alene districts many faults appear to have provided access for hydrothermal solutions. An example is shown in Figure 12. In some places the hydrothermal alteration is confined to only one side of a fault, indicating ponding and diversion of the solutions.

In summary, faults commonly are conduits for water and in the past have served as channels for hydrothermal solutions. However, in many mining districts, they may not be the only permeable zones and even may be far less permeable than less conspicuous fractures and pervasive joint systems having little displacement.

Faulting versus foliation

Foliated rocks such as phyllites and schists are common in the Coeur d'Alene district and, certainly, some have an origin related to shearing and faulting. Sheared argillites in and near fault zones almost everywhere have a schistose structure. The Deadman Shear zone, which is several hundred meters wide, is a good example of a diffuse zone of shearing (Fig. 13). The body of foliated rocks itself is believed to represent large-scale, strike-slip shear distributed relatively uniformly across a broad zone. But later disruption of the foliated body by shear, represented by concentrated zones of gouge a few centimeters across, also has taken place. In addition, kink banding has developed in the highly foliated body. The planes of kink banding dip at various low angles, and the intersections of kink-band planes are subhorizontal, suggesting predominantly vertical displacement as the last phase of deformation within the zone of shearing originally produced by strike slip.

Distributed shear is also indicated in the steep flanks of large anticlines, those having amplitudes and wavelengths of a kilometer or more. Beds of quartzite are broken in places by closely spaced shear zones, into thin tablets at most a few centimeters across, although the entire bed may not be offset appreciably (Fig. 9). A common error in geologic mapping has been to consider the vertical shear planes as bedding, because in detail they are lithologic boundaries. The true bedding planes (see dark unit) cross the foliation or shear planes at an angle.

Discussion and significance for seismology

These descriptions of faults in three dimensions suggest several features that would have a bearing on the earthquake process. The fact that no faults are perfectly planar and that large irregularities characterize many of them provides a simple, geometric model for visualizing asperities or locked zones along faults. The irregularities are at many scales, and slip on faults seems to be difficult because of the bends, branches, and buckles. Many large faults are not persistent to depths as coherent structures.

The similarity of density and complexity of faulting, whatever the scale of mapping, shows that the crust of the earth, in a region such as the Coeur d'Alene mining

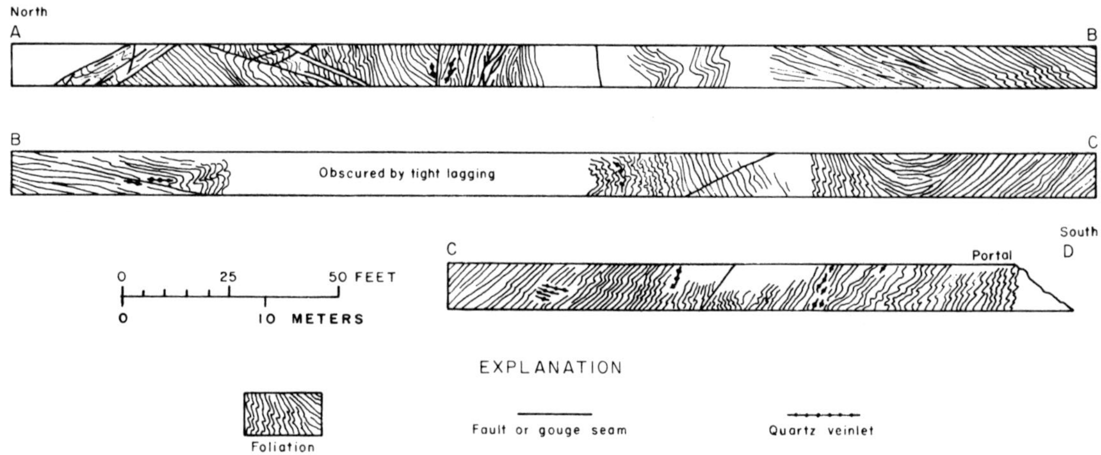

Figure 13
The Deadman shear zone, Coeur d'Alene mining district, is composed of well-foliated rocks, highly sheared and folded. Folds range from tens of meters across to kink bands a few centimeters across, representing planar shears of a few centimeters of displacement. Folds and kink bands represent secondary deformation. From HOBBS et al., 1964, p. 110.

district, Idaho, or along the San Andreas fault in California, is not a solid, unbroken block but, rather, a composite of blocks of a wide range of sizes, separated from each other by faults. Deformation of such a cluster of 'grains' must be different from an isotropic, unbroken, virgin piece of rock.

The characteristic complexity and network of faults of many sizes makes the number and wide distribution of aftershocks seem very expectable. The network of faults could adjust only gradually to a new state of grain packing, and faults of such a range in size could account for the range in size of aftershocks.

The size and plasticity of wet bodies of fault gouge must introduce an element of ductility into the deformational system. Wet gouge must be rather mobile under tectonic stress, and it could be intruded into positions of extension or lowest compressional stress.

The exploration and testing of hypocentral regions by deep drilling is apt to be fraught with difficulties because of the inherent complexities and irregularities of fault zones.

References

Charlewood, G. H. (1964), *Geology of deep developments of the main ore zone at Kirkland Lake: Compilation.* Ontario Dept. Mines Geol., *Circular 11*, 49 pp.

Fryklund, V. C., Jr. (1964), *Ore deposits of the Coeur d'Alene district, Shoshone Country, Idaho.* U.S. Geol. Surv. Prof. Paper 445, 103 pp.

Hobbs, S. W., Griggs, A. B., Wallace, R. E. and Campbell, A. B. (1965), *Geology of the Coeur d'Alene district, Shoshone County, Idaho.* U.S. Geol. Surv. Prof. Paper 478, 139 pp.

Hopkins, H. (1950), *Geology of the Wright-Hargreaves mine.* Ontario Dept. Mines Ann. Rept., 1948, 57, Pt. 5, pp. 161–176.

Mills, J. W. (1950), *Structural control of ore bodies as illustrated by the use of vein contours at the O'Brien gold mine, Cadillac, Quebec.* Econ. Geol. *45*, 786–807.

Morris, H. T. (1968), 'The main Tintic mining district, Utah', in *Ore Deposits of the United States, 1933–1967.* Am. Inst. Min. Met. Pet. Engrs., II, pp. 1043–1073.

Morris, H. T. and Lovering, T. S. (1979), *General geology and mines of the East Tintic mining district, Utah and Juab Counties, Utah.* U.S. Geol. Surv. Prof. Paper 1024, 203 pp.

Ransome, F. L. and Calkins, F. C. (1908), *Geology and ore deposits of the Coeur d'Alene district, Idaho.* U.S. Geol. Surv. Prof. Paper 62, 203 pp.

Wallace, R. E., Griggs, A. B. Campbell, A. B. and Hobbs, S. W. (1960), 'Tectonic settings of the Coeur d'Alene district, Idaho', in *Short Papers in the Geological Sciences.* U.S. Geol. Surv. Prof. Paper 400-B, pp. B25–B27.

Wallace, R. E. and Morris, H. T. (1979), 'Characteristics of faults and shear zones as seen in mines at depths as much as 2.5 km below the surface', in *Proc. Conf. VIII, Analysis of Actual Fault Zones in Bedrock* J. Evernden (ed.). U.S. Geol. Surv. Open-File Rept. 79-1239, p. 79–100.

(Received 25th September 1985, revised 14th March 1986, accepted 17th March 1986)

PAGEOPH, Vol. 124, Nos. 1/2 (1986) 0033-4533/86/020127-14$1.50+0.20/0

Gravity Anomaly and Density Structure of the San Andreas Fault Zone

CHI-YUEN WANG,[1] FENG RUI,[2] YAO ZHENGSHENG[3] and SHI XINGJUE[4]

Abstract—A densely spaced gravity survey across the San Andreas fault zone was conducted near Bear Valley, about 180 km south of San Francisco, along a cross-section where a detailed seismic reflection profile was previously made by McEVILLY (1981). With FENG and McEVILLY's velocity structure (1983) of the fault zone at this cross-section as a constraint, the density structure of the fault zone is obtained through inversion of the gravity data by a method used by PARKER (1973) and OLDENBURG (1974). Although the resulting density picture cannot be unique, it is better constrained and contains more detailed information about the structure of the fault than was previously possible. The most striking feature of the resulting density structure is a deeply seated tongue of low-density material within the fault zone, probably representing a wedge of fault gouge between the two moving plates, which projects from the surface to the base of the seismogenic zone. From reasonable assumptions concerning the density of the solid grains and the state of saturation of the fault zone the average porosity of this low-density fault gouge is estimated as about 12%. Stress-induced cracks are not expected to create so much porosity under the pressures in the deep fault zone. Large-scaled removal of fault-zone material by hydrothermal alteration, dissolution, and subsequent fluid transport may have occurred to produce this pronounced density deficiency. In addition, a broad, funnel-shaped belt of low density appears about the upper part of the fault zone, which probably represents a belt of extensively shattered wall rocks.

Key words: Density structure of the San Andreas fault zone.

Introduction

Density in a fault zone reveals important information about its composition and its constitutive properties. For example, fracture porosity and thus the amount of pore fluids in a fault zone, which play a primary role in determining the fault zone's constitutive properties (BOOKER, 1974; RICE and RUDNICKI, 1979; LACHENBRUCH, 1980), can be evaluated if the density structure of the fault zone is determined. Previous modellings of the density structure of the San Andreas fault zone (PAVINI, 1973; WANG et al., 1978; ROBBINS, 1982; STIERMAN, 1984) have pointed out that the

[1] Department of Geology and Geophysics, University of California, Berkeley, California 94720.
[2] Institute of Geophysics, State Seismological Bureau, Beijing, China.
[3] Institute of Seismology, State Seismological Bureau, Beijing, China.
[4] Department of Geophysics, Chinese University of Science and Technology, Hefei, China.

density in the fault zone may be relatively low in comparison to the densities of the wall rocks; but owing to the inherent non uniqueness in the inversion and in the interpretation of gravity data, and to the lack of independent constraints, these studies can provide only grossly averaged pictures of the density structure of the fault zone.

In 1978 a detailed seismic reflection profiling across the San Andreas fault zone was carried out near Bear Valley (McEvilly, 1981), about 180 km south of San Francisco, California. Inverting this data set, Feng and McEvilly (1983) obtained a detailed velocity structure of the fault zone and the surrounding crust. This velocity structure provides a most useful constraint for density inversion from gravity data.

We carried out, in 1983, a detailed gravity survey across the San Andreas fault near Bear Valley along the same route of the seismic profiling. The results of this survey have yielded a dense and evenly distributed set of gravity data which provides an excellent control on the lateral gradient of the gravity anomaly, which turns out to be an important criterion in constraining density models for the subvertical structures in the fault zone. On the basis of this new data set and the seismic results we are able to provide a picture of the deep density structures of the San Andreas fault zone, one which is better constrained and contains more detailed information than was previously possible.

Experiment

1. Gravity survey

Figure 1 shows the regional location of our gravity survey. Beginning in the Gabilan Range of the Salinian Block, it extends northeast, crossing the San Andreas fault zone at a latitude of about 36° 34′ and ending in the Diablo Range to the northeast. This section of the San Andreas fault is among the most seismically active sections in northern California. Figure 2 shows the relative position of our survey line with respect to the local gravity field (Bishop and Chapman, 1967). Owing to the sparse station density, this gravity map does not provide sufficient constraint for precise determination of the lateral gradient of the gravity anomalies.

Comparing Figures 1 and 2, one sees a broad correlation between the regional geology and the gravity anomalies. Whereas the granitic and metamorphic terrains of the Gabilan Range show a relative gravity high, the sedimentary section in the Diablo Range is characterized by lower gravity. A steep gradient appears on the west side of the San Andreas fault zone, which separates the two ranges.

Two identical LaCoste–Romberg gravity meters (Model G) were used in our survey. One was used for routine measurements in the field; the other was kept at a sheltered base station throughout the period of the field experiment for continuous

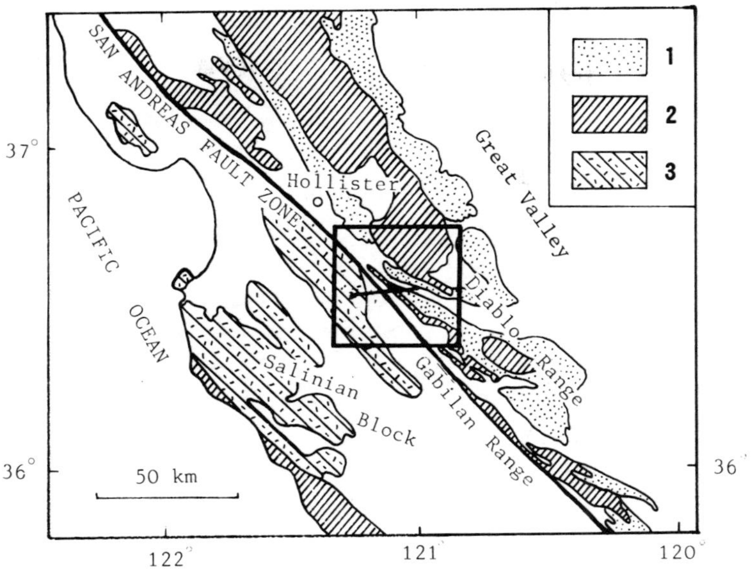

Figure 1
Generalized geoligic map of the studied area (modified from Walter and Mooney, 1982). The position of the present gravity survey is indicated by the straight line across the fault. The small box at the center of the map shows the position of Figure 2. Index 1 is the Great Valley Sequence, index 2 is the Franciscan Assemblage, and index 3 is the granitic and metamorphic terrains.

records of tides. Both gravimeters had been calibrated by the manufacturer just before our field work. A U.S. Geological Survey reference gravity station at the nearby Hollister airport provided the tie to the global gravity network.

By means of a large-scaled topographic map giving precise locations of the seismic vibrator points used during the reflection survey (McEvilly, 1981) we were able to reoccupy the vibrator points during our field work. A total of 162 evenly spaced gravity stations along a profile about 21 km in length was occupied in this way. Some stations were repeatedly occupied during the course of each day (and throughout the surveying period) for the purpose of monitoring instrumental drift. Sketches of the local land masses and measurements of their slopes were made at each station for close-in terrain corrections, to be done at a later time.

2. *Data reduction*

The measured gravity values were first corrected for tidal variations through the continuous digital records of the stationary gravimeter at the base station. The secular instrumental drift was corrected by repeated readings at certain stations during fixed hours of a day throughout the field experiment.

The elevations of our stations had been determined previously by a surveying

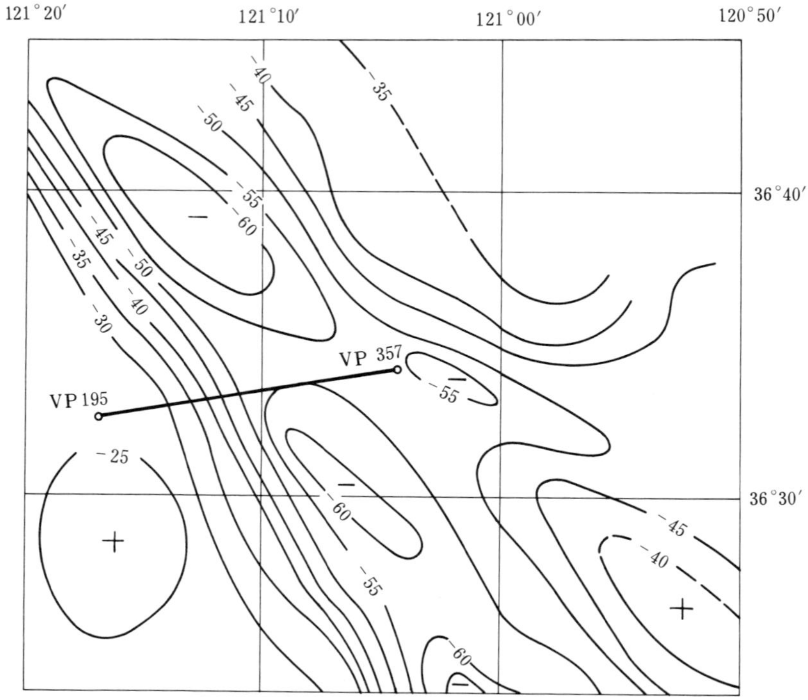

Figure 2
Bouguer gravity map of the studied area (modified from Bishop and Chapman, 1967); the location of this map is given in Figure 1, and the position of the present survey is given by the heavy line. VP, vibrator point.

team as part of the seismic experiment, and they are accurate to within 0.3. Corrections for the free-air effect are thus accurate to 0.1 mgal. A Bouguer density of 2 670 kg/m^3 was used in the single Bouguer correction. Considering the general low-density sediment coverage in the studied area, we judge that this density may be too high. A different choice of Bouguer density, however, will not alter the lateral differences in the gravity anomaly, which are the essential information for the present density modelling.

The latitudes and longitudes of the gravity stations were read directly from 7.5 min topographic maps. These data were input to a VAX computer at the U.S. Geological Survey, where digitized terrain data were used for the terrain correction at each gravity station to an outer radius of 166.7 km; the magnitudes of these corrections range from 2 to 7 mgal. Finally, manual corrections for the effects of land masses around the gravity stations were made to an outer radius of 50 m at about every ten stations by use of the field notes regarding the local distribution of land masses. The manual corrections were simplified by selecting stations where the local topography on the two sides of the survey route can be approximated by two

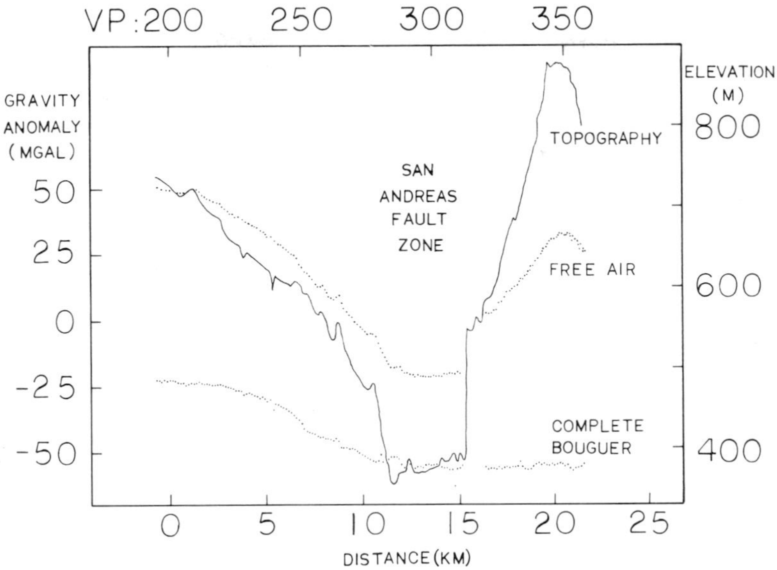

Figure 3
Free-air gravity anomaly, complete Bouguer anomaly, and topography along the route of the present survey across the San Andreas fault zone. CP, vibrator point.

half-planes; an 'inclined plane' solution from OLIVER and ROBINS (unpublished results) was used in estimating the magnitude of this effect, which turns out to be less than 0.3 mgal for all our stations where such corrections were made. This 'inclined plane' solution is similar to the solution of SANDBERG (1958) for an infinite plane, except that a 'kink' inserted in the inclined plane represents the survey route.

3. *Gravity anomalies*

The profiles of the free-air gravity anomalies, the complete Bouguer anomalies, and the topography along the route of our survey are given in Figure 3. In order to tie the gravity profile to the seismic profile we have marked at the top of this figure the numbering of the vibrator points (VP). The Bouguer anomaly shows a relative high of about -20 mgal at the southwest end of our profile, decreasing to a minimum of -53 mgal in the San Andreas fault zone. Thus the transition from the relative gravity high on the west of the San Andreas fault zone to the relative gravity low across the fault zone takes place over a broad region, about 10 km in width. The gradient of the gravity anomalies decreases eastward, first gently then steeply, reaching an absolute magnitude of 9 ± 1 mgal/km near VP 250, and then increases gradually until it becomes zero near VP 300, where the Bouguer anomaly reaches its minimum. Then the gradient becomes positive but remains close to zero within the surveyed area. This refined control on the lateral gradient of gravity anomaly turns out to be quite important in a discrimination between different density models for the subvertical fault structures.

Density Inversion

A rapid computation scheme was given by PARKER (1973) for forward calculation of the gravity attraction of a two-dimensional, uneven layer of constant density. OLDENBURG (1974) rearranged Parker's equation into a form suitable for iterative inversion procedure. We summarize their main results as follows.

Take the x–y plane to represent a horizontal surface, and let the z-axis point upwards. If the lower and the upper boundaries of a layer are specified by $z = 0$ and $z = h(x)$ respectively, the Fourier transform of the gravity anomalies is related to that of the topography of the layer by the following equation (PARKER, 1973),

$$F[g(x)] = -2\pi G\,\Delta\rho \exp\{-|k|z_0\} \sum_{n=1}^{\infty} \frac{|k|^{n-1}}{n!} F[h^n(x)]$$

where k is the wavenumber of the transformed function, z_0 is the average depth of the upper boundary of the layer below the observational plane, $\Delta\rho$ is the given density contrast between the two layers, and G is Newton's gravitational constant. Transposing the $n = 1$ term from the summation and rearranging terms, OLDENBURG (1974) obtained:

$$F[h(x)] = -\frac{F[\Delta g(x)] \exp\{|k|z_0\}}{2\pi G\,\Delta\rho} - \sum_{n=2}^{\infty} \frac{|k|^{n-1}}{n!} F[h^n(x)]$$

As OLDENBURG (1974) showed, this equation can be used iteratively for calculating $h(x)$ when both $\Delta\rho$ and z_0 are given. A first-guess solution (i.e., the initial model discussed in the next section) is used in evaluating the right-hand side of the equation. The inverse Fourier transform of this quantity then gives an updated position of the density boundary. The iterative procedure was continued until the RMS value of the differences between the computed positions in two consecutive iterations became smaller than or equal to 1 m.

Besides taking the 162 gravity measurements along the 21 km profile, we extrapolated on the basis of the regional gravity map of BISHOP and CHAPMAN (1967) outward from both ends of our gravity profile to a total length of about 40 km. One

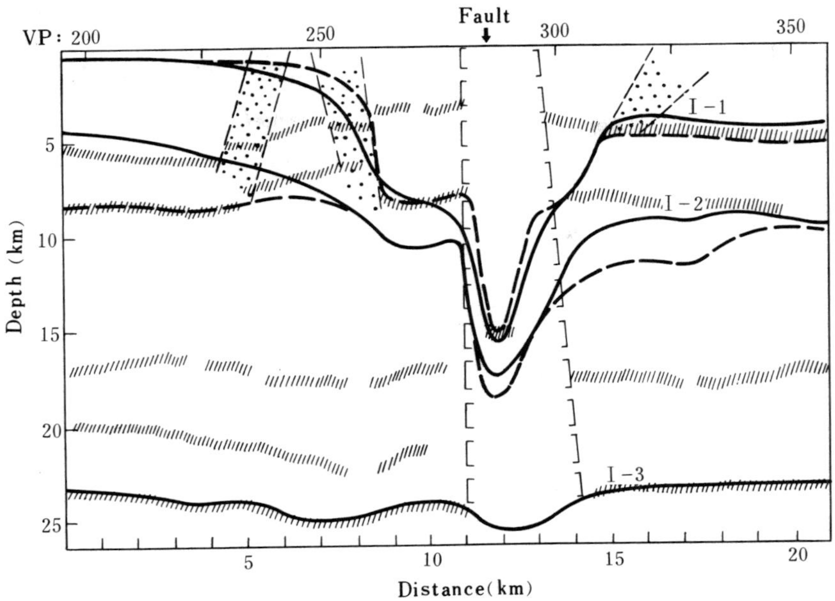

Figure 4
The initial density model (dashed curves) and the final density model (solid curves); note the close similarity between the two. VP, vibrator point. Curves I–1, I–2, and I–3 show the positions of the density interfaces, with density contrasts of 200, 170, and 410 kg/m^3 across the respective interfaces. Inclined slashes show positions of seismic reflectors inferred from reflection study (Feng and McEvilly, 1983). Dotted areas represent inferred faults or fractured rocks. The two near-vertical clumns of brackets indicate the inferred width of the San Andreas fault zone by Feng and McEvilly (1983).

reason for this extrapolation of profile length, as pointed out by OLDENBURG (1974), is to ensure that the two ends of the profile have approximately the same gravity values, as required in the computation of a finite transform. Another reason is to minimize the so-called edge effect in the computation. Before the inversion is made, the contribution due to the undulations of the Moho boundary (see Figure 4) is removed; this contribution amounts to no more than 0.1 mgal, given 400 kg/m^3 as the density contrast across the Moho. The inversion is then carried out by an iterative refining of the positions of the two crustal boundaries, one after the other.

The velocity structure shown in Figure 6 of FENG and MCEVILLY (1983) is used for defining the initial model of our density structure. One reasonable assumption for constructing the initial model would be that the seismic reflections inferred by FENG and MCEVILLY (1983), which are represented in Figure 4 by the subhorizontal bands of inclined slashes, also represent the boundaries between layers of different densities. A difficulty of this association is that the apparently coherent seismic reflectors shown by FENG and MCEVILLY (1983) cut across contours of equal siesmic velocity. From experimental studies of rocks and minerals one expects the changes in density to be associated with changes in seismic velocity. Therefore, choosing the seismic reflectors to represent the boundaries of distinct density layers would violate the empirical velocity–density relationship. For this reason we decided instead to choose two contours of about equal seismic velocity, 5.0 and 5.5 km/sec, which lie fairly closely to the upper seismic reflectors, as the boundaries between distinct density layers in our initial model; these are shown by the dashed curves in Figure 4. The Moho, naturally, is taken to be the density boundary between the crust and the upper mantle.

An average seismic velocity for each crust layer is then obtained from the velocity structure of FENG and MCEVILLY (1983), and the corresponding average density is obtained from an empirical velocity–density relation of NAFE and DRAKE (1957). The precise values of velocities and densities are not important for the calculation of a gravity anomaly; the important parameter is, instead, the density contrasts between adjacent layers, which are determined, by the procedure outlined above, to be 200, 170, and 410 kg/m^3, respectively, across the two crustal boundaries and the Moho.

Figure 4 shows the final model of the present inversion. It is seen that the final model does not differ substantially from the initial model. The most striking feature of this density model is a deeply seated low-density region inside the fault zone. This low-density region is in the shape of a wedge, extending to the base of the seismogenic zone at a depth of about 15 km. In addition, a broad belt of low density, in the shape of a funnel, occurs around the upper 8 km of the fault zone. The lower boundary of this funnel-shaped low-density belt follows closely the 5 km/sec velocity contour in the velocity structure shown in Figure 6 of FENG and MCEVILLY (1983).

Figure 5 shows the computed gravity anomalies for our final density model and compares them with the observational data. It shows that the agreement between the computed and the observed gravity anomalies is very good: the standard error of fitting is less than 1 mgal. A large part of this error arises from some high-frequency components in the observed field (Fig. 3 and 4) which are not reproduced in the computed field. Such high-frequency variations can only be due to surficial changes in mass distribution and may in part reflect the fact that the terrain corrections in the present study were made only at every ten points. No further refined adjustment of the model is made, because the gross fitting to the observed field is considered satisfactory.

An important result of the present study is that only minor modifications of the initial density model are required to yield a close agreement between the computed and the observed gravity anomalies. This demonstrates that our choice of the initial density model, based upon the given seismic velocity structure and the empirical velocity–density relation, is quite reasonable.

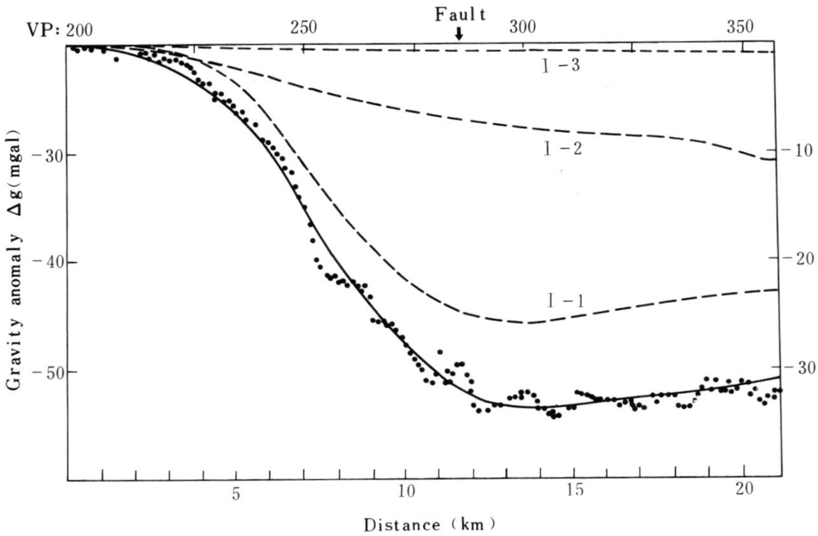

Figure 5
The computed gravity anomalies from our final density model (solid curve). Curves I–1, I–2, and I–3 show the gravity anomalies contributed by the three corresponding density interfaces given in Figure 4. Dots indicate the 'observed' Bouguer anomalies, given for comparison with the computed results.

Discussion

Although the density structure as obtained above cannot be unique and remains rather coarse in resolution, it is constructed on the basis of a densely and evenly distributed gravity data base, a detailed seismic velocity structure of the San Andreas fault zone along the same cross-section, and a well-accepted empirical relation between seismic velocity and density. It is therefore believed to be better constrained and contains more detailed information than was previously possible.

As expected, the lateral inhomogeneities in the surficial density distribution in the studied region play an important role in effecting the observed gravity anomalies. According to WALTER and MOONEY (1982), the sediment thickness on the southwestern side of the fault it the Gabilan Range is less than 2 km; on the northeastern side in the Diablo Range it varies between 0.2 and 3.4 km. Our density model (Fig. 4) put the depths of the basement at 1 to 2 km on the southwestern side of the fault and 5 km on its northeastern side. This variation in the thickness of the surficial layer accounts for about 75% of the observed lateral variation in the gravity anomalies (Fig. 5). The contribution of the second 'layer' to the observed gravity anomalies is only 24%, even though the vertical amplitudes in the variations of its position are as great as those of the first 'layer'. The small undulations of the Moho boundary in the studied area contribute less than 1% to the observed gravity anomalies.

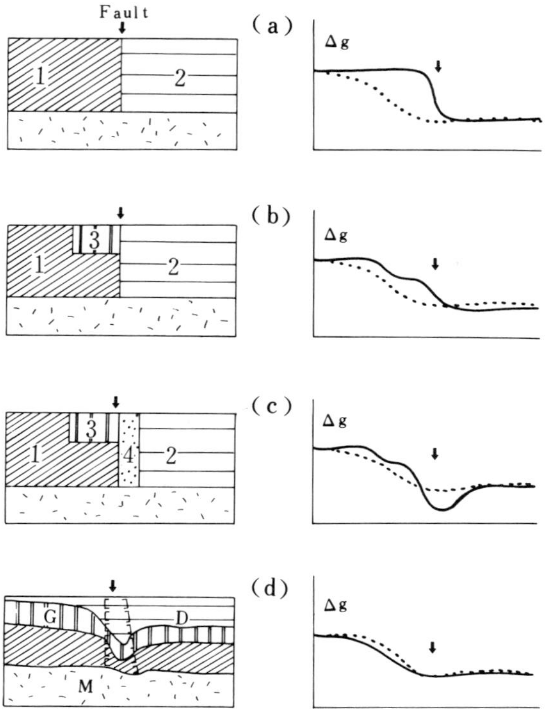

Figure 6
Sketch comparisons of computed and observed anomalies for some fault-zone models. In the diagrams on the right dotted curves show the observed anomalies, solid curves show the computed results for (a) a simple vertical fault, (b) a vertical fault with a fracture zone (low density) added on the left side of the fault, (c) same as in (b) except that the simple fault is broadened into a zone of finite thickness, and (d) a funnel-shaped low-density fault zone (our final density model).

From their reflection study FENG and MCEVILLY (1983) suggest that the width of the San Andreas fault zone in the studied area is about 3 km at the surface, widening to about 4 km at the bottom of the crust. The gravity anomaly profile, on the other hand, shows a broad transition zone of almost 10 km in width, from about -20 mgal on the southwestern end of the profile to about -53 mgal within the fault zone (Fig. 3), greatly exceeding the width of the fault zone suggested by FENG and MCEVILLY. In modelling the density structure we have tried various fault boundaries. Figure 6 shows the computed gravity profiles for several representative models and compares them with the observed gravity profile. For models with single vertical density boundaries the lateral variation in the computed gravity anomalies near the fault is invariably too sharp for the models, to be admissible. To fit the observed gradient in the computed gravity anomaly we find it necessary to include in the density model, in addition to the deeply seated low-density region within the fault zone, a broad funnel-shaped belt of low density about the upper section of the fault (Fig. 4 and 6d). We infer that this belt may represent extensively shattered wall rocks with high fracture porosity. This interpretation is consistent with the observation of CHESTER and LOGAN (1986) on the Punchbowl fault zone in southern California, and with the suggestion of STIERMAN and KOVACH (1979) that fracturing and weathering of the wall rocks about the San Andreas fault in central California may have an important effect on the local gravity anomalies. In addition, sediments washed from the adjacent hilly areas may have contributed to the total thickness of this funnel-shaped low-density belt.

In our view, the most interesting result of the present study is the deeply seated low-density region inside the fault zone, which extends to the base of the seismogenic zone at a depth of about 15 km. We infer that this low-density wedge may represent a zone of fault gouge with a significant amount of porosity.

It may be of interest to estimate the porosity of this gouge zone, since porosity, as noted previously, plays an important part in determining the fault-zone constitutive properties. In order to make this estimate we need to know the density of the solid grains in the fault gouge and the state of saturation in the fault zone. In central California the electrical resistivity of the San Andreas fault zone to a depth of 12 km or more is found to be exceedingly low, of the order of 10 ohm·m (PHILLIPS and KUCKES, 1983; MAZZELLA, 1976). Laboratory studies at elevated pressures (WANG, 1984; LOCKNER and BYERLEE, 1985) have shown that, for the electrical resistivity of fault gouges collected from the San Andreas fault zone to have such low resistivity, the gouge must be saturated with aqueous fluids with certain amounts of salinity. This conclusion is not likely to be changed even if the deeper part of the fault zone is composed of materials different from those collected from its surficial exposures. The average porosity of the fault gouge can then be estimated from the following elementary relation

$$n = \frac{\Delta\rho}{\rho_s - \rho_w}$$

where $\Delta\rho$ is the density contrast between the fault gouge and the wall rock and ρ_w and ρ_s are the densities of the aqueous fluid and the solid grains, respectively, in the fault gouge. Taking $\Delta\rho = 200$ kg/m^3, $\rho_s = 2{,}700$ kg/m^3, and $\rho_w = 1{,}000$ kg/m^3, we obtain a porosity of 12%. Since the value of $\Delta\rho$ used in this calculation is the lower bound in our model, the above-mentioned result on the average porosity of the low-density wedge inside the faul-t zone should be taken as a conservative estimate. Any reasonable choice of values for ρ_s and ρ_w is not likely to alter this result substantially.

Since high effective pressure tends to eliminate porosity, the presence of such high porosity in the deep fault zone is somewhat surprising. The mechanism responsible for the generation and maintenance of this deeply seated low-density region is not clear. Dilatant volumetric strain in crystalline rocks due to shearing rarely exceeds 1% under high pressure (e.g., BRACE et al., 1966; ZOBACK and BYERLEE, 1975; HADLEY, 1976). The amount of fracture porosity that can be supported in rocks at elevated pressures is not well understood. On the other hand, one can be quite certain that enhanced fracture porosity would increase permeability and thus facilitate the processes of weathering and hydrothermal alteration in the fault zone. The extent of weathering may be limited to the upper few hundred meters (STIERMAN and WILLIAMS, 1985). WANG (1984) argued on the basis of the molar volumes of minerals that hydrothermal alteration of intensely fractured rocks can lead to the low density inferred to exist inside the San Andreas fault zone in central California. STIERMAN (1984), on the other hand, suggested that pervasive dissolution of fault-zone material and subsequent removal by fluid transport may be responsible. These suggestions are compatible with the results from studies of fluid infiltration in ancient fault zones (Kerrich, this volume).

Irrespective of the mechanism responsible for the creation of the low-density saturated fault gouge, we may expect that the constitutive properties of this fault gouge will have important effects on the deformation behavior of this segment of the San Andreas fault zone, which is characterized by continuous and episodic fault creep (e.g., BURFORD et al., 1973; NASON, 1973). MOONEY and GINZBURG (1986) point out that the wide zone of low seismic velocity inside the San Andreas fault zone in central California may be restricted to the creeping segment, and may be either very narrow or even absent along the locked segments north of Watsonville and south of Parkfield. If so, the density structure below the locked segments of this fault may also be quite different from the one shown in this study. On the other hand, it appears that low density in fault zones may not be restricted to this particular segment of the San Andreas. STIERMAN (1984) showed that a steep decrease in gravity and a corresponding decrease in density occur across the San Jacinto fault in southern California. HORN et al. (1984) showed that both the seismic velocities and the gravity are abnormally low across the southern Queen Charlotte transform fault zone, off western Canada. Further studies are needed to more fully characterize the density structures of major transform fault zones and their relationship to the constitutive properties and fault behaviors.

Acknowledgments

We thank Thomas McEvilly and Lane Johnson for the use of their gravimeters and the Berkeley Seismographic Station for the use of the Stone Canyon facilities, during our field work. We are grateful to Rich Clymer for his thoughtful and detailed arrangements for our field operations. Phil Cummings helped in the use of a digital recorder for the continuous recording of tides. Russel Seal helped in arranging the field vehicle. We thank Carter Roberts and Don Plouff for permitting us to use the digitized terrain data and the computing facility at the U.S. Geological Survey, Menlo Park. Carter Roberts provided information on the reference gravity stations in the Hollister area. Don Plouff showed us how to use the U.S. Geological Survey facilities and other effective means for terrain corrections. Don Stierman and an anonymous reviewer carefully reviewed this article and offered helpful comments. This research was supported in part by the U.S. Department of Interior, Contract 14-08-0001-21352.

References

Bishop, C. C. and Chapman, R. H. (1967), *Bouguer gravity map of California, Santa Cruz sheet.* Calif. Div. Mines Geol.

Booker, J. R. (1974), *Time dependent strain following faulting of a porous medium.* J. Geophys. Res., *79*, 2037–2044.

Brace, W. F., Paulding, Jr. and Scholz, C. (1966), *Dilatancy in the fracture of crystalline rocks.* J. Geophys Res. *71*, 3939–3953.

Burford, R. O., Allen, S. S., Lamson, R. J. and Goodreau, D. D. (1973), 'Accelerated fault creep along the central San Andreas fault after moderate earthquakes during 1971–1973,' in *Proc. Conf. Tectonic Problems of the San Andreas Fault System*, edited by R. L. Kovach and A. Nur (eds.), Stanford Univ. Publ., Geol. Sci. *13*, 268–274.

Chester, F. M. and Logan, J. M. (1986), *Implications for Mechanical Properties of Brittle Faults from Observations of the Punchbowl Fault Zone, California*, PAGEOPH 124, 79–106

Feng, R. and McEvilly 1983), *Interpretation of seismic reflection profiling data for the structure of the San Andreas fault zone.* Bull. Seism. Soc. Am. *73*, 1701–1720.

Hadley, K. (1976), *The effect of cyclic stress on dilatancy: Another look.* J. Geophys. Res. *81*, 2471–2474.

Horn, J. R., Clowes, R. M., Ellis, R. M. and Bird, D. N. (1984), *The seismic structure across an active oceanic/continental transform fault zone.* J. Geophys. Res. 89, 3107–3120.

Kerrich, R. (1986), *Fluid Infiltration into Fault Zones: Chemical, Isotopic and Mechanical Effects*, PAGEOPH, 124, 225–268.

Lachenbruch, A. H. (1980), *Frictional heating, fluid pressure, and the resistance to fault motion.* J. Geophys. Res. *85*, 6097–6112.

Lochner, D. A. and Byerlee, J. D. (1985), *Complex resistivity of fault gouge and its significance for earthquake lights and induced polarization.* Geophys. Res. Lett. *12*, 211–214.

Mazzella, A. T. (1976), *Resistivity study across the San Andreas fault zone.* Ph.D. Thesis, Univ. California, Berkeley.

McEvilly, T. V. (1981), *Extended reflection survey of the San Andreas fault zone.* U.S. Geol. Surv. Open-file Rept. No. 81–388.

Mooney, W. D. and Ginzburg, A. (1986), *Seismic measurements of the internal properties of fault zones.* PAGEOPH, 124, 141–157.

Nafe, J. and Drake, C. L. (1957), *Variation with depth in shallow and deep water marine sediments of porosity, density and the velocities of compression and shear waves.* Geophys. 22, 523–552.

Nason, R. D. (1973), 'Fault creep and earthquakes on the San Andreas Fault,' in *Proc. Conf. Tectonic Problems of the San Andreas Fault System.* R. L. Kovach and A. Nur (eds.), Stanford Univ. Publ., Geol. Sci. *13*, 275–285.

Oldenburg, D. W. (1974), *The inversion and interpretation of gravity anomalies.* Geophysics *39*, 526–536.

Parker, R. L. (1973), *The rapid calculation of potential anomalies.* Geophys. J. R. Astr. Soc. *31*, 447–455.

Pavoni, N. (1973), 'A structural model for the San Andreas fault zone along the northeast side of the

Gabilan Range,' in *Proc. Conf. Tectonic Problems of the San Andreas Fault System.* R. L. Kovach and A. Nur (eds), Standford Univ. Publ. Geol. Sci. *13.*

PHILLIPS, W. J. and KUCKES, A. F. (1983), *Electrical conductivity structure of the San Andreas fault in central California.* J. Geophys. Res. *88*, 7467–7474.

RICE, J. R. and RUDNICKI, J. W. (1979), *Earthquake precursory effects due to pore fluid stabilization of a weakening fault zone.* J. Geophys. Res. *84*, 2177–2193.

ROBBINS, S. L. (1982), *Complete Bouguer gravity, aeromagnetic and generalized geologic map of the Hollister 15-minute quadrangle, California, scale 1:62,500.* U.S. Geol. Surv. Geophys. Invest. Map GP-945.

SANDBERG, C. H. (1958), *Terrain corrections for an inclined plane in gravity computation.* Geophys. *23*, 701–711.

STIERMAN, D. J., (1984), *Geophysical and geological evidence for fracturing, water circulation and chemical alteration in granitic rocks adjacent to major strike-slip faults.* J. Geophys. Res. *89*, 5849–5857.

STIERMAN, D. J. and KOVACH, R. L. (1979), *An in situ velocity study: The Stone Canyon well.* J. Geophys. Res. *84*, 672–678.

STIERMAN, D. J. and WILLIAMS, A. E. (1985), *Hydrologic and geochemical properties of the San Andreas fault at the Stone Canyon well.* Pure Appl. Geophys. *122*, 403–424.

WALTER, A. W. and MOONEY, W. D. (1982), *Crustal structure of Diablo and Gabilan ranges, central California: A reinterpretation of existing data.* Bull. Seism. Soc. Am. *72*, 1567–1590.

WANG, C. Y. (1984), *On the constitution of the San Andreas fault zone in central California.* J. Geophys. Res. *89*, 5858–5866.

WANG, C. Y., LIN, W. and WU, F. T. (1978), *Constitution of the San Andreas fault zone at depth.* Geophys. Res. Lett. *5*, 741–744.

ZOBACK, M. D. and BYERLEE, J. D. (1975), *The effect of cyclic effective stress on dilatancy in Westerly granite under uniaxial and triaxial conditions.* J. Geophys. Res. *80*, 1526–1530.

(Received 20th September 1985, revised 15th January 1986, accepted 18th January 1986)

PAGEOPH, Vol. 124, Nos. 1/2 (1986)

0033-4533/86/020141-17$1.50+0.20/0

Seismic Measurements of the Internal Properties of Fault Zones

WALTER D. MOONEY[1] and AVIHU GINZBURG[1,2]

Abstract—The internal properties within and adjacent to fault zones are reviewed, principally on the basis of laboratory, borehole, and seismic refraction and reflection data. The deformation of rocks by faulting ranges from intragrain microcracking to severe alteration. Saturated microcracked and mildly fractured rocks do not exhibit a significant reduction in velocity, but, from borehole measurements, densely fractured rocks do show significantly reduced velocities, the amount of reduction generally proportional to the fracture density. Highly fractured rock and thick fault gouge along the creeping portion of the San Andreas fault are evidenced by a pronounced seismic low-velocity zone (LVZ), which is either very thin or absent along locked portions of the fault. Thus there is a correlation between fault slip behavior and seismic velocity structure within the fault zone; high pore pressure within the pronounced LVZ may be conducive to fault creep. Deep seismic reflection data indicate that crustal faults sometimes extend through the entire crust. Models of these data and geologic evidence are consistent with a composition of deep faults consisting of highly foliated, seismically anisotropic mylonites.

Key words: Seismic refraction, mylonites, fractures, seismic reflection, low velocity zone, microcracks.

Introduction

The deformation of rocks due to faulting causes changes in lithology, pore pressure, and seismic velocity, all of which are evident from geophysical measurements. The internal properties of fault zones may provide information on the manner in which deformation takes place, the slip behavior of faults, and variations in the depth of seismogenic zones. Fault zones are likely regions for successful earthquake-prediction measurements, since they are expected to exhibit premonitory phenomena. Thus, an understanding of the internal properties of fault zones contributes to an understanding of the faulting process.

In this paper we review seismic measurements of fault zones and adjacent rocks. We discuss principally the in-situ compressional-wave velocity (V_p) structure of fault zones, since this parameter has been the most commonly measured; where data are available, we discuss the shear-wave velocity (V_s) properties, and we relate both

[1] U.S. Geological Survey, 345 Middlefield Road, MS 977, Menlo Park, California 94025.

[2] On leave from Department of Geophysics and Planetary Sciences, Tel Aviv University, Tel Aviv, Israel.

these velocity determinations to laboratory measurements.

We proceed from the effects of intragrain microcracks and rock fractures associated with minor faults to fault gouge along major faults and, finally, to the seismic evidence of the internal constitution of faults in the lower crust. We present no new data, but we critically reëvaluate previously published results. We present evidence that, in the San Andreas fault of California, the internal properties of the continuously creeping sections of the fault differ markedly from the locked sections of the fault that fail catastrophically in great earthquakes. In the final section we use recent seismic reflection data to evaluate SIBSON's (1977) conceptual model that deep faults consist of mylonite zones.

Seismic velocities of fractured rocks

Microcracks

The effect of microcracks on the wave velocity of crystalline rocks has been studied in the laboratory and in boreholes. BIRCH (1960, 1961) showed in laboratory measurements that dry crystalline rocks' seismic velocities at low confining pressure are much lower than their intrinsic velocities (uncracked, high pressure) and that the closure of microcracks with increased confining pressure causes a rapid increase in P-wave velocity. DORTMAN and MAGID (1969) and NUR and SIMMONS (1969) later demonstrated that the seismic velocity of saturated granitic rocks is higher than that of unsaturated (dry) rocks, and that the rate of velocity increase with pressure is diminished, owing to the low compressibility of water. The important effect of water on the velocity of cracked rocks was also emphasized by SIMMONS and NUR (1968), who found higher velocities in situ at two wet boreholes than in laboratory measurements of dry samples. Water is available in most cases in situ, and WANG and SIMMONS (1978) found the in situ velocities in gabbroic rocks at a 5.3 km depth in the Michigan Basin to be about equal to the velocity of saturated laboratory samples at the appropriate pressure.

Macrofractures

A measurable velocity reduction is expected theoretically and observed experimentally in densely cracked and fractured media. O'CONNELL and BUDIANSKY (1974) considered the effects of cracks on elastic moduli for the case in which crack interaction is important. Their theoretical model predicts that for saturated cracks an increase in spatial crack density decreases both V_p and V_s and increases V_p/V_s. The lower velocity of densely cracked and fractured rocks has also been measured in situ. STIERMAN and KOVACH (1979) reported the results of logging a 600 m borehole in Stone Canyon, California, 1.2 km from the active trace of the San Andreas fault.

The well penetrates wet quartz diorite, ranging in condition from extremely weathered and friable to hard and unweathered. Their logs showed velocities that are 30 to 500 percent lower than the velocities commonly measured in saturated granite. They concluded that the low velocities are due to the fact that the macrofractures in the well are held open by tectonic shear stress associated with the San Andreas fault. They contend that their conclusion is supported by the occurrence of microearthquakes near the borehole, indicating that the rocks are near failure.

The relationship between macrofracture density and V_p and V_s was further investigated by MOOS and ZOBACK (1983), who reported sonic logs from four deep boreholes (0.6 to 1.2 km) drilled in granitic rocks. In wells with a relatively low density of macroscopic fractures the velocity in situ is similar to that of a saturated core measured in the laboratory (about 5.9 km/sec.). With dense fracturing, however, they found that V_p and V_s are very much reduced and V_p/V_s increased relative to laboratory measurements. Significantly, of the two boreholes with low fracture density the hole with the higher density exhibited a higher sonic velocity. This observation demonstrates that velocity is not directly proportional to fracture density, but that other factors, such as rock composition, play important roles as well. MOOS and ZOBACK (1983) also found lower velocities in boreholes from unfractured rocks adjacent to the macrofractures, which they interpreted as indicating chemical or mechanical alteration of these rocks. In those intervals of the borehole that are highly fractured it was not possible to measure the S-wave velocity, because of the attenuation of the waveform.

CHRISTENSEN (1984) and CHRISTENSEN and WANG (1985) reported laboratory measurements of the effects of pore pressure on basalts and on Berea sandstone. They demonstrated that, at a given confining pressure, the shear-wave velocity is more sensitive to increased pore pressure than is the compressional-wave velocity, thereby causing increases in V_p/V_s and Poisson's ratio. Thus, high pore-pressure zones may potentially be identified on the basis of Poisson's ratio, in the crust, where both shear and compressional waves can be measured.

Besides depending on crack density and degree of saturation, velocity variations in fractured rocks will depend on the preferred fracture orientation. If oriented fractures remain open, seismic velocities will be anisotropic, with lower P-wave velocities perpendicular to the cracks (CRAMPIN, 1978; HUDSON, 1981); if fractures are completely saturated, theory predicts a 4θ dependence, with slow P waves parallel and perpendicular to the fractures. CRAMPIN (1984a and 1984b) discussed in detail the key theoretical and observational aspects of the propagation of waves in anisotropic media. Velocity anisotropy for P waves around a borehole drilled in crystalline rock in a geothermal field was reported by LEARY and HENYEY (1985). They interpreted the anisotropy as arising from the existence of fractures in the basement having an orientation subparallel to the latest episode of local faulting. BAMFORD and NUNN (1979) presented additional examples of seismic methods of locating fracture zones, and CRAMPIN et al. (1980) estimated crack parameters by using synthetic seismograms to model shear-wave polarization anomalies.

Fault zones: the San Andreas

Whereas cracked and fractured rocks are found in association with minor faults and adjacent to major faults, the near-surface portion of a major fault zone is often observed to be filled with 10 to 100 m of fault gouge due to severe alteration and hydration of the rocks. This is the case within the San Andreas fault zone, which in some locations is filled with severely altered rocks containing a significant amount of clay minerals. Since it is known from laboratory studies that fault gouge has a lower velocity than its protoliths (WU, 1978; WU et al., 1975; WANG et al., 1978), seismic methods are effective in delineating the extent of alteration within the fault zone.

Along the linear stretch of the San Andreas fault system (Fig. 1) south of Hollister, California, a substantial seismic low-velocity zone (LVZ) has been determined (HEALY and PEAKE, 1975; AKI and LEE, 1976; FENG and MCEVILLY, 1983). In the model of HEALY and PEAKE (1975) the LVZ is 3 km wide and 12 km deep, with velocities ranging from 3.0 to 4.5 km/sec, compared to 5.2 to 6.5 km/sec outside the fault zone. The detailed nature of the velocity distribution within this zone is indeterminate. The low average velocities determined for this zone may actually represent the velocity throughout the zone. Alternatively, it is possible that very high-resolution refraction measurements (e.g., SJOGREN, 1984) would reveal that the LVZ medium with an average velocity of 4.0 to 4.5 km/sec actually consists of narrow zones with velocities as low as 3.0 km/sec alternating with largely unfractured rocks with velocities approaching 6.0 km/sec. This model is appealing, since gouge zones at the surface are always much thinner than the 3 km LVZ we have discussed.

Some discrepancies exist in the depths to which fault-zone LVZ's penetrate the crust. Along the portion of the San Andreas fault system discussed above, AKI and LEE (1976) found no evidence of low velocities at depths below 5 km, whereas HEALY and PEAKE (1975) report them at depths of 12 km at the same location, and FENG and MCEVILLY (1983) extend the LVZ to the Moho.

From seismic refraction measurements there is general agreement that along the Calaveras fault zone (Fig. 1) an LVZ exists within the fault, with velocities of 3.0 to 4.0 km/sec extending to a depth of 6 km (MAYER-ROSA, 1973; MOONEY and LUETGERT, 1982; BLÜMLING et al., 1985; MOONEY and COLBURN, 1985). Fan profiles recorded across the fault indicate that the seismic low-velocity zone is 1 to 2 km wide at the near-surface (BLÜMLING et al., 1985). At this same location, however, THURBER (1983) found no evidence of low velocities in an inversion of earthquake travel times (Fig. 2). To resolve the discrepancy between the results of THURBER (1983) and the seismic refraction results, CORMIER and SPUDICH (1984) calculated synthetic seismograms for earthquakes from a model of the Calaveras fault which included a deep wedge-shaped LVZ (Fig. 3). Their calculations successfully modeled

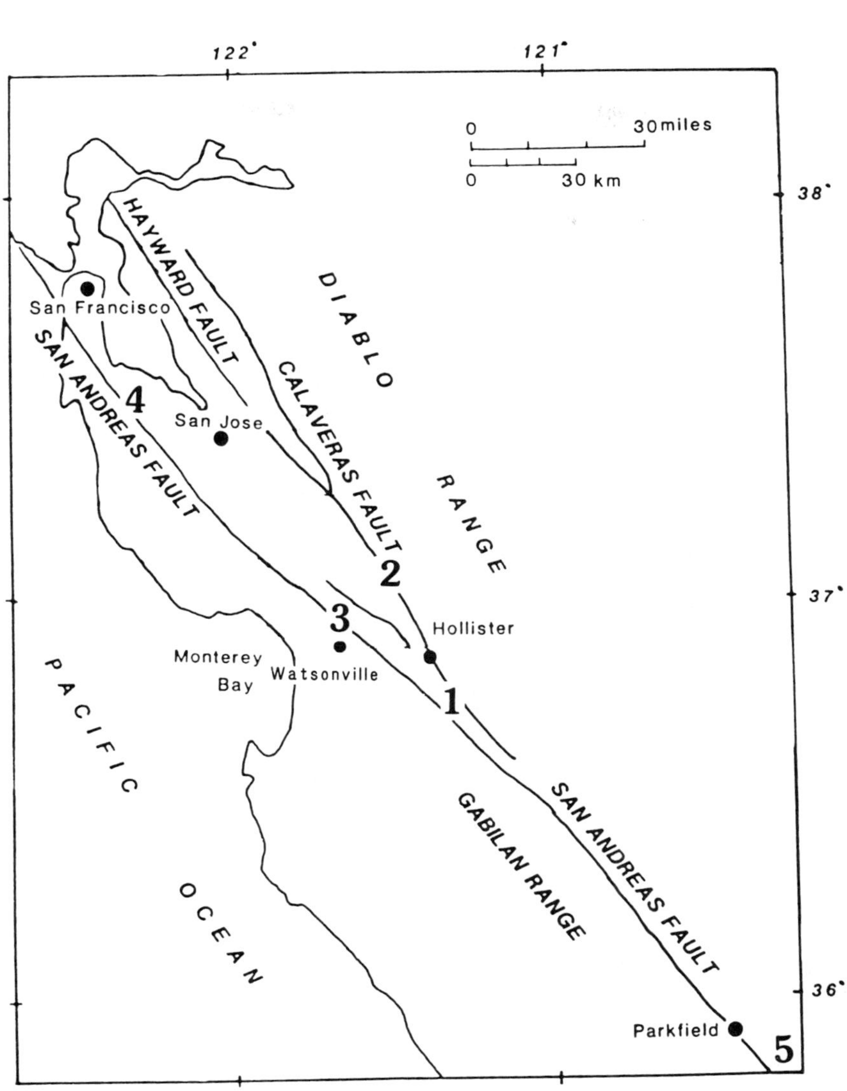

Figure 1
Location map of studies of fault-zone structure in central California. The San Andreas fault is seismically active (creeping) from Parkfield to Watsonville. The Calaveras fault is active from Hollister to the latitude of San Francisco. Numbers next to faults refer to seismic studies discussed in text. 1: AKI and LEE (1976), HEALY and PEAKE (1975), FENG and MCEVILLY (1983). 2: MAYER-ROSA (1973), MOONEY and LUETGERT (1982), THURBER (1983), CORMIER and SPUDICH (1984), MOONEY and COLBURN (1985), BLÜMLING et al. (1985). 3: THURBER (1983), MOONEY and COLBURN (1985). 4: BOKEN and MOONEY (1982). 5: TREHU and WHEELER (1986), MEREU (1986).

the seismic energy trapped and focused within the wedge, leading to locally high amplitudes directly over the fault (Fig. 3). Thus, their model contradicts Thurber's

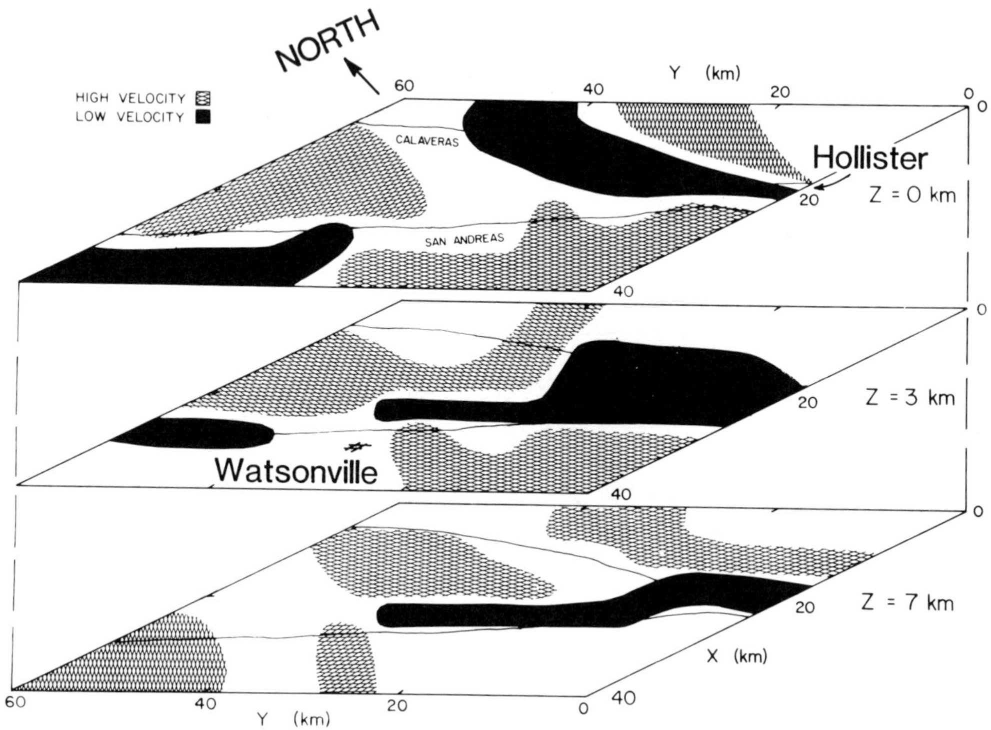

Figure 2

Seismic-velocity inversion results for an area in central California, from Thurber (1983). The Calaveras and San Andreas faults are labeled, as are Hollister and Watsonville. Areas of high and low velocity are indicated at three depths: 0, 3, and 7 km. At 7 km a low-velocity zone follows the San Andreas fault as far north as Watsonville; such a zone is absent along the Calaveras fault at that depth. Along the San Andreas fault the low-velocity zone is consistent with other seismic studies; the lack of one on the Calaveras fault is inconsistent with these studies. See text for discussion.

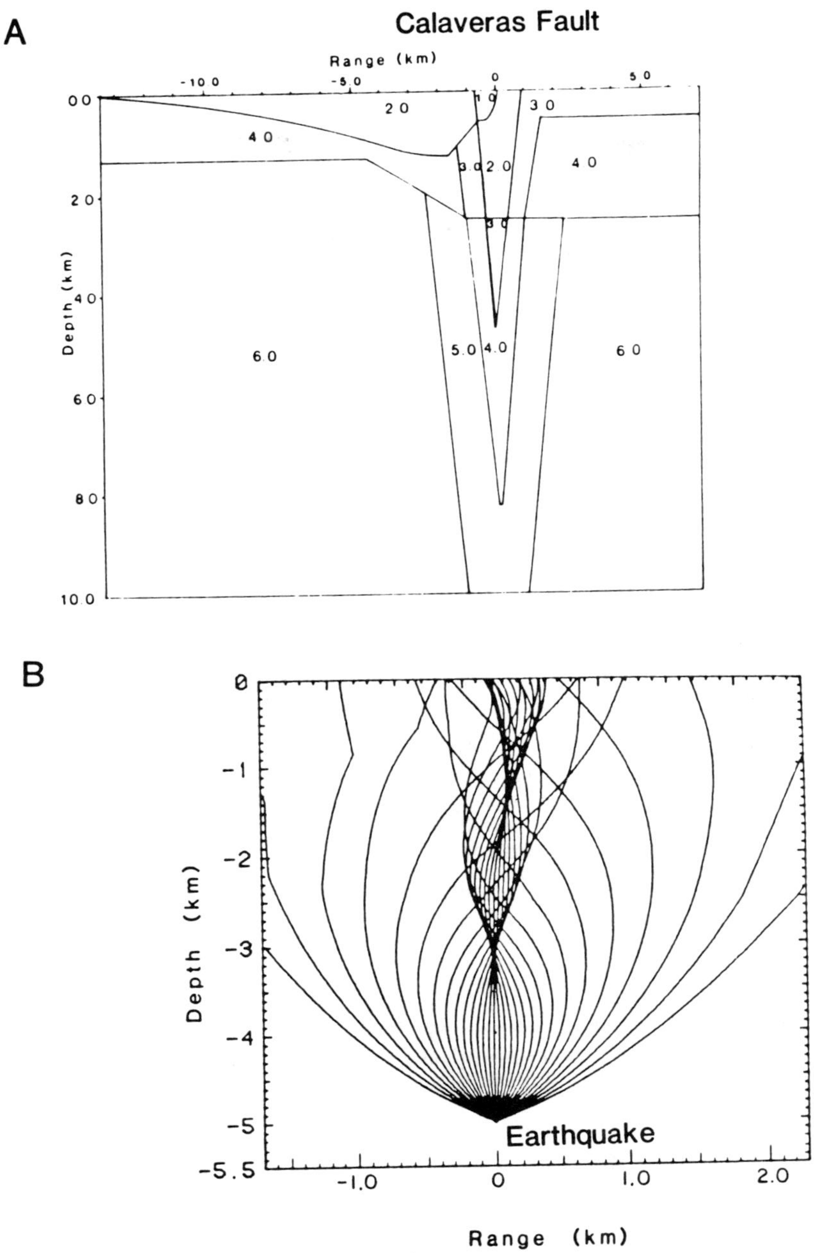

Figure 3

(A) Velocity model across the Calaveras fault, used by CROMIER and SPUDICH (1984) to model earthquake seismograms recorded over the fault; the fault model consists of a pronounced low-velocity zone to a depth of 10 km. (b) Diagram of ray paths propagating from a depth of 5 km through model in (A); note ray multipathing and high ray density at the fault, correctly predicting local site amplification. This modelling supports the existence of a substantial low-velocity zone along the Calaveras fault.

conclusion that an LVZ is lacking along this portion of the Calaveras fault. An important aspect of their study was that it indicated the potential for deriving the properties of the fault zones from dense array recordings of microearthquakes.

The discrepancies among the seismic interpretations discussed above are due largely to the resolutions inherent in the methods and to the varying spatial densities in the data. The studies of AKI and LEE (1976) and THURBER (1983) were based on the inversion of earthquake travel times to a permanent array of seismograms. The resolution of these inversions is controlled by several factors, particularly station and hypocentral distribution. In general, features less than 2 km in width will not be resolved by this method.

The seismological evidence summarized above indicates that the creeping sections of the San Andreas fault system may be characterized by a pronounced seismic LVZ. WANG (1984) considers geologic, geophysical, and geochemical evidence for the composition of the LVZ material. He shows that the seismic velocity models which show a deep LVZ along the central San Andreas fault are consistent with laboratory measurements of fault-gouge materials at elevated pressures. Electrical resistivity values for the fault zone are consistent with a gouge composition. Citing the laboratory results of CHRISTENSEN (1966, 1978), WANG (1984) concludes that the seismic velocity of serpentine is higher than the velocities reported for the fault zone. This contrasts with earlier suggestions that the San Andreas fault in central California is rich in serpentine at depth (ALLEN, 1968). WANG (1984) concludes from this and additional geologic and geochemical evidence that throughout its seismogenic regime (0 to 15 km) the creeping portion of the fault in central California is composed of saturated fault gouge. WU et al. (1975) and WANG et al. (1978) reached the same conclusion from a smaller amount of evidence. WANG (1984) is cautious about extending his deep fault-gouge model to outside the creeping section of the San Andreas fault, but he admits the possibility that it is applicable in those other regions. We examine this possibility below.

Several studies indicate that the San Andreas fault does not have a substantial LVZ along the locked portions of the fault. Although the inversion results of THURBER (1983) do show a significant deep LVZ running just northeast of the San Andreas fault in the area of Hollister (Fig. 2), further northwest along the San Andreas fault the LVZ terminates approximately where creep and seismic activity cease, near Watsonville (Fig. 2). The lack of an LVZ on the fault north of Watsonville is confirmed by two seismic refraction profiles across the fault: neither a profile just north of Watsonville (MOONEY and COLBURN, 1985) nor one 40 km further north (BOKEN and MOONEY, 1982) detected an LVZ at the fault.

Across the locked portion of the San Andreas fault south of Parkfield, California, a seismic refraction profile has been variously interpreted as being with (TREHU and WHEELER, 1986) and without an upper crustal LVZ (MEREU, 1986). Thus the available evidence may indicate that substantial LVZ's do not occur along locked sections of the fault, and accordingly we hypothesize that *substantial* LVZ's occur only on creeping sections of the fault. It appears that LVZ's are conducive to fault creep—for example, through high pore pressures due to the high water content.

Two observations contradict our hypothesis that substantial LVZ's occur only along the creeping section of the fault. First, due to the volumetric increase (and density decrease) that accompanies its formation, fault gouge often upwells along the fault. The resultant weltlike ridge is readily visible along many portions of the fault, such as the locked section in the Carrizo Plain (WALLACE, 1949; ALLEN, 1968, 1981). Small amounts (30 m) of surficial fault gouge are also reported along the locked portion of the fault north of Watsonville (HALL, 1984). These observations, however, do not demonstrate that a thick zone of fault gouge exists at depth, only that some gouge exists in the near-surface.

The second observation indicating thick fault gouge at depth is the narrow, elongate Bouguer gravity low which is nearly coincident with much of the fault trace. Several investigators have modelled this gravity low as due to a low-density zone (presumably fault gouge) 10 km deep and between one and several kilometers wide (PAVONI, 1973; WANG et al., 1978; STIERMAN, 1984). Of these studies, only that of STIERMAN (1984) concerns a locked portion of the fault.

STIERMAN (1984) considered the Bouguer gravity data along with borehole and other geophysical evidence and concluded that fracturing and chemical alteration exist in a wide zone along the fault that extends to seismogenic depths. He demonstrates that the altered zone is not limited to a single narrow trace a few tens or even hundreds of meters wide, that rocks a few kilometers from the primary dislocation surface are also modified to some extent by the earthquake process.

To evaluate the properties within the fault zone, STIERMAN (1984) modeled gravity data across the San Jacinto fault of southern California at locations where the observed lows are clearly not attributable to surficial basins. He modeled these lows with a series of crustal prisms 10 km deep by 1 to 2 km wide with density reductions of 0.1 to 0.025 gm/cm^3 on either side of the fault, and a zone 200 m wide with a density reduction of 0.4 gm/cm^3 directly at the fault. This model, with several kilometers of low-density prisms, appears to be consistent with a pronounced LVZ at this locked portion of the fault, and STIERMAN (1984) concluded that a fault-zone LVZ is probably not confined to the creeping portion of the San Andreas fault. To test this hypothesis we convert the density model to a velocity model, using an approximation to a velocity–density curve (LUDWIG et al., 1970):

$$\Delta V_p = 3.33\ \Delta\rho$$

This relationship yields a velocity reduction of only 0.33 km/sec corresponding to a density reduction of 0.1 gm/cm^3, and a velocity reduction of 1.33 km/sec corresponding to a density reduction of 0.4 gm/cm^3. Thus, if a *pronounced* LVZ is indicated by a velocity reduction of 0.5 to 1.0 km/sec or greater (such as found along creeping faults), then STIERMAN's model (1984) implies an LVZ that is only 200 m wide across the San Jacinto fault, as compared to one that is 3 to 4 km wide along the creeping

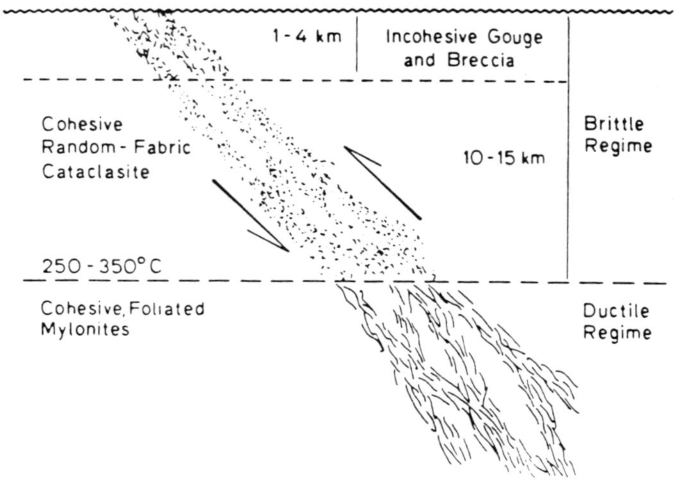

Figure 4
Geologic model of a deep continental fault (Sibson, 1977). In the ductile regime, cohesive, foliated mylonites are expected.

segment of the San Andreas fault in central California.

We note that our hypothesis of a substantial LVZ along creeping sections of the San Andreas fault is not inconsistent with Stierman's hypothesis (1984) that rocks adjacent to locked portions of the fault are fractured and locally altered; it is merely that the velocity reduction there will be significantly less than within the pronounced LVZ found on the creeping central San Andreas fault.

In conclusion, a substantial body of geologic and geophysical evidence is consistent with an LVZ as wide as 3 to 4 km along the creeping section of the San Andreas fault, and a narrow LVZ (200 m or much less) along the locked sections. Thus, fault slip behavior correlates with the internal properties of the fault zone, and we speculate that high pore pressure within thick LVZ's is conducive to fault creep. Important questions raised by this hypothesis are whether locked faults evolve into creeping faults as gouge is developed and, indeed, whether fault creep creates fault gouge or fault gouge leads to creep.

Deep structure of faults

On the basis of geological and physical considerations Sibson (1977) proposed a conceptual model of the deep structure of faults. In his model (Fig. 4) fault gouge and breccia are formed in the upper 1 to 4 km of the crust, where brittle processes dominate and water is available to alter the crushed rocks into clay-rich gouge. At depths greater than 4 km, low-porosity cataclasites are formed as a result of elasticofrictional deformation. These rocks generally have random fabrics and reduced grain size. At depths greater than about 15 km and temperatures of about 300 °C ductile processes dominate and mylonites are formed. The mylonites in this quasi-plastic zone are well foliated and show tectonically induced grain-size reduction. The depth of the boundary between the elastic and the quasi-plastic styles of deformation will be determined by several factors, including heat flow, presence of fluids, and composition of the rocks. Numerous seismic reflection and refraction profiles completed in the course of crustal studies and exploration activities have detected faults, and in this study we summarize the seismological evidence for the internal structure of fault zones and use these to evaluate SIBSON's model.

SIBSON (1977) hypothesizes that both vertical and low-angle faults should in some instances penetrate to the lower crust. The primary evidence he cites for deep faults is the existence of mylonites in ancient fault zones that have been uplifted and exposed. In many cases the mineral assemblages and metamorphic grades of the rocks within and enclosing the fault zone give unquestionable evidence that they come from deep crustal levels. Given this geologic evidence, we consider how deeply into the crust faults can be traced on seismic reflection profiles. A profile recorded northwest of Scotland shows a reflection that projects through the crust nearly to the Moho (Fig. 5). PEDDY (1984) shows that after correction for a near-surface sediment wedge the apparently reactivated thrust penetrates through the entire lower crust and into the uppermost mantle. Another reflection profile across the Wind River mountains of Wyoming shows a low-angle fault that can be traced from the surface to depths approaching 32 km (SMITHSON et al., 1979; LYNN, 1979). Given these examples, the question remains whether these seismic observations can be explained by the presence of mylonites (SIBSON, 1977) within faults.

The origin of reflections from deep faults was considered by JONES and NUR (1984) in a combined analysis of field observations of ductile fault zones, laboratory measurements of the physical properties of mylonites, and seismic modelling of fault-zone velocity structures. Their field observations show that the mylonites are characterized by both compositional banding and strong foliation parallel to the boundaries of the shear zone. Velocities were measured in mylonite samples recovered from the fault. No significant difference in average velocity and density, between the mylonites and adjacent rocks, was found, but a seismic anisotropy of 7 per cent or greater was found in several mylonites. In particular, quartz and phyllosilicates are readily deformed ductilely to effect a strong preferred orientation of crystallographic axes. Phyllosilicates have a much greater single-crystal anisotropy than does quartz, so the phyllosilicate composition and fabric will control the seismic anisotropy. JONES and NUR (1984) showed, using synthetic seismograms, that the anisotropy and finely laminated structure in the lower crust could produce strong reflections from mylonite zones.

FOUNTAIN et al. (1984) also assessed the evidence that mylonite zones are the cause of deep crustal reflections. They note that the shallow dips (less than 30°) and lateral continuity of these zones are conducive to producing laterally continuous,

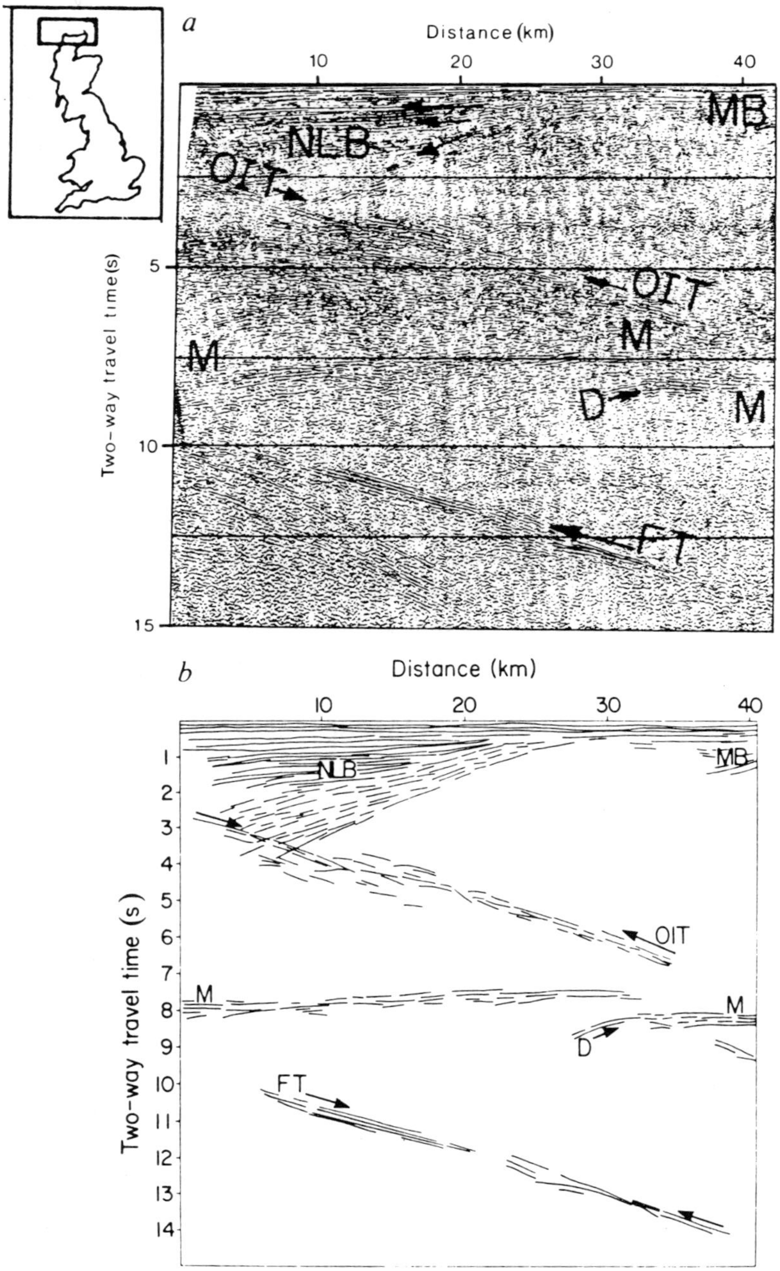

Figure 5

(a) Deep seismic reflection data from a marine survey north of Scotland (Brewer et al., 1983): NLB, North Lewis Basin; MB, Minch Basin; OIT, Outer Island Thrust; M, Moho; D, diffraction; FT, Flannel thrust. (b) Line drawing of main events in (a) above. The OIT appears to reach almost to the Moho; PEDDY (1984) shows that it does in fact reach and cross the Moho, after data corrections are made for the surficial basins. The Flannel thrust indicates that faults exist as well in the mantle.

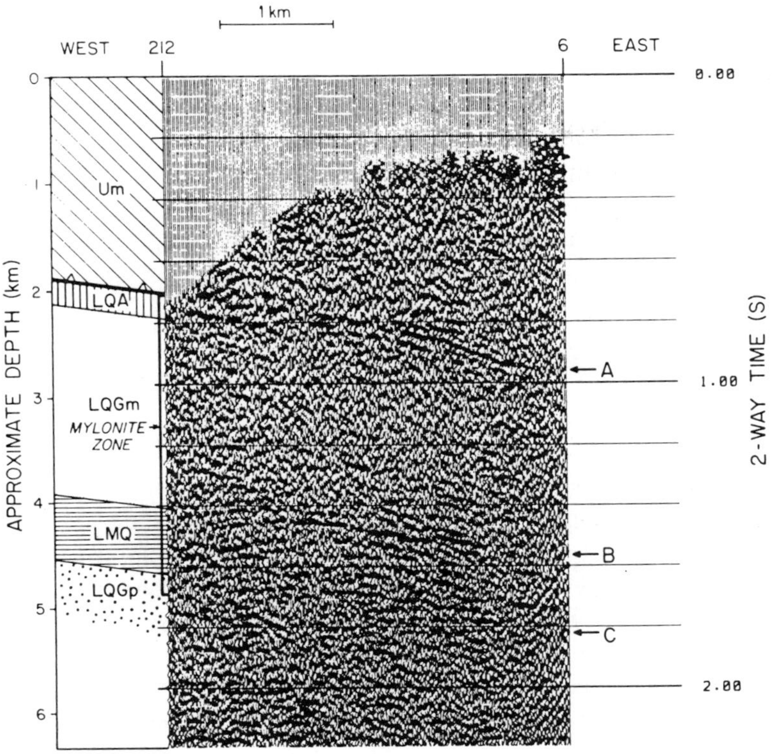

Figure 6
Seismic reflection section showing a correlation between the downdip projection of an exposed mylonite zone and seismic data (HURICH et al., 1985), 1:1 scale: A, reflection from fault and upper part of mylonite zone; B, reflections from interlayered metamorphic rocks; C, reflection from top of crystalline basement; Um, upper-plate marble, phyllite; LQA, lower-plate quartzite, ampphibolite; LQGm, lower-plate quartzofeldspathic gneiss; LMQ, lower-plate marble, quartzite, schist; LQGp, lower-plate porphyroclastic gneiss.

correlatable reflections. They also note that many mylonite zones exhibit either a laminated or anastomosing geometry. Both geometries produce relatively high-amplitude seismic reflections where constructive interference occurs. Furthermore, field and petrographic studies demonstrate that mylonitic zones commonly exhibit strong retrograde metamorphic effects, compared to undeformed zones. The effect of this retrogression will be to lower the velocity of mylonitic zones relative to neighboring rocks, or to enhance the seismic anisotropy by the synkinematic growth of new phyllosilicates (FOUNTAIN et al., 1984).

A direct demonstration of the seismic response of a mylonite zone was obtained by HURICH et al. (1985), who recorded a seismic reflection profile over an exhumed zone in eastern Washington State. The presence of sillimanite within the zone indicated a minimal depth of about 15 km for the deformation in the fault zone. A downdip projection of the exposed mylonite zone into the seismic profile indicated that the compositionally layered mylonite is reflective and traceable to a depth of at least 5 km, as a complex zone of multicyclic reflections (Fig. 6).

In summary, deep seismic reflection data sometimes show pronounced reflection events that penetrate the entire crust. These events consist of multicyclic reflections with relatively high amplitudes and good lateral continuity. Modelling of these events and field tests confirm that fault zones consisting of mylonite zones 0.5 to 2 km thick in the crust can explain these observations. The properties of mylonite zones that appear to make them reflective include velocity anisotropy, laminated or anastomosing geometry, and retrograde metamorphism.

Summary

The deformation of upper crustal rocks due to faulting may be described in three levels, corresponding to successive reductions in the seismic velocities of the rocks. Faulting at the lowest level leads to microcracking and limited macrofracturing of the rocks and does not have a significant effect on their seismic velocities, particularly if the rocks are saturated, as is generally the case in situ. The middle level, densely fractured rocks, such as are found adjacent to major strike-slip fault zones, show greatly reduced V_p and V_s and increased V_p/V_s. If the fractures are oriented by a deviatoric stress field, velocity anisotropy results and can be determined with precise velocity measurements and shear-wave polarization analysis. At the third level, large amounts of continuous deformation produce fault gouge which is generally 30% to 40% lower in seismic velocity than is the protolith, at a given confining pressure. Fault gouge therefore is easily detected within the fault zone by seismic measurements.

Seismic studies in central California have identified densely fractured rocks adjacent to the San Andreas fault and a pronounced low-velocity zone (LVZ) extending to seismogenic depths along the creeping segment of the fault. This LVZ has been attributed to deep fault gouge, and WANG (1984) presents geologic and geochemical evidence that fault gouge is stable to depths of 10 to 12 km in the crust, which lends credence to this hypothesis. Evidence is presented here that the creeping segments of the San Andreas fault in central California contain thick LVZ's, whereas the locked segments of the fault have either no LVZ or very thin ones (200 m or much less). Thus the slip behavior along segments of the fault appears to be related to the presence or absence of an LVZ along the fault, and it is possible that high pore pressure within thick LVZ's is conducive to fault creep.

Faults may be traced locally on seismic reflection sections into the lower crust. According to SIBSON's model (1977) for continental fault zones, deep faults consist of mylonitic rocks. Recent seismic reflection data and computer modelling of these data are consistent with this model.

The discussion of the observations of fault-zone properties given here indicates that the study of the internal properties of fault zones is at a rather rudimentary stage. Particular attention needs to be paid to shear-wave velocities and the seismic quality factor (*Q*). Progress may be expected with continued interpretation of seismic reflection and refraction data, borehole measurements, and array recordings made with three-component sensors deployed directly across and along fault zones.

Acknowledgments

Critical reviews of an early draft by M. C. Andrews, S. Klemperer, and A. W. Walter significantly improved this paper. Discussions with S. Peacock, J. McCarthy, and E. L. Ambos were helpful. Review comments by N. I. Christensen, A. Lindh, T. V. McEvilly, and R.-S. Zeng (Beijing) are greatly appreciated.

References

Aki, K and Lee, W. H. K. (1976), *Determination of three-dimensional velocity anomalies under a seismic array using first P arrival times from local earthquakes: 1. A homogeneous initial model.* J. Geophys. Res. *81*, 4381–4399.

Allen, C. R. (1968), 'The tectonic environments of seismically active and inactive areas along the San Andreas fault system', in Proc. Conf. Geol. Problems San Andreas Fault System, W. R. Dickinson and A. Grantz (eds.). Stanford Univ. Publ. Geol. Sci. *11*, 70–80.

Allen, C. R. (1981), 'The modern San Andreas fault', in *The Geotectonic Development of California*, W. G. Ernst (ed.). Prentice-Hall, Engelwood Cliffs, New Jersey, p. 512–534.

Bamford, D. and Nunn, K. R. (1979), *In situ measurement of crack anisotropy in the Carboniferous limestone in northwest England.* Geophys. Prosp. *27*, 322–338.

Birch, F. (1960), *The velocity of compressional waves in rocks to 10 kilobars: Pt. 1.* J. Geophys. Res. *65*, 1083–1102.

Birch, F. (1961), *The velocity of compressional waves in rocks to 10 kilobars: Pt. 2.* J. Geophys. Res. *66*, 2199–2224.

Blümling, P., Mooney, W. D., and Lee, W. H. K. (1985), *Crustal structure of the southern Calaveras fault zone, central California, from seismic refraction investigations Bull. Seis. Soc. Am.* *75*, 193–209.

Boken, A. and Mooney, W. D. (1982), *A refraction study of the Santa Cruz Mountains, west-central California* (abstr.). Trans. Am. Geophys. Union, EOS *63*, 1036.

Brewer, J. A., Matthews, D. M., Warner, M. R., Hall, J., Smythe, and Wittington, R. J. (1983), BIRPS deep seismic reflection studies of the British Caledonides. Nature *305*, 206–210.

Christensen, N. I. (1966), *Elasticity of ultramafic rocks.* J. Geophys. Res. *71*, 5921–5931.

Christensen, N. I. (1978), *Ophiolites, seismic velocities and oceanic crustal structure.* Tectonophys. *47*, 131–157.

Christensen, N. I. (1984), *Pore pressure and oceanic crustal seismic structure.* Geophy. J. Roy. Astr. Soc. *79*, 411–424.

Christensen, N. I. and Wang, H. F. (1985), *The influence of pore pressure and confining pressure on dynamic elastic properties of Berea sandstone.* Geophysics *50*, 207–213.

Cormier, V. F. and Spudich, P. (1984), *Amplification of ground motion and waveform complexity in fault zones: Examples from the San Andreas and Calaveras Faults.* Geophys. J. Roy. Astr. Soc. *79*, 135–152.

Crampin, S. (1978), *Seismic wave propagation through a cracked solid: Polarization as a possible dilatancy diagnostic.* Geophys. J. Roy. Astr. Soc. *53*, 467–496.

CRAMPIN, S. (1984a), *An introduction to wave propagation in anisotropic media.* Geophys. J. Roy. Astr. Soc. *76*, 17–28.

CRAMPIN, S. (1984b), *Effective anisotropic elastic constants for wave propagation through cracked solids.* Geophys. J. Roy. Astr. Soc. *76*, 135–145.

CRAMPIN, S., MCGONIGLE, R. and BAMFORD, D. (1980), *Estimating crack parameters from observation of P-wave velocity anisotropy.* Geophysics *45*, 345–360.

DORTMAN, N. B. and MAGID, M. S. (1969), *New data on velocities of elastic waves in crystalline rocks as a function of moisture.* Int. Geol. Rev. *11*, 517–523.

FENG, R. and MCEVILLY, T. V. (1983), *Interpretation of seismic reflection profiling data for the structure of the San Andreas fault zone.* Bull. Seis. Soc. Am. *73*, 1701–1720.

FOUNTAIN, D. M., HURICH, C. A. and SMITHSON, S. B. (1984), *Seismic reflectivity of mylonite zones in the crust.* Geology *12*, 195–198.

HALL, N. T. (1984), *Holocene history of the San Andreas fault between Crystal Springs Reservoir and San Andreas Dam, San Mateo County, California.* Bull. Seis. Soc. Am. *74*, 281–299.

HEALY, J. H. and PEAKE, L. G. (1975), *Seismic velocity structure along a section of the San Andreas fault near Bear Valley, California.* Bull. Seis. Soc. Am. *65*, 1177–1197.

HUDSON, J. A. (1981), *Wave speeds and attenuation of elastic waves in material containing cracks.* Geophys. J. Roy. Astr. Soc. *64*, 133–150.

HURICH, C. A., SMITHSON, S. B., FOUNTAIN, D. M. and HUMPHREYS, M. C. (1985), *Seismic evidence of mylonite reflectivity and deep structure in the Kettle dome metamorphic core complex, Washington.* Geology. *13*, 577–580.

JONES, T. D. and NUR, A. (1984), *The Nature of seismic reflection from deep crustal fault zones.* J. Geophys. Res. *89*, 3153–3173.

LEARY, P. C. and HENYEY, T. L. (1985), *Anisotropy and fracture zones about a geothermal well from P-wave velocity profiles.* Geophysics *50*, 25–36.

LYNN, H. G. (1979), *Migration and interpretation of deep crustal seismic reflection data.* Ph.D. Thesis, 158 pp. Stanford Univ., Stanford, Calif.

LUDWIG, W. J., NAFE, K. E. and DRAKE, C. L., 'Seismic refraction', in *The Sea*, Vol. 4, Pt. I, A. Maxwell (ed.). Wiley, New York, 1970, p. 53–84.

MAYER-ROSA, D. (1973), *Traveltime anomalies and distribution of earthquakes along the Calaveras fault zone. California.* Bull. Seis. Soc. Am. *63*, 713–729.

MEREU, R. F. (1986), *An interpretation of the central California profiles for the CCSS Shizooka workshop,* submitted to USGS Open-File Rep., 7 pp.

MOONEY, W. D. and COLBURN, R. (1985), *A seismic refraction profile across the San Andreas, Sargent, and Calaveras faults, west-central California.* Bull. Seis. Soc. Am. *75*, 175–191.

MOONEY, W. D. and LUETGERT, J. H. (1982), *A seismic refraction study of the Santa Clara Valley and southern Santa Cruz Mountains, west-central California.* Bull. Seis. Soc. Am. *72*, 901–909.

MOOS, D. and ZOBACK, M. D. (1983), *In situ studies of velocity in fractured crystalline rocks.* J. Geophys. Res. *88*, 2345–2358.

NUR, A. and SIMMONS, G. (1969), *The effect of saturation on velocity in low porosity rocks.* Earth Planet. Sci. Lett. *7*, 183–193.

O'CONNELL, R. J. and BUDIANSKY, B. (1974), *Seismic velocities in dry and saturated cracked solids.* J Geophys. Res. *79*, 5412–5426.

PAVONI, N. (1973), 'A structural model for the San Andreas fault zone along the northeast side of the Gabilan Range', in *Proc. Conf. Tectonic Problems San Andreas Fault System*, R. L. Kovach and A. Nur (eds.). Stanford Univ. Publ. Geol. Sci. *13*, 259–267.

PEDDY, C. P. (1984), *Displacement of the Moho by the Outer Isles thrust shown by seismic modeling.* Nature *312*, 628–630.

SIBSON, R. H. (1977), *Fault rocks and fault mechanism.* J. Geol. Soc. Lond. *133*, 191–213.

SIMMONS, G. and NUR, A. (1968), *Granites: Relation of properties in situ to laboratory measurements.* Science *162*, 789–791.

SJOGREN, B. (1984), *Shallow Refraction Seismics*, Chapman and Hall, London, 268 pp.

SMITHSON, S. B., BREWER, J. A., KAUFMAN, S., OLIVER, J. E. and HURICH, C. (1979), *Structure of the*

Laramide Wind River uplift, Wyoming, from COCORP deep reflection data and from gravity data. J. Geophys. Res. *84*, 5955–5972.

STIERMAN, D. J. (1984), *Geophysical and geological evidence for fracturing, water circulation, and chemical alteration in granitic rocks adjacent to major strike-slip faults.* J. Geophys. Res. *89*, 5849–5857.

STIERMAN, D. J. and KOVACH, R. L. (1979), *An in situ velocity study: The Stone Canyon well.* J. Geophys. Res. *84*, 672–678.

THURBER, C. H. (1983), *Earthquake locations and three-dimensional crustal structure in the Coyote Lake Area, central California.* J. Geophys. Res. *88*, 8226–8236.

TREHU, A. M. and WHEELER, W. H. (1986), *Possible evidence for subducted sediments beneath central California.* Geology (submitted).

WALLACE, R. E. (1949), *Structure of a portion of the San Andreas rift in southern California.* Geol. Soc. Am. Bull. *60*, 781–806.

WANG, C. Y. (1984), *On the constitution of the San Andreas fault zone in central California.* J. Geophys. Res. *89*, 5858–5866.

WANG, C. Y., LIN, W. N. and WU, F. T. (1978), *Constitution of the San Andreas Fault Zone at depth.* Geophys. Res. Lett. *5*, 741–744.

WANG, H. F. and SIMMONS, G. (1978), *Microcracks in crystalline rock from 5.3 km depth in the Michigan Basin.* J. Geophys. Res. *83*, 5849–5856.

WU, F. T. (1978), *Mineralogy and physical nature of clay gouge.* Pure Appl. Geophys. *116*, 655–689.

WU, F. T., BLATTER, L. and ROBERTSON, H. (1975), *Clay gouges in the San Andreas fault system and their possible implications.* Pure Appl. Geophys. *113*, 87–96.

(Received 5th November 1985, revised 20th March 1986, accepted 23rd March 1986)

PAGEOPH, Vol. 124, Nos. 1/2 (1986) 0033-4533/86/020159-17$1.50+0.20/0

Brecciation Processes in Fault Zones: Inferences from Earthquake Rupturing

RICHARD H. SIBSON [1]

Abstract—Surface-rupture patterns and aftershock distributions accompanying moderate to large shallow earthquakes reveal a residual brittle infrastructure for established crustal fault zones, the complexity of which is likely to be largely scale-invariant. In relation to such an infrastructure, continued displacement along a particular master fault may involve three dominant mechanical processes of rock brecciation: (a) *attrition brecciation*, from progressive frictional wear along principal slip surfaces during both seismic and aseismic sliding, (b) *distributed crush brecciation*, involving microfracturing over broad regions when slip on the principal slip surfaces is impeded by antidilational jogs or other obstructions, and (c) *implosion brecciation*, associated with the sudden creation of void space and fluid-pressure differentials at dilational fault jogs during earthquake rupture propagation. These last, high-dilation breccias are particularly favorable sites for hydrothermal mineral deposition, forming transitory low-pressure channels for the rapid passage of hydrothermal fluids. Long-lived fault zones often contain an intermingling of breccias derived from all three processes.

Key words: Breccias, faults, earthquakes, mineralization.

1. *Introduction*

Fault breccias group with gouge and cataclasite-series material as the largely random-fabric rock products of friction-dominated faulting within the seismogenic regime, which typically occupies the top 10–15 km of actively deforming continental crust (SIBSON, 1977, 1983). Despite their widespread occurrence and their frequently noted role as sites for hydrothermal mineralization (e.g., HULIN, 1929; NEWHOUSE, 1942; MITCHAM, 1974; WILKINS and HEIDRICK, 1982), comparatively little attention has been paid to the mechanics of their origin, though the range of textures, matrix-clast relationships, and structural associations observed implies a variety of formative processes. There is a common supposition (e.g., ROBERTSON, 1982) that fault breccias develop by progressive frictional attrition of sidewalls as fault displacement increases. This assumption, however, takes no account of the self-similar structural complexities inherent in natural fault systems (KING, 1983) nor of important dynamic effects which accompany earthquake rupturing, the dominant mechanism of fault slip at

[1] Department of Geological Sciences, University of California, Santa Barbara, California 93106.

high crustal levels. Moreover, this theory of progressive frictional wear does not explain the existence of isolated but locally extensive patches of distributed cataclasis and brecciation, which have been described in the general vicinity of master faults (e.g., BROCK and ENGELDER, 1977; FLINN, 1977; DAVIS et al., 1980), but seemingly bear no direct relationship to the major slip surfaces.

This paper therefore addresses the mechanical processes accompanying slip that may induce rock brecciation within fault zones, taking into consideration their brittle infrastructure and, especially, the likely dynamic effects accompanying earthquake rupturing. Particular attention is paid to aftershock distributions following major earthquakes, which serve to outline patterns of subsidiary brittle deformation related to irregularities in the main rupture. To at least some extent, the complexity of natural fault-fracture systems appears scale-invariant (TCHALENKO, 1970; KING, 1983), so these deformation patterns related to fault irregularities may mimic those directly observable, generally on lesser scales, in ancient fault zones. No account is taken of other breccias that may, in a broad sense, be considered fault-related, such as those arising from metasomatic replacement along fault zones or from structurally controlled igneous intrusive activity.

2. *Development of brittle infrastructure*

Field studies of the initiation and development of brittle fault zones (AYDIN and JOHNSON, 1978; GAY and ORTLEPP, 1979; SEGALL and POLLARD, 1983) show that they rarely evolve by the propagation of a single isolated shear fracture. If a single slip surface is dominant, it is generally far from planar initially, refracting across lithologic contacts (e.g., MURAOKA and KAMATA, 1983) and deviating along preëxisting planes of weakness. More commonly, a brittle fault zone develops by the progressive incorporation and amalgamation of newly formed and preëxisting subsidiary shears and extension fractures of varying size and orientation. Subsidiary shears may sometimes be arranged regularly in echelon Riedel-type arrays (TCHALENKO, 1970), but often the step sense of the echelon shears is quite nonsystematic, especially over extended distances.

Consider, then, the problems of accommodating large shear displacements across such an irregular, newly formed brittle infrastructure under an applied shear stress τ and an effective compressive normal stress, $\sigma_n' = (\sigma_n - P)$, where P is the fluid pressure (Fig. 1). It is apparent that some form of residual through-going shear system must develop in a tabular zone of fractured and brecciated rock whose thickness is largely determined by the amplitude of the initial irregularities (cf. JACKSON and DUNN, 1974). Inevitably, the development of such a *fault initiation breccia* must involve some dilation, the amount depending inversely on σ_n'. For low values of σ_n' or in the (rare?) situation when $\sigma_n' < 0$, dilation may be extreme.

The evolution to a residual infrastructure involves the development of one or

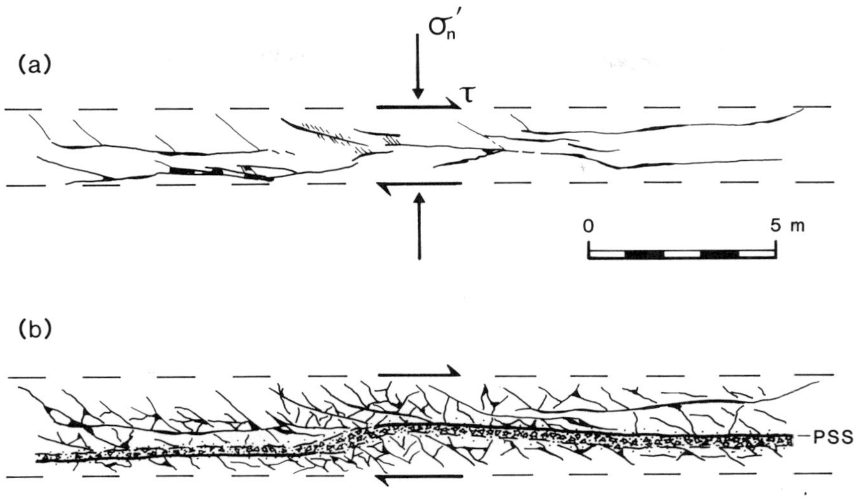

Figure 1
Brecciation accompanying the development of residual PSS infrastructure under shear stress τ and effective normal stress $\sigma_n{}'$ in intact rock: (a) initial irregular shear fracturing (example taken from the initiation infrastructure of a mining-induced normal fault in intact quartzites, from GAY and ORTLEPP, 1979); (b) imagined evolution with continuing displacement into a *fault initiation breccia*, with development of through-going PSS.

more relatively planar and discrete through-going principal slip surfaces (PSS) which persist in a particular configuration for extended time periods and accommodate most subsequent shearing. Commonly these PSS are localized at one or other, or sometimes both, of the margins to the tabular breccia zones (BROCK and ENGELDER, 1977; AYDIN and JOHNSON, 1978; GROCOTT, 1981; SEGALL and POLLARD, 1983). Analogies may be drawn with experimental shearing of artificial gouge zones where stick-slip sliding likewise tends to be restricted to discrete surfaces at the gouge–rock interface (ENGELDER et al., 1975; BYERLEE, et al., 1978). This broad-scale infrastructure of PSS, which may be irregular and discontinuous on scales up to the width of the tabular fault zone, appears to determine the location, style, and intensity of subsequent brittle deformation accompanying displacement.

Dilational and antidilational jogs

Broad-scale PSS infrastructure is revealed in map view along transcurrent fault zones. Studies of brittle deformation in exhumed fault zones (FLINN, 1977) and of both individual earthquake ruptures and rupture patterns arising from multiple episodes of faulting along active faults (CLARK, 1972; TCHALENKO and BERBERIAN, 1975; VEDDER and WALLACE, 1970) reveal irregularities on scales extending up to and sometimes beyond the topographically defined widths of the fault zones, which

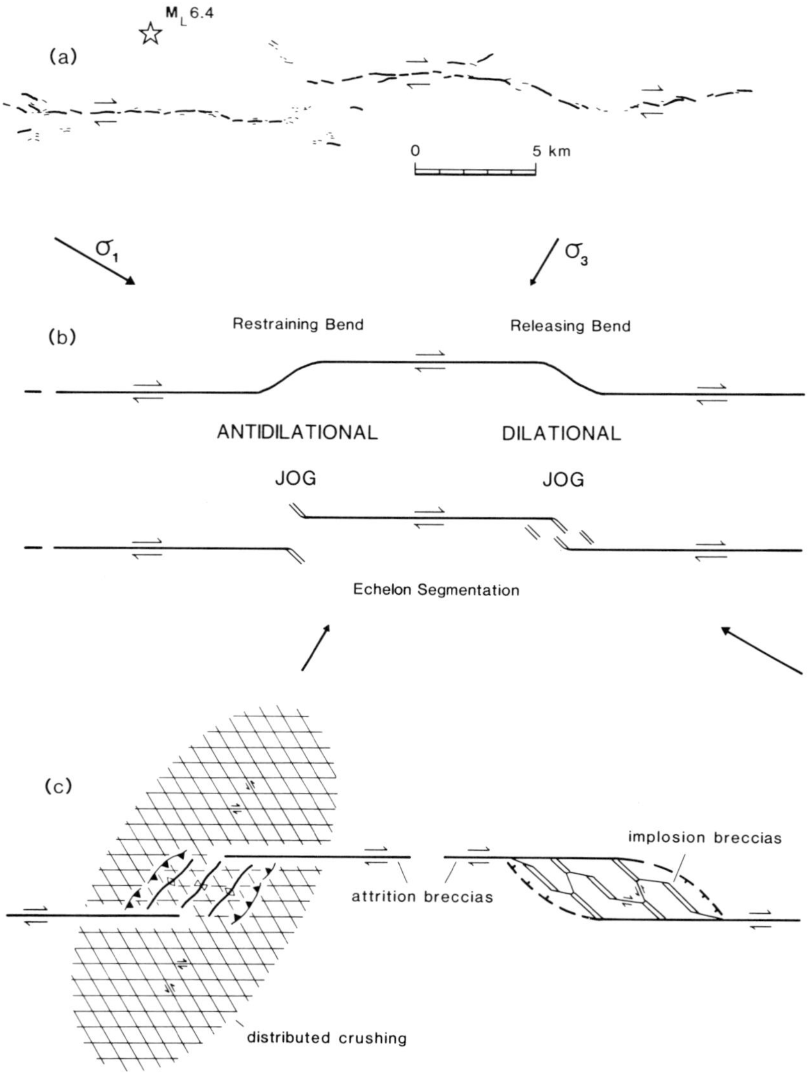

Figure 2

Effect of brittle PSS infrastructure on brecciation processes in fault zones: (a) surface-rupture trace of the 1968 M_L 6.4 Borrego Mountain earthquake (star represents epicenter), showing broad-scale infrastructure (after CLARK, 1972); (b) classification of infrastructure into dilational and antidilational jogs with respect to slip sense and far-field principal compressive stresses ($\sigma_1 > \sigma_2 > \sigma_3$); (c) inferred internal structure of strike-slip fault jogs, showing association with different brecciation processes (thrusts represented by sawtooth lines, fold axial traces by thick wavy lines, subsidiary strike-slip faults by criss-cross lines, vertical extension fractures by thin parallel lines, and normal faults by thick dashed lines with tick on downthrown side).

may range up to a kilometer or so (Fig. 2a). While the full three-dimensional infrastructure of the PSS may be exceedingly complex, we restrict our attention here to irregularities as they appear in the plane defined by the fault slip vector and the pole to planar PSS segments. Two basic classes of irregularity can then be recognized: echelon segmentation and local bends. Individual linear or gently curving PSS strands may be traced for distances of up to 10 km or so, often along one or other of the fault-zone margins, before slip is transferred laterally over perhaps 10^2 to 10^3 m to an echelon strand. It is comparatively rare for an individual rupture strand to curve continuously in a sigmoidal manner across such steps, but subsidiary fracturing often becomes intense in the stepover regions. High-precision microearthquake studies demonstrate that this broad-scale brittle infrastructure often extends throughout the seismogenic regime to depths of 10 km or more (EATON et al., 1970; REASENBERG and ELLSWORTH, 1982). Abrupt changes in the amount of slip accompanying individual earthquakes are commonly associated with the larger stepovers (CLARK, 1972; TCHALENKO and BERBERIAN, 1975; SIEH, 1978). Geomorphic evidence suggests this broad fault infrastructure often remains fixed in one particular configuration for lengthy time periods, certainly well in excess of 10^5 years in some instances (SIBSON, 1986).

The PSS irregularities occur in two alternative asymmetries with respect to the overall sense of shear (Fig. 2b). Depending on the tendency for areal increase or reduction as a consequence of slip transfer across the stepovers, the irregularities are usefully classified as dilational or antidilational fault jogs respectively, independent of faulting mode or sense of shear.

3. *Incremental deformation at fault jogs*

Clearly, jog structures in the PSS form impediments to rigid-body displacement and may induce deformation in and adjacent to fault zones. To evaluate the character of this deformation, it is necessary to consider not only quasi-static elastic interactions between echelon fault segments and their implications for the internal structure of jogs, but also dynamic effects associated with rupture–jog interaction in fluid-saturated crust.

Quasi-static analysis

Stress-field perturbations arising from elastic interaction between echelon fault segments in the two alternative asymmetries have been investigated by SEGALL and POLLARD (1980) in a two-dimensional quasi-static analysis. In the case of a dilational jog, elastic interaction between the fault segments decreases frictional resistance at the segment tips, facilitating slip, and also lowers mean stress throughout the intersegment region, to the extent that the fault tips tend to propagate into extension fractures, and a linking fracture mesh may develop, involving further extension fracturing and subsidiary faulting. However, for an antidilational jog, both the frictional resistance at the segment tips and the mean stress in the intersegment region increase, inhibiting slip transfer across the step. Fault segment tips tend to propagate away from each other, and deformation in the form of subsidiary faulting is predicted over a broad swathe, extending beyond the stepover into regions where mean stress has been somewhat reduced (Fig. 2c). Somewhat similar situations should arise when PSS traces occur in sigmoidal curves to form releasing and restraining bends (CROMWELL, 1974).

Thus, in comparison with antidilational fault jogs which clearly form major obstructions, this quasi-static analysis suggests that dilational jogs present comparatively little hindrance to slip transfer, owing to the ease of linkage between echelon fault segments.

Internal structure of jogs

Detailed analysis of aftershock focal mechanisms following moderate to large strike-slip earthquake ruptures (EATON et al., 1970; HAMILTON, 1972; REASENBERG and ELLSWORTH, 1982) shows that, while minor components of normal and reverse slip are sometimes associated, respectively, with dilational and antidilational fault jogs, the dominant faulting mechanism in the immediate vicinity of both types of jog remains strike-slip on subvertical planes. Aftershock distributions and mechanisms are therefore broadly in accord with expectations from the SEGALL and POLLARD (1980) analysis (Figs. 2c and 3). Slip transfer around an antidilational jog involves mostly distributed strike-slip microfaulting over a broad region in the vicinity of the jog, perhaps with some subsidiary thrusting. This may be viewed as a form of large-scale cataclastic flow around the obstruction. By contrast, deformation associated with a dilational fault jog tends to be localized within the stepover. Structural comparisons with small-scale dilational jogs, the dominance of strike-slip aftershocks, and the inference of induced tensile stresses within the stepover, all suggest that an appropriate structural model for a major dilational jog is that proposed by HILL (1977), whereby the echelon fault segments are linked by small vertical extension fractures and shears arranged in a 'honeycomb' mesh or stockwork. An important consequence of this model is that slip transfer across a dilational fault jog involves extensional opening of the same order in the direction of fault motion.

Dynamic effects

Directly contrary to expectations from the quasi-static analysis, dilational fault jogs appear to form equal or greater impediments to the passage of earthquake ruptures than antidilational jogs (SIBSON, 1985, 1986). It is noteworthy, however, that rupture arrest at a dilational jog is often followed by slow post-seismic slip transfer across the stepover. This time-dependent behavior has been attributed to

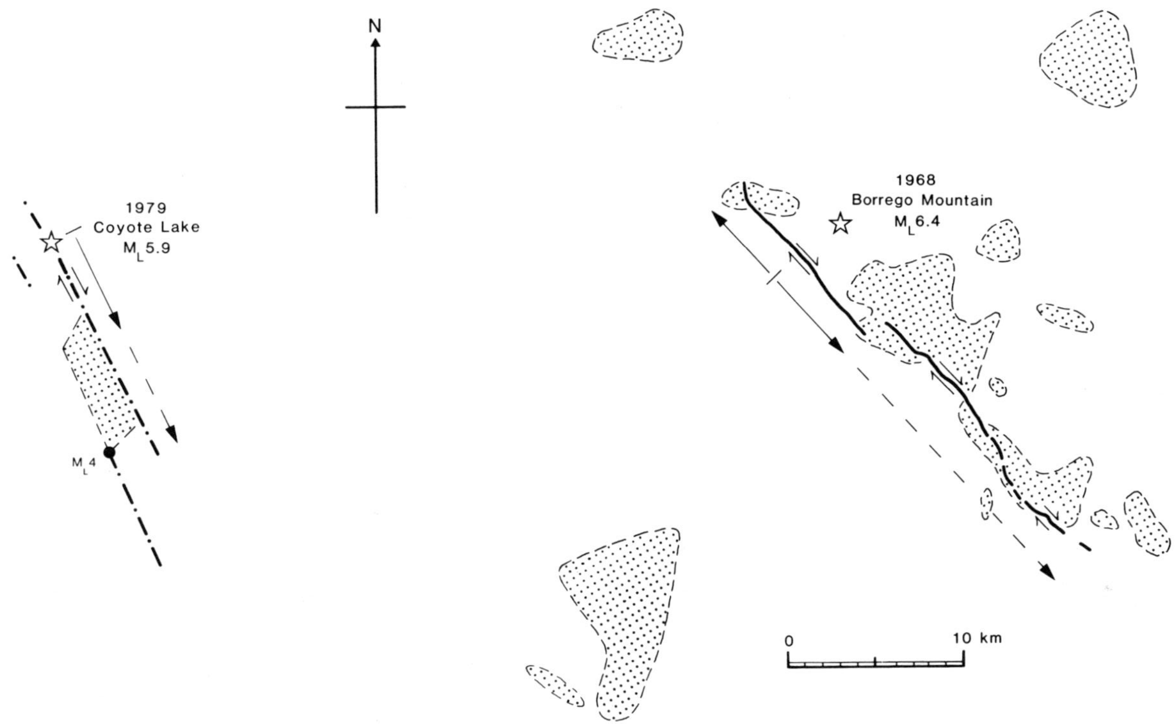

Figure 3
Aftershock distributions associated with two dextral strike-slip ruptures in the San Andreas fault system. Instrumental epicenters represented by stars, surface-rupture traces by bold lines, microearthquake lineaments by dash-dot lines, aftershock clusters by stippling, propagation directions and extent of mainshock ruptures by arrows.

the difficulty in opening linking extensional fracture systems in fluid-saturated crust within periods comparable to the slip durations of moderate to large earthquakes, typically a few seconds. Over a wide range of rock permeabilities, the rapid incipient opening of extension fractures is sufficient to induce fluid-pressure differentials between wall-rock and crack interiors, comparable to initial hydrostatic pressures at any depth throughout the seismogenic regime. Further extensional opening is then opposed by an induced suctional force which scales with the width of the dilational jog. Depending on the magnitude of the suction, rupture propagation may be perturbed or arrested by the jog. If arrested, delayed slip transfer across the linking fracture system may occur as fluid pressures reëquilibrate by diffusion.

Antidilational jogs, by contrast, appear to act as barriers equally to both rapid and long-term slip transfer.

4. *Strike-slip ruptures and aftershock distributions*

Contrasting patterns of deformation arising from earthquake rupture interaction with dilational and antidilational fault jogs are revealed by aftershock epicentral distributions in strike-slip fault systems. As examples we consider two dextral strike-slip ruptures in California: the 1979 M_L 5.9 Coyote Lake earthquake on the Calaveras fault and the 1968 M_L 6.4 Borrego Mountain earthquake within the San Jacinto fault zone (Fig. 3). Information on these events is variously derived from seismic waveform analyses, surface-rupture patterns, and high-precision aftershock studies (under favorable circumstances microearthquake hypocenters may be located to epicentral and focal depth uncertainties of under ± 1 km).

In the Coyote Lake earthquake, mainshock slip averaging ≈ 0.2 m occurred on a subvertical plane delineated by subsequent aftershocks, the rupture extending from 10 to 3 km in depth and propagating unilaterally southeastwards from the focus for 6 to 14 km, to terminate adjacent to a dilational jog defined by a rhomboidal area of diffuse aftershock activity (BOUCHON, 1982; LIU and HELMBERGER, 1983). Delayed slip transfer to an echelon fault segment ≈ 2 km to the southwest began some five hours after the main earthquake with an M_L 4.0 shock, after which activity progressively migrated southeastwards for ≈ 8 km, defining a sharp microearthquake lineament (REASENBERG and ELLSWORTH, 1982). Aftershock focal mechanisms were almost uniformly consistent with strike slip on subvertical planes parallel to the main rupture, except that within the dilational jog a slight clockwise swing was noted in the strike of probable fault planes. Very similar behavior has been noted for other strike-slip ruptures which terminate in dilational fault jogs (SIBSON, 1985).

The rupturing that accompanied the Borrego Mountain earthquake was rather more complex. A segmented surface break was mapped over a distance of ≈ 31 km and consisted of three principal rupture strands separated by antidilational and dilational jogs with stepovers of ≈ 1.5 km (CLARK, 1972). Measured surface slip reached ≈ 0.4 m along the northern segment, ≈ 0.3 m along the central, and ≈ 0.1 m along the southern, but the last two estimates include substantial post-seismic afterslip. Seismological analyses of the main shock suggest that most of the radiated energy was derived from the northernmost segment, along which subsequent aftershock activity was a minimum (BURDICK and MELLMAN, 1976; EBEL and HELMBERGER, 1982). It seems probable, therefore, that the main rupture was at least severely perturbed, and possibly temporarily arrested, by the antidilational jog. Well-located aftershocks were broadly distributed over a region extending well outside the subvertical rupture zone at depths down to 12 km (HAMILTON, 1972). Some significant correlations with the surface-rupture geometry are, however, apparent. Two remote aftershock clusters, symmetrically located roughly equidistant from the rupture trace, have been attributed to off-fault increases in shear stress as a consequence of the elastic response of the surrounding crust to the main rupture (DAS and SCHOLZ, 1981). The most intense aftershock concentration occurred northeast of the central rupture segment adjacent to the antidilational jog. By constrast, aftershocks associated with the southern dilational jog mostly lay between the echelon rupture segments.

Observed aftershock distributions are therefore broadly in accord with inferences from the SEGALL and POLLARD (1980) analysis of elastic interaction between echelon fault segments, subsidiary deformation being concentrated within dilational jogs, but outside and adjacent to the Borrego Mountain antidilational jog on the side opposite the rupture segment where relief of shear stress is likely to have been greatest. In both situations the greatest concentrations occur in regions where mean stress is likely to have been reduced as a consequence of the main rupture.

Aftershocks and fluid flow

This last observation—that aftershocks associated with both varieties of jog tend to concentrate in regions of lowered mean stress—lends credence to the hypothesis (NUR and BOOKER, 1972; BOOKER, 1974) that the location and time-dependent decay of aftershock activity on subsidiary fractures is controlled by redistribution of aqueous fluids around rupture terminations. Shear resistance of such fractures may be approximated by a criterion of Coulomb form,

$$\tau_f = C + \mu_s \sigma_{n'} = C + \mu_s(\sigma_n - P) \tag{1}$$

where C and μ_s are respectively the cohesive strength and the static frictional coefficient. Changes in fluid pressure immediately following the main rupture should reflect induced changes in mean stress, so that intiially there is little variation in the effective normal stress across subsidiary fractures or in their frictional strength. The disequilibrium in fluid pressure thus established, however, is slowly dissipated as fluids diffuse from the regions of increased mean stress to regions of lowered mean stress, where pressure levels are progressively restored back up to their premainshock values. In turn, this causes time-dependent reductions in frictional strength and shear failure along fractures within the region of reduced mean stress.

One may therefore infer a concentrated fluid inflow into dilational jogs from surrounding areas following rupture arrest. In the case of an antidilational jog, a rather more diffuse flow into the broad surrounding regions of reduced mean stress should occur.

5. *Mechanical processes of brecciation*

It is apparent from the foregoing that much of the brittle deformation induced by an increment of slip on a master fault may occur well away from the PSS. Three dominant processes of rock breakage, *attritional wear*, *distributed crushing*, and *implosion brecciation*, are inferred at different sites in relation to the brittle infrastructure (Fig. 2c). Varieties of intermediate mixed behavior may also occur, so that these three mechanisms should be considered as end-member processes.

Attrition brecciation

Development of sidewall breccias by frictional wear with accompanying grain cataclasis is to be expected during both seismic and aseismic slip on the PSS. Wear processes include brittle shearing of asperities on all scales, asperity indentation and ploughing with the possibility of some asperity tip plasticity and even melting, sidewall cracking and plucking, and progressive rock fragmentation and grain comminution by inter- and intra-granular cracking. Rotation of entrained grains and rock fragments is integral to the process. In general, both grain size and the degree of sorting tend to decrease with increasing displacement. These processes have been documented experimentally (ENGELDER, 1974, 1978; JACKSON and DUNN, 1974) and inferred from field-based microstructural studies (BROCK and ENGELDER, 1977; HOUSE and GRAY, 1982). MOORE and SIBSON (1978) suggest that sidewall fracturing leading to brecciation may also be enhanced by transient thermal stresses generated during seismic slip.

Experimentally it is found that the thickness of gouge-breccia zones generally increases with displacement; the same is believed to hold true for natural fault zones (ROBERTSON, 1982). The rate of thickening depends primarily on the level of effective normal stress to the fault, σ_n' (ENGELDER et al. 1975), which also determines the degree of dilation accompanying progressive gouge-breccia development. Other factors, such as the inclination to faults of existing planar anisotropies (bedding, foliation, etc.), may also affect the rate of gouge-breccia generation (JACKSON and DUNN, 1974).

There are major difficulties in discriminating between the cataclastic products of seismic and aseismic slip (SIBSON, 1977), but the development of locally strong shape fabrics within fault gouge or cataclasite (e.g., HOUSE and GRAY, 1982; CHESTER et al. 1985) seems more likely to be indicative of relatively steady aseismic shearing flow.

Distributed crush brecciation

Where displacement along the PSS is impeded by antidilational jogs or other such obstructions, distributed cataclastic crushing (microcracking and microfaulting) of varying intensity may occur over broad areas, corresponding to the off-fault aftershock clusters described previously. While it is true that much of this deformation may result from attritional processes on subsidiary faults, the shearing displacements are likely to be minute in comparison with slip on the master fault. On experimental evidence, the pervasiveness of the microfaulting can be expected to increase with increasing effective overburden pressure (PATERSON, 1978).

Rock products are likely to be structureless microbreccias and cataclasites of the kind described by FLINN (1977), adjacent to the major transcurrent faults in Shetland. Hydrothermal alteration (and sometimes neomineralization) is often widespread in such rocks (DAVIS et al. 1980), perhaps arising from the intermittent influxes of fluid postulated to accompany aftershock activity.

Implosion brecciation

Fault-related brecciation may also occur by the implosion of wall rock into cavity space generated during rapid slip, most commonly at dilational fault jogs. As previously discussed, the deviatoric stress state in such structures tends to favor the development of linking extension fracture systems, and field descriptions of rhomboidal to lensoidal cavity infillings associated with dilational jogs are widespread, especially in the mining literature (e.g., SPURR, 1925; HULIN, 1929; GAMOND, 1983; SEGALL and POLLARD, 1983). Dilational fault jogs containing a breccia of wall-rock clasts set in a matrix of pseudotachylyte friction melt, the product of localized seismic slip in dry crystalline rocks (SIBSON, 1975; GROCOTT, 1981), are of particular significance. They provide direct evidence that substantial cavity opening may accompany seismic slip in strong rocks at depths of several kilometers.

Most commonly, however, the infilling is of hydrothermal minerals (e.g., quartz, calcite, sulphides) occurring either as extensional vein systems or as the matrix to high-dilation breccias of angular wall-rock fragments. A frequently noted feature of such breccias is that they possess an *exploded jigsaw* texture, with the clasts showing little evidence of frictional attrition (PHILLIPS, 1972). Textures often record a history of incremental extensional opening across the jog, or of multiple-episode brecciation and hydrothermal recementation. This strongly suggests that, while implosion of wall rocks into cavities at dilational jogs may sometimes occur solely from the surrounding lithostatic pressure, as suggested by MITCHAM (1974), the more usual mechanism involves hydraulic implosion, as a consequence of the sudden fluid-pressure differentials generated in dilational jogs during rupture arrest (SIBSON, 1985). The transitory low fluid pressures induced within dilational jogs, coupled with enhanced fracture permeability, clearly make them particularly favorable sites for rapid deposition of hydrothermal minerals (SHARP, 1965; TOULMIN and CLARK, 1979). In association with subsidiary shears and their attrition breccias, the extensional fracture system within a dilational jog may therefore resemble a vein stockwork (Fig. 2), while the driving echelon PSS segments themselves remain barren of hydrothermal mineralization and relatively inconspicuous. In some instances, however, brecciation by hydraulic implosion may occur without subsequent hydrothermal deposition. Discrimination from other breccia types may then be difficult.

Somewhat similar processes of brecciation by hydraulic fracturing have been envisaged by MASSON (1972), PHILLIPS (1972), REDWINE (1981), and ROEHL (1981).

Conditions for hydraulic implosion. The difference in fluid pressures between the wall rock (P_w) and the interior of an opening extension fracture (P_i) in a dilational jog may be written,

$$\Delta P = P_w - P_i \tag{2}$$

and the condition for hydraulic implosion of wall rock of tensile strength T is simply

$$\Delta P \geqslant T \tag{3}$$

Given the incompressibility of aqueous fluids, it can be shown that for a broad range of likely rock permeabilities the incipient opening of extension fractures over periods comparable to earthquake slip durations is sufficient to generate values of ΔP approaching initial fluid-pressure levels at any depth within the seismogenic regime (SIBSON, 1985). On the assumption that initial fluid pressures are hydrostatic, the maximum potential pressure differential therefore increases with depth at ≈ 100 bar/km. Thus, taking a reasonable value of tensile strength for the wall rock of, say, 50 bars (ETHERIDGE, 1983), one may infer a likely depth control for brecciation by hydraulic implosion. For depths $z < 0.5$ km no implosion of fracture sidewalls should occur and, given the availability of appropriate hydrothermal solutions, linking extension fractures in a dilational jog should develop as extensional vein fills. At $z > 0.5$ km sidewall implosion is likely to occur with each forced increment of extensional opening, giving rise to the characteristic multiply recemented breccias.

Hydraulic implosion breccias of similar character may also develop when shear occurs on a fault across which the effective normal stress is tensile (i.e., $\sigma_n' < 0$) at failure. For this probably rare situation to arise the fault must possess significant cohesive strength, perhaps from hydrothermal cementation.

6. *Discussion*

A first point to emphasize is that the concepts developed here are equally applicable to all modes of faulting; the discussion hitherto has been restricted to strike-slip faulting because brittle fault infrastructure is then simply revealed in map view and outlined by aftershock epicentral distributions. More generally, local brecciation zones associated with fault jogs should occur as crudely linear structures within the plane of the fault zone, lying orthogonal to the slip vector. In the case of pure normal and reverse dip slip, extension fracturing within dilational jogs should be, respectively, subvertical and subhorizontal. It should also be borne in mind that our attention has been concentrated on the simplest fault infrastructure, as seen in two dimensions. Much more complex structures and patterns of subsidiary deformation may arise in three dimensions from crustal inhomogeneities, splay fault intersections, etc.

Distinguishing breccia types

It clearly would be desirable to have a simple means of distinguishing the fault breccias arising from the different mechanisms, but the problem is made complicated by crustal compositional heterogeneity, and also by the finite lifetime of a fault infrastructure in one particular configuration. For large-displacement faults a complex intermingling of breccias derived from all three processes is likely to arise, with one mode of brittle disruption overprinting another. It might be supposed that the high-dilation character of breccias formed by hydraulic implosion, coupled with low internal deformation of the clasts, renders them the most easily recognized. Indeed, it seems probable that many of the *crackle* and *shatter* breccias described in the mining literature, where fault-related, may have originated by this mechanism, but where hydraulic implosion occurs without significant hydrothermal deposition, distinction from other breccia types may be problematic. The difficulties may be compounded by other common processes such as hydrothermal mineral replacement. Thus, in attempting to distinguish different breccia types it is usually necessary to consider a range of likely textual, microstructural, and structural association characteristics, as summarized in Table 1. Even then, certain discrimination may not always be possible.

Implications for hydrothermal mineralization

Because of their high permeability, the importance of fault breccias as channelways for hydrothermal fluids and sites favorable to mineral deposition has long been recognized (HULIN, 1929; NEWHOUSE, 1942). The mechanisms of breccia development outlined here, though, suggest a more intimate relationship between intermittent seismic slip on a master fault, the development of permeable breccia

Table 1

Possible Distinguishing Characteristics and Structural Associations for Fault Breccias Derived from the Different Brecciation Mechanisms

		BRECCIATION PROCESSES		
		ATTRITION	DISTRIBUTING CRUSHING	IMPLOSION
	STRUCTURAL ASSOCIATION	Tabular PSS zones Major shear separations	Vicinity antidilational jogs Minor shear separations	Dilational jogs Subsidiary extension fractures
		? —— ? —— ? —— ? —— ? ——→ increasing dilation		
DISTINGUISHING CHARACTERISTICS	Textural Features	Extensively rolled clasts	Widespread pervasive micro-fracturing	Jigsaw puzzle textures
	Internal Clast Deformation	High	Moderate→High	Low
	Clast Size Distribution	Poorly to moderately sorted	Poorly sorted	Often well sorted
	Clast Composition	Often bimodal or polymict	Often unimodal	Often unimodal
	Matrix Composition	Similar to clasts	Similar to clasts (minor hydrothermal)	Often exotic hydrothermal

zones in regions of subsidiary brittle deformation, and the inflow of hydrothermal fluids. Significant dilation may accompany the development of off-fault crush breccias in the broad regions of lowered mean stress induced by antidilational jogs. Hydrothermal mineralization in such situations is likely to be of a rather diffuse nature, perhaps similar to that occurring in the broad zones of brecciation associated with detachment faults (WILKINS and HEIDRICK. 1982). However, hydrothermal fluid inflow and mineral deposition are clearly likely to be most intense in the high-dilation extensional fracture systems and implosion breccias associated with dilational jogs. Examples of strongly mineralized breccias in dilational jogs are widespread in the literature (e.g., SPURR, 1925; NEWHOUSE, 1942; MCKINSTRY, 1948). At the present day it is interesting to speculate on the likely consequences for mineral deposition of a large strike-slip rupture on the southern San Andreas fault, terminating in one of the major, geothermally active, dilational fault jogs in the Salton Sea area (cf. MCKIBBEN and ELDERS, 1985).

Acknowledgments

I thank Jim Andrews, Barbara John, and Steve Richard for comments on various aspects of this paper. Research leading to this publication was supported by the National Science Foundation, Grant Number EAR83-05876.

References

AYDIN, A. and JOHNSON, A. M. (1978), *Development of faults as zones of deformation bands and as slip surfaces in sandstone.* Pure Appl. Geophys. *116*, 931–942.

BOOKER, J. (1974), *Time-dependent strain following faulting of a porous medium.* J. Geophys. Res. *79*, 2037–2043.

BOUCHON, M. (1982), *The rupture mechanism of the Coyote Lake earthquake of 6 August 1979 inferred from near-field data.* Bull. Seism. Soc. Am. *72*, 745–757.

BROCK, W. G. and ENGELDER, J. T. (1977), *Deformation associated with the movement of the Muddy Mountain overthrust in the Buffington window, southeastern Nevada.* Geol. Soc. Am. Bull. *88*, 1667–1677.

BURDICK. L. J. and MELLMAN, G. R. (1976), *Inversion of body waves from Borrego Mountain earthquake to the source mechanism.* Bull. Seism. Soc. Am. *66*, 1485–1499.

BYERLEE, J., MJACHKIN, V., SUMMERS, R. and VOEVODA, O. (1978), *Structures developed in fault gouge during stable sliding and stick-slip.* Tectonophys. *44*, 161–171.

CHESTER, F. M., FRIEDMAN, M., and LOGAN. J. M. (1985), *Foliated cataclasite.* Tectonophys. *111*, 134–146.

CLARK, M. M. (1972), *Surface rupture along the Coyote Creek fault.* U.S. Geol. Surv. Prof. Paper *787*, 55–86.

CROWELL, J. C. (1974), *Origin of Late Cenozoic basins in southern California.* Soc. Econ. Pal. Mineral. Spec. Publ. *22*, 190–203.

DAS, S. and SCHOLZ, C. H. (1981), *Off-fault aftershock clusters caused by shear stress increase.* Bull. Seism. Soc. Am. *71*, 1669–1675.

DAVIS, G. A., ANDERSON, J. L., FROST, E. G. and SHACKELFORD, T. J. (1980), *Mylonitization and detachment faulting in the Whipple-Buckskin-Rawhide Mountains terrane, southeastern California and western Arizona.* Geol. Soc. Am. Mem. *153*, 79–129.

EATON, J. P., O'NEILL, M. E., and MURDOCK, J. N. (1970), *Aftershocks of the 1966 Parkfield-Cholame, California, earthquake: A detailed study.* Bull. Seism. Soc. Am. *60*, 1151–1197.

EBEL, J. E. and HELMBERGER, D. V. (1982), *P-wave complexity and fault asperities: The Borrego Mountain, California, earthquake of 1968.* Bull. Seism. Soc. Am. *72*, 413–437.

ENGELDER, J. T. (1974), *Cataclasis and the generation of fault gouge.* Geol. Soc. Am. Bull. *85*, 1515–1522.

ENGELDER, J. T. (1978), *Aspects of asperity-surface interaction and surface damage of rocks during experimental frictional sliding.* Pure Appl. Geophys. *116*, 705–716.

ENGELDER, J. T., LOGAN, J. M., and HANDIN, J. (1975), *The sliding characteristics of sandstone on quartz fault-gouge.* Pure Appl. Geophys. *113*, 69–86.

ETHERIDGE, M. (1983), *Differential stress magnitudes during regional deformation and metamorphism: Upper bound imposed by tensile fracturing.* Geology *11*, 231–234.

FLINN, D. (1977), *Transcurrent faults and associated cataclasis in Shetland.* J. Geol. Soc. Lond. *133*, 231–248.

GAMOND, J. F. (1983), *Displacement features associated with fault zones: A comparison between observed examples and experimental models.* J. Struct. Geol. *5*, 33–46.

GAY, N. C. and ORTLEPP, W. D. (1979), *Anatomy of a mining-induced fault zone.* Geol. Soc. Am. Bull. *90*, 47–58.

GROCOTT, J. (1981), *Fracture geometry of pseudotachylyte generation zones: A study of shear fractures formed during seismic events.* J. Struct. Geol. *3*, 169–178.

HAMILTON, R. M. (1972), *Aftershocks of the Borrego Mountain earthquake from April 12 to June 12, 1968.* U.S. Geol. Surv. Prof. Paper *787*, 31–54.

HILL, D. P. (1977), *A model for earthquake swarms*. J. Geophys. Res. *82*, 1347–1352.

HOUSE, W. M. and GRAY, D. R. (1982), *Cataclasites along the Saltville thrust, U.S.A., and their implications for thrust-sheet emplacement*. J. Struct. Geol. *4*, 257–269.

HULIN, C. D. (1929), *Structural control of ore deposition*. Econ. Geol. *24*, 15–49.

JACKSON, R. E. and DUNN, D. E. (1974), *Experimental sliding friction and cataclasis of foliated rocks*. Int. J. Rock Mech. Min. Sci. *14*, 235–249.

KING, G. C. P. (1983), *The accommodation of large strains in the upper lithosphere of the earth and other solids by self-similar fault systems: The geometrical origin of b-value*. Pure Appl. Geophys. *121*, 762–815.

LIU, H.-L. and HELMBERGER, D. V. (1983), *The near-source ground motion of the 6 August 1979 Coyote Lake, California, earthquake*. Bull. Seism. Soc. Am. *73*, 201–218.

MASSON, H. (1972), *Sur l'origine de la cornieule par fracturation hydraulique*. Eclogae geol. Helv. *65*, 27–41.

MCKIBBEN, M. A. and ELDERS, W. A. (1985), *Fe–Zn–Cu–Pb mineralization in the Salton Sea geothermal system, Imperial Valley, California*. Econ. Geol. *80*, 539–559.

MCKINSTRY, H. E., *Mining Geology*. Prentice-Hall, New Jersey, 1948, 677 pp.

MITCHAM, T. W. (1974), *Origin of breccia pipes*. Econ. Geol. *69*, 412–413.

MOORE, H. E. and SIBSON, R. H. (1978), *Experimental thermal fragmentation in relation to seismic faulting*. Tectonophys. *49*, T9–T17.

MURAOKA, H. and KAMATA, H. (1983), *Displacement distributions along minor fault traces*. J. Struct. Geol. *5*, 483–495.

NEWHOUSE, W. H., *Ore Deposits as Related to Structural Features*. Princeton Univ. Press, New Jersey, 1942, 280 pp.

NUR, A., and BOOKER, J. (1972), *Aftershocks caused by pore fluid flow?*. Science *175*, 885–887.

PATERSON, M. S., *Experimental Rock Deformation—The Brittle Field*. Springer-Verlag, Berlin, 1978, 254 pp.

PHILLIPS, W. J. (1972), *Hydraulic fracturing and mineralization*. J. Geol. Soc. Lond. *128*, 337–359.

REASENBERG, P. and ELLSWORTH, W. L. (1982), *Aftershocks of the Coyote Lake, California, earthquake of August 6, 1979: A detailed study*. J. Geophys. Res. *87*, 10, 637–10, 665.

REDWINE, L (1981), *Hypothesis containing dilation, natural hydraulic fracturing and dolomitisation to explain petroleum reservoirs in Monterey Shale, Santa Maria area, California*, in *The Monterey Formation and Related Siliceous Rocks of California*. R. E. Garrison, R. G. Douglas, K. E. Pisciotto, C. M. Isaacs and J. C. Ingle (eds.), Spec. Publ. Soc. Econ. Pal. Mineral., Los Angeles, p. 221–248.

ROBERTSON, E. C. (1982), *Continuous formation of gouge and breccia during fault displacement*, in *Issues in Rock Mechanics*. Proc. 23rd Symp. Rock Mechanics, R. E. Goodman and F. E. Heuse (eds.), Am. Inst. Mining Metall. Petrol. Eng., New York, 397–403.

ROEHL, P. O. (1981), 'Dilation brecciation—*A proposed mechanism of fracturing, petroleum expulsion and dolomitization in the Monterey Formation*, California', in *The Monterey Formation and Related Siliceous Rocks of California*. R. E. Garrison, R. G. Douglas, K. E. Pisciotto, C. M. Isaaca, and J. C. Ingle (eds.), Spec. Publ. Soc. Econ. Pal. Mineral., Los Angeles, p. 285–315.

SEGALL, P. and POLLARD, D. D. (1980), *Mechanics of discontinuous faulting*. J. Geophys. Res. *85*, 4337–4350.

SEGALL, P. and POLLARD, D. D. (1983), *Nucleation and growth of strike-slip faults in granite*. J. Geophys. Res. *88*, 555–568.

SHARP, W. E. (1965), *The deposition of hydrothermal quartz and calcite*. Econ. Geol. *60*, 1635–1644.

SIBSON, R. H. (1975), *Generation of pseudotachylyte by ancient seismic faulting*. Geophys. J. R. Astr. Soc. *43*, 775–794.

SIBSON, R. H. (1977), *Fault rocks and fault mechanisms*. J. Geol. Soc. Lond. *133*, 191–213.

SIBSON, R. H. (1983), *Continental fault structure and the shallow earthquake source*. J. Geol. Soc. Lond. *140*, 741–767.

SIBSON, R. H. (1985), *Stopping of earthquake ruptures at dilational fault jogs*. Nature *316*, 248–251.

SIBSON, R. H. (1986), *Rupture interaction with fault jogs*, in *Earthquake Source Mechanics*. S. Das, J. Boatwright and C. H. Scholz (eds.), Maurice Ewing Ser. 6, Am. Geophys. Union Mon. 37, 157–168.

SIEH, K. E. (1978), *Slip along the San Andreas fault associated with the great 1857 earthquake*. Bull. Seism. Soc. Am. *68*, 1421–1448.

SPURR, J. E. (1925), *The Camp Bird compound veindike*. Econ. Geol. *20*, 115–152.

TCHALENKO, J. S. (1970), *Similarities between shear zones of different magnitudes*. Geol. Soc. Am. Bull. *81*,

1625–1640.

TCHALENKO, J. S. and BERBERIAN, M. (1975), *Dasht-e Bayaz Fault, Iran: Earthquake and earlier related structures in bed rock.* Geol. Soc. Am. Bull. *86*, 703–709.

TOULMIN, P. and CLARK, S. P. (1979), *Thermal aspects of ore formation*, in *Geochemistry of Hydrothermal Ore Deposits.* H. L. Barnes (ed.), Holt, Rinehart, New York, 437–464.

VEDDER, J. G. and WALLACE, R. E. (1970), *Map showing recently active fault breaks along the San Andreas and related faults between Cholame Valley and Tejon Pass, California.* U.S. Geol. Surv. Misc. Invest. Map I-574, scale 1:24,000.

WILKINS, J. and HEIDRICK, T. L. (1982), *Base and precious metal mineralization related to low-angle tectonic features in the Whipple Mountains, California, and Buckskin Mountains, Arizona*, in *Mesozoic-Cenozoic Tectonic Evolution of the Colorado River Region, California, Arizona and Nevada.* E. G. Frost and D. L. Martin (eds.), Cordilleran Publishers, San Diego, p. 182–203.

(Received 10th October 1985, revised 20th February 1986, accepted 23rd February 1986)

PAGEOPH, Vol. 124, Nos. 1/2 (1986) 0033-4533/86/020177-25$1.50 + 0.20/0

Relations between Left-Lateral Strike-Slip Faults and Right-Lateral Monoclinal Kink Bands in Granodiorite, Mt. Abbot Quadrangle, Sierra Nevada, California

RUSSELL K. DAVIES[1] and DAVID D. POLLARD[2]

Abstract—Small left-lateral strike-slip faults and right-lateral monoclinal kink bands with subvertical fold axes may be related to the formation of a very large right-lateral kink band (Bear Creek kink band), about 8 km wide and at least 15 km long, trending N27W along Bear Creek Valley in the Mt. Abbot quadrangle, Sierra Nevada, California.

A foliation within Bear Creek Valley is defined by vertical slabs of granodiorite bounded by joints and faults. Small strike-slip faults and larger fault zones have nucleated along preëxisting joints and accommodated shearing between granodiorite slabs. The orientations of small cracks that occur near the tips of faults or connect adjacent fault segments indicate that the direction of maximum compression was about 20° counterclockwise from traces of joints at the time the faults nucleated. In some places where faults are closely spaced there are small, right-lateral kink bands with widths of 1 to 20 m. The slabs of granodiorite are gently curved through the kink bands, and analysis of the orientations of slabs in the limbs of the small kink bands indicates that the direction of maximum compression during kink-band formation was 15° to 20° counterclockwise from the traces of faults outside the kink bands. The orientation of the maximum compression for the formation of the small cracks at tips of many strike-slip faults and for the formation of the small kink bands, relative to the orientation of the maximum compression inferred from the joints on the limb of Bear Creek kink band, suggests that the foliation within the Bear Creek Valley has reoriented a maximum of 40° to 60° clockwise. Although the various orientations of joints, faults, and kink bands could be explained in terms of different regional compression directions at different places and at different times in the Mt. Abbot quadrangle, a much simpler interpretation, based on analysis of large and small structures in the granodiorite in Bear Creek Valley, is that they all formed in response to one maximum regional compression in the direction N25E.

Key words: Strike-slip faults, kink bands, Sierra Nevada, stress orientation.

Introduction

Major lineaments, representing zones of closely spaced joints and faults (LOCKWOOD and LYDON, 1975), change trend with a definite pattern (Fig. 1) across the Mt. Abbot quadrangle, Sierra Nevada, California (Fig. 2). Near Florence Lake

[1] University of Cincinnati, Cincinnati, Ohio 45220. Present address: Center for Tectonophysics, Texas A and M University, College Station, TX 77843.

[2] Stanford University, Stanford, CA 94305.

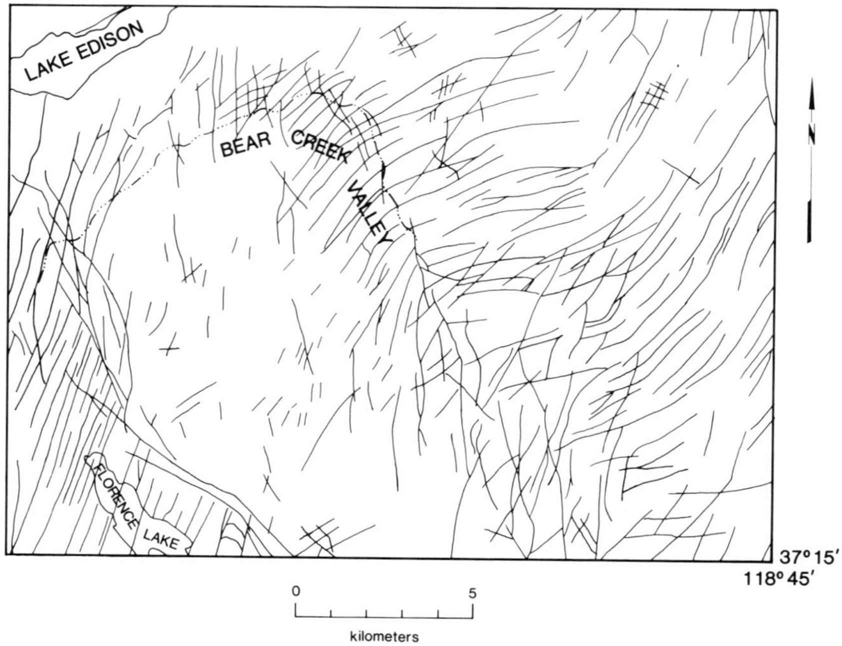

Figure 1
Traces of major lineaments, or zones of closely spaced joints and faults, in southern half of Mt. Abbot quadrangle. Modified from LOCKWOOD and LYDON, 1975.

(Fig. 1) the lineaments are subparallel, straight, and trend approximately N20E. Within a wide band, about 8 km wide and trending northwest along Bear Creek Valley, the lineaments have distinctly different orientations, dominated by traces of lineaments that curve from N30E to N75E and back to N30E from southwest to northeast across the band. The lineaments in the area northeast of the band are subparallel to those near Florence Lake.

An investigation made by SEGALL and POLLARD (1983a) has shown that the lineaments near Florence Lake (Fig. 1) are subparallel to numerous small, closely spaced, vertical fractures that are predominantly joints, across which displacements are extensional. According to detailed mapping by SEGALL and POLLARD (1983a), the traces of the joints are straight or slightly curved and range in length from about 1 to 100 m, and the joints are irregularly spaced, the spacings ranging between 20 cm and 25 cm. The surfaces of joints trend N20E and dip 70° to 90° east. The detailed study of SEGALL and POLLARD (1983a, 1984) suggests that the lineaments near Florence Lake are predominantly joints with little or no shear displacement, reflecting maximum compression in the direction N20E and extension in the direction N70W.

According to another study made by SEGALL and POLLARD (1983b), the linea-

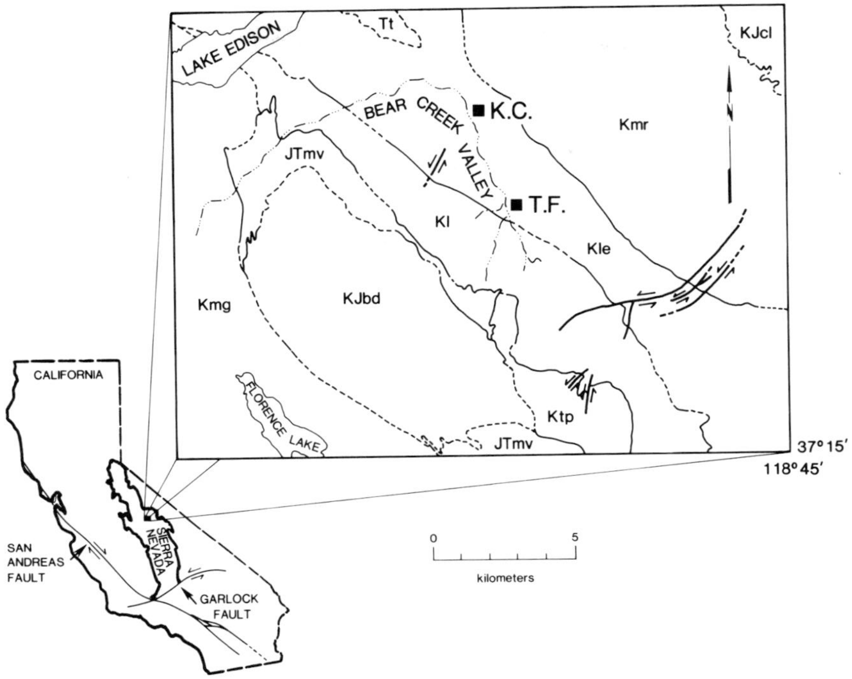

Figure 2

Geological map of southern half of Mt. Abbot quadrangle, simplified from LOCKWOOD and LYDON (1975), showing locations of two map areas, Kip Camp and Trail Fork: Kmg, Cretaceous Mount Givens granodiorite; KJbd, Jurassic or Cretaceous quartz monzonite of Bear Dome; Ktp, Cretaceous quartz monzonite of Turret Peak; Kl, Cretaceous granodiorite of Lake Edison; Kmr, Cretaceous quartz monzonite of Mono Recesses; KJcl, Jurassic or Cretaceous granodiorite of Chickenfoot Lake; Tt, Tertiary olivine trachybasalt; JTmv, Triassic or Jurassic metamorphic rocks.

ments within the wide band along Bear Creek Valley are subparallel to joints and many small, subparallel faults with left-lateral, oblique slip. SEGALL and POLLARD (1983b) demonstrated that the faults had formed after the joints, and that the faults had nucleated along the joints. Locally, granodiorite slabs bounded by the faults and joints have been folded into monoclinal kink bands. Figure 3 shows traces of many of the longer faults and joints and a kink band at Kip Camp in Bear Creek Valley (see also Fig. 2). The fault traces are straight or gently curved and range in length from 5 to 80 m, the spacings of adjacent faults ranging from several centimeters to several meters. The faults consist of overlapping segments that step to the left or to the right. (To define left-stepping and right-stepping traces, consider two overlapping fault segments; as viewed along the trace of one, the other is left-stepping if it is offset to the left of the first segment and right-stepping if it is offset to the right of it.) The strike component of the slip along faults in Bear Creek Valley is left-lateral and ranges from a few mm up to several m. Slickenlines on the fault surfaces

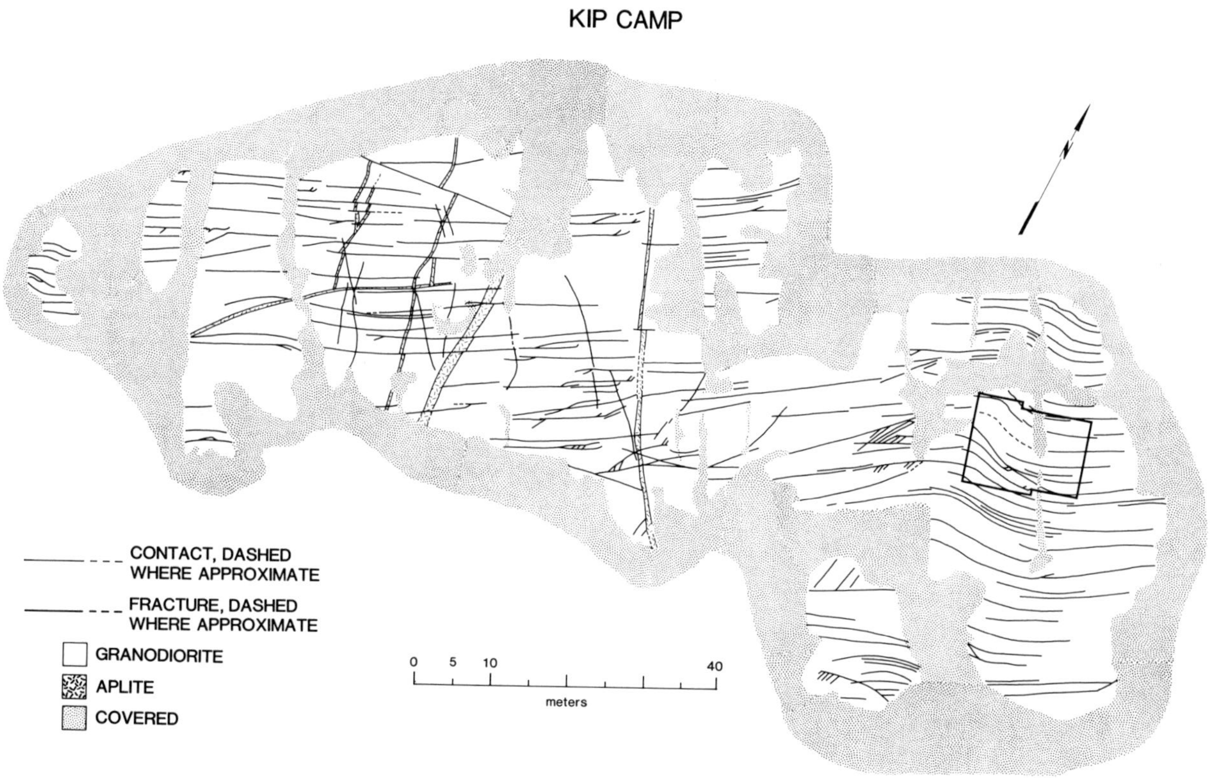

Figure 3
Map at Kip Camp outcrop showing geometry and orientations of fault and joint traces and kink bands. Box: location of detailed map in Figure 9. Modified from SEGALL and POLLARD, 1983b, p. 557.

plunge 0° to 25° NE showing that the last motion on the faults was oblique left-lateral slip. The results of the investigation show that the gently curved lineaments in the wide band striking northwest through the quadrangle are both faults and joints and that the faults show a left-lateral shear displacement (SEGALL and POLLARD, 1984).

This paper contains a structural synthesis of the kink bands and cracks associated with the faults, including stress directions that may have been responsible for the geometry of the joints and faults along Bear Creek Valley. Two field areas, Kip Camp and Trail Fork (Fig. 2) were chosen for detailed mapping of small kink bands, left-lateral faults, and cracks associated with the faults. The results are synthesized into possible mechanisms in the formation of the major and minor structures within the granodiorite in the southern half of the Mt. Abbot quadrangle.

Minor Structures Within Bear Creek Valley

Extension cracks

Extension cracks with orientations of 15° to 65° counterclockwise (CCW) from the trend of traces of prominent faults are associated with the faults throughout Bear Creek Valley (Fig. 3). Some of the geometric relations between faults and these extension cracks are shown schematically in Figure 4.

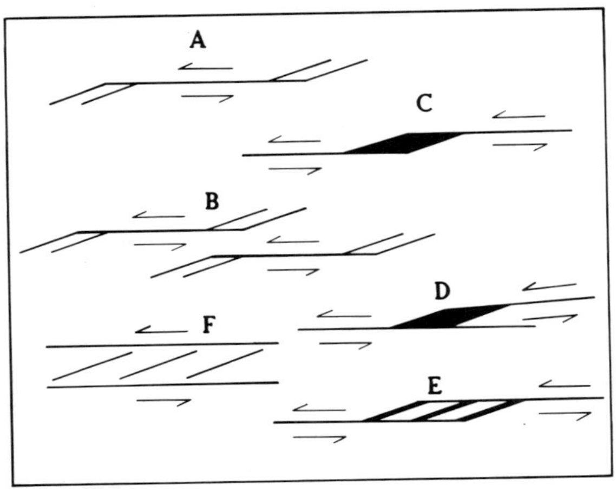

Figure 4
Schematic showing relations between left-lateral strike-slip faults and associated extension cracks oriented at an acute angle counterclockwise from the fault trace: A, cracks at ends of isolated fault segment are similar to cracks (see B) at ends of two right-stepping overlapping fault segments; C, D, and E, filled cracks between two left-stepping overlapping fault segments with various geometries; F, extension cracks between mid-sections of two left-lateral fault segments.

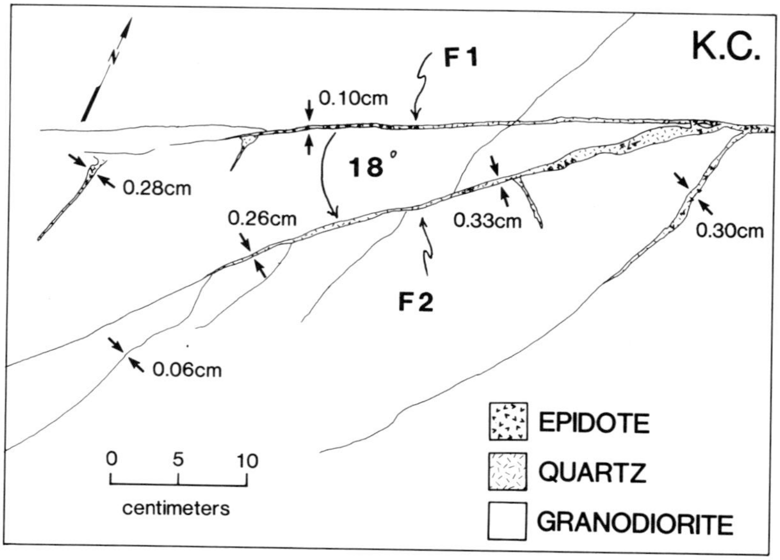

Figure 5
Map from photographic base showing two branches, F1 and F2, of end of left-lateral strike-slip fault and associated extension cracks oriented 50° to 55° counterclockwise from horizontal fault trace: single line, crack trace; location of figure indicated at upper right; K.C., Kip Camp (Fig. 9) and T. F., Trail Fork (Fig. 10); numbers indicate specific localities.

At the ends of single, isolated fault segments are extension cracks branching out of the plane of the slip surface. These cracks range in length from several centimeters up to several meters and are oriented 15° to 50° CCW from the traces of the left-lateral faults (Fig. 4A). Figure 5 shows such extension cracks near the termination of a fault with two branches. One branch of the fault, F1, terminates near the left-hand edge of the figure, where its terminus bifurcates into two subparallel, hairline fractures. Two wide, short cracks oriented 50° to 55° CCW with respect to the trace of the fault occur at its terminus. The second branch, F2, radiates downward from the lower edge of the first at an orientation of 18°. Oriented at an acute angle counterclockwise to its trace are several short hairline cracks and one wide crack almost normal to its trace. A long crack to the right of the intersection of the two fault branches has an orientation about 55° CCW with respect to F1.

Near the terminations of right-stepping left-lateral, faults, with little or no overlap, the pattern of extension cracks is the same as that near the terminations of isolated faults (Fig. 4B).

At the tips of two left-stepping left-lateral fault segments, cracks connect the segments, and rhomb-shaped cavities, filled chiefly with quartz, occur between the segments. A common geometry is a single rhomb (Fig. 4C). Single rhombs also occur at the ends of fault segments with large overlap (Fig. 4D). For example, Figure 6A shows a quartz-filled rhomb at the termination of a fault, F1, connecting a second

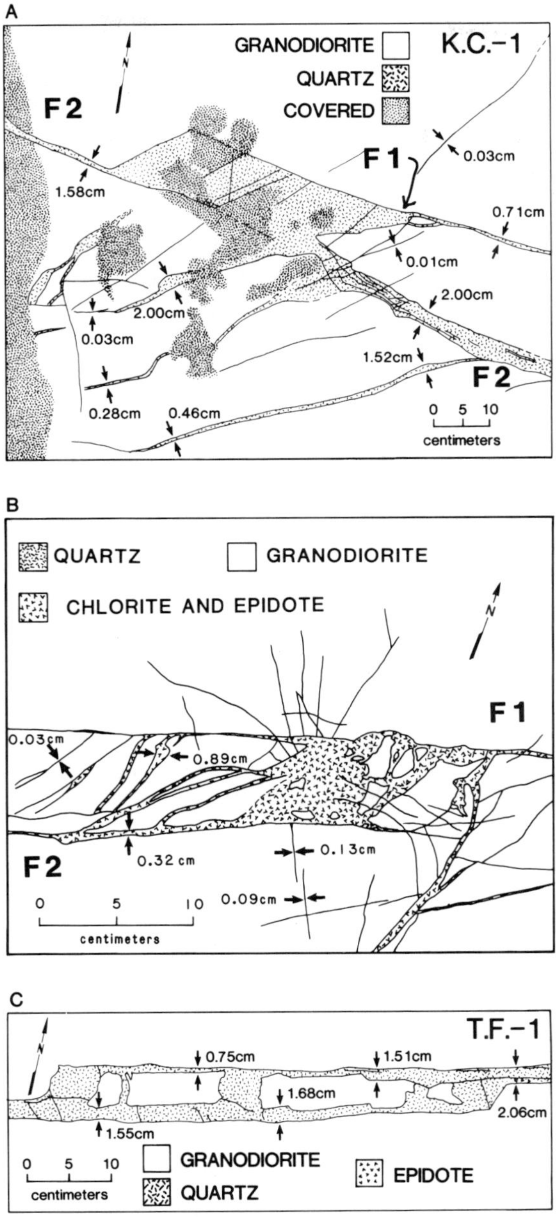

Figure 6

Various geometries of filled cavities between two left-stepping overlapping left-lateral faults. Mapped from photographic base: single line, crack trace, dashed where approximate; location of figure indicated at upper right. (A) Rhomb-shaped cavity connecting end of upper fault, F1, to mid-section of lower fault, F2. (B) Complex web pattern of predominantly epidote-filled cracks connecting two left-stepping left-lateral faults (F1 and F2). (C) Regular pattern of quartz-filled cracks connecting two left-stepping left-lateral faults; fault traces are horizontal to page.

fault, F2, the trace of which passes through the figure. The ends of the rhomb are straight and are oriented 45° to 50° CCW with respect to the trends of the faults.

Several filled cracks may occur between two overlapping, left-stepping fault segments, and connect them (Fig. 4E). The filled cracks vary in thickness, orientation, and length. In some cases they form a complex web. For example, Figure 6B shows two faults, F1 and F2, parallel to the base of the figure, connected by a series of cracks oriented about 30° to 65° CCW with respect to the traces of the faults. In other cases the filled cracks are regularly distributed between the overlapping fault segments. Figure 6C shows two narrow, and three wide, quartz-filled cracks connecting fault segments.

Filled cracks connect left-stepping fault segments, so that segments behave as a single slip surface; the slip is transferred from one segment to an adjacent segment through the connecting filled cracks. Where the filled crack is a single rhomb (Fig. 4C and 4D), the distance between the sides of the rhomb, along a line parallel to the fault trace, is a measure of the strike component of slip. Where there is a series of filled cracks (Fig. 4E and 6C), the sum of the widths parallel to the fault traces approximates the strike component of slip.

Closely spaced, subparallel cracks commonly occur with an en-echelon pattern between parallel fault segments (Fig. 4F). An example is shown in Figure 7, where

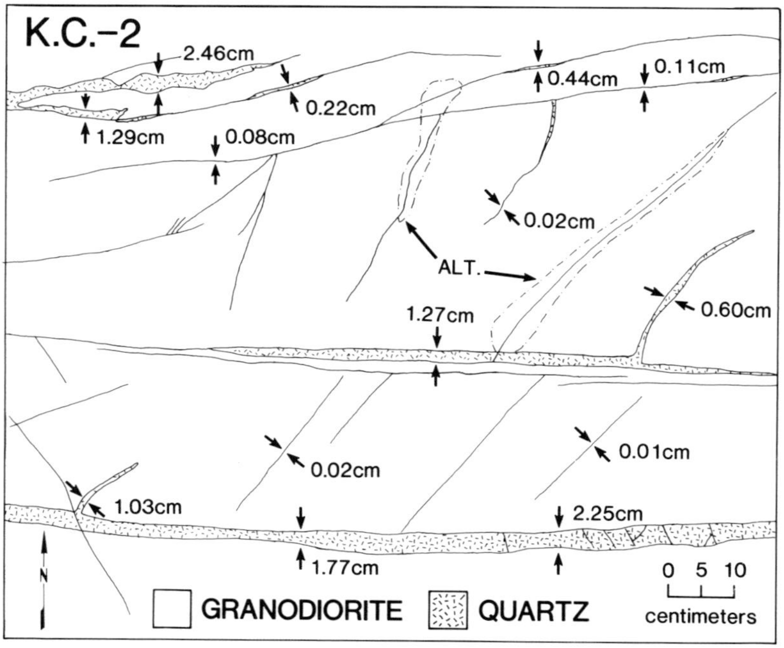

Figure 7
Echelon extension cracks between mid-section of two quartz-filled left-lateral faults: traces of faults are horizontal; single line, crack trace; ALT, alteration; location of figure indicated at upper left.

cracks extend between two thick faults in the lower part of the view. The cracks tend to split the granodiorite into rhombohedral blocks, although the lengths of the cracks may be less than the space between the bounding faults.

Fault zones

SEGALL and POLLARD (1983b) have suggested that the formation of en-echelon cracks between parallel fault segments (Fig. 7) is an early step in the formation of fault zones in granodiorite. The en-echelon fractures break the intervening granodiorite into rhombohedral blocks and accommodate strike slip. Further slip along the bounding faults breaks the rhombohedral blocks into irregular fragments. Fully developed fault zones are septae of intensely deformed granodiorite between two parallel fault segments which accommodate up to 10 or more meters of slip. Most of the slip within the fault zones occurs on the bounding faults. Figure 8 is a map of Trail Fork that shows aplite dikes crosscutting fault zones and the left-lateral offset across the zones. Slickenlines on the surfaces of the parallel bounding faults plunge eastward up to 10°, like those on isolated faults. The fault zones are spaced approximately 100 m apart in Bear Creek Valley and erode into wide gullies (SEGALL and POLLARD, 1983b, 1984); the gullies are the lineaments shown in Figure 1.

Kink bands

Most of the faults in Bear Creek Valley are left-lateral oblique slip faults; conjugate right-lateral shear, however, was accommodated locally by a right-lateral deflection of granodiorite slabs into monoclinal kink bands with subvertical fold axes and right-lateral slip along small, isolated cracks oriented 45° to 55° CCW from the left-lateral faults. Kink bands have been mapped at Kip Camp and Trail Fork and have been observed in granodiorite in other parts of Bear Creek Valley.

Most of the kink bands in Bear Creek Valley are right-lateral and are defined in outcrops by gentle, open, monoclinal flexuring of granodiorite slabs. (Right-lateral kink bands are those in which the relative displacement of the layers outside the band is right-lateral.) Figure 3, for example, shows a large kink band east of an area of nonfolded slabs at Kip Camp. Table 1 documents the geometrical elements of the kink band.

Figure 9 is a detailed map of the central part of the kink band at Kip Camp (location shown in Fig. 3). At this scale many more faults are shown, with wide openings between some slabs and small subparallel cracks at an acute angle counterclockwise from the traces of the major fault set. The slabs are tightly curved in the western hinge and gently curved in the eastern hinge. The curvature also changes along each slab because of irregularities in orientations of faults and joints bounding

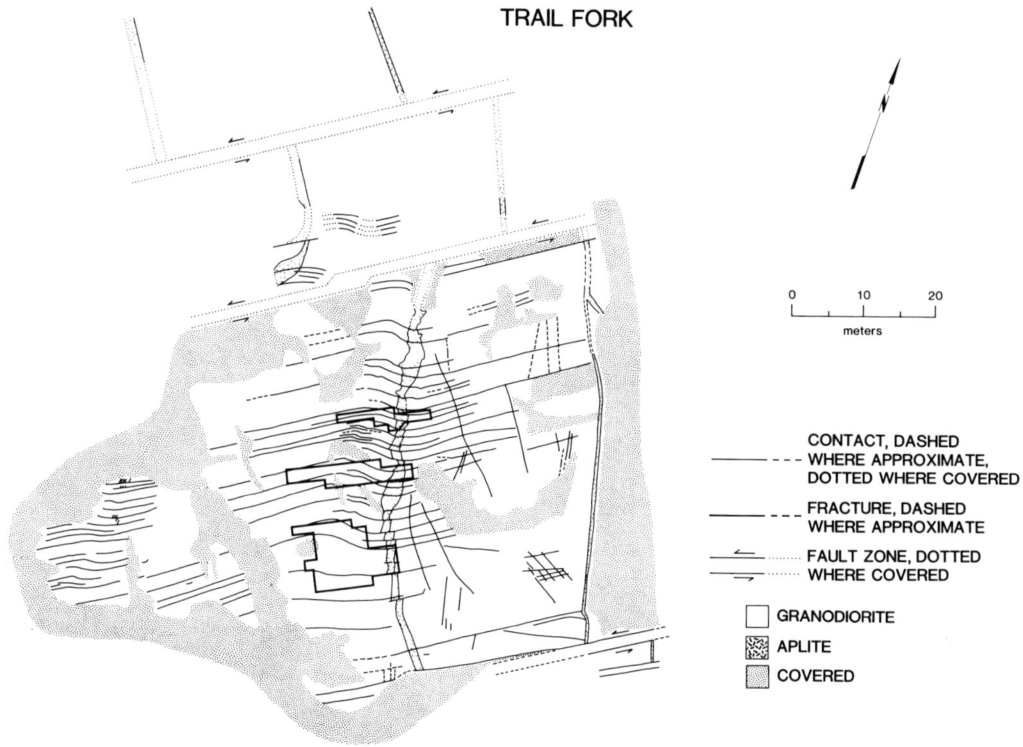

Figure 8
Map at Trail Fork outcrop showing geometry and orientations of fault and joint traces, fault zones, and kink bands: boxes, locations of detailed maps in Figure 10. Modified from SEGALL and POLLARD, 1983b, p. 563.

Table 1

*Data on Large Kink Bands**

	Kip Camp	Trail Fork
Length	>40 m	>42 m
Width	12–20 m	7–9 m
Trend of hinges	N0E ± 5°	N02E ± 5°
Reorientation of slabs in band	45° CW	40° CW
Displacement of slabs across band	4–8 m	1.5–3.5 m

* CW, clockwise.

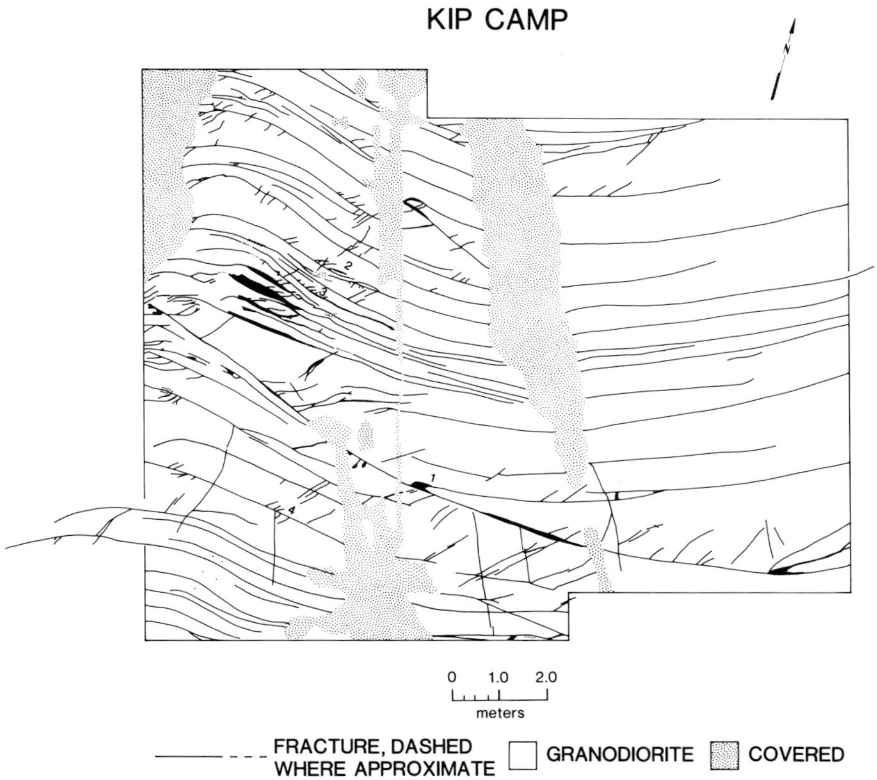

Figure 9

Detailed map of central part of kink band at Kip Camp showing fault orientations and geometries, wide fillings between granodiorite slabs, and small subparallel cracks at acute angle counterclockwise from trace of major fault set. For location see Figure 3.

the slabs. The spacings of the faults are smaller inside the band than outside: approximately 40 to 80 cm apart outside the band and 35 to 40 cm inside the band.

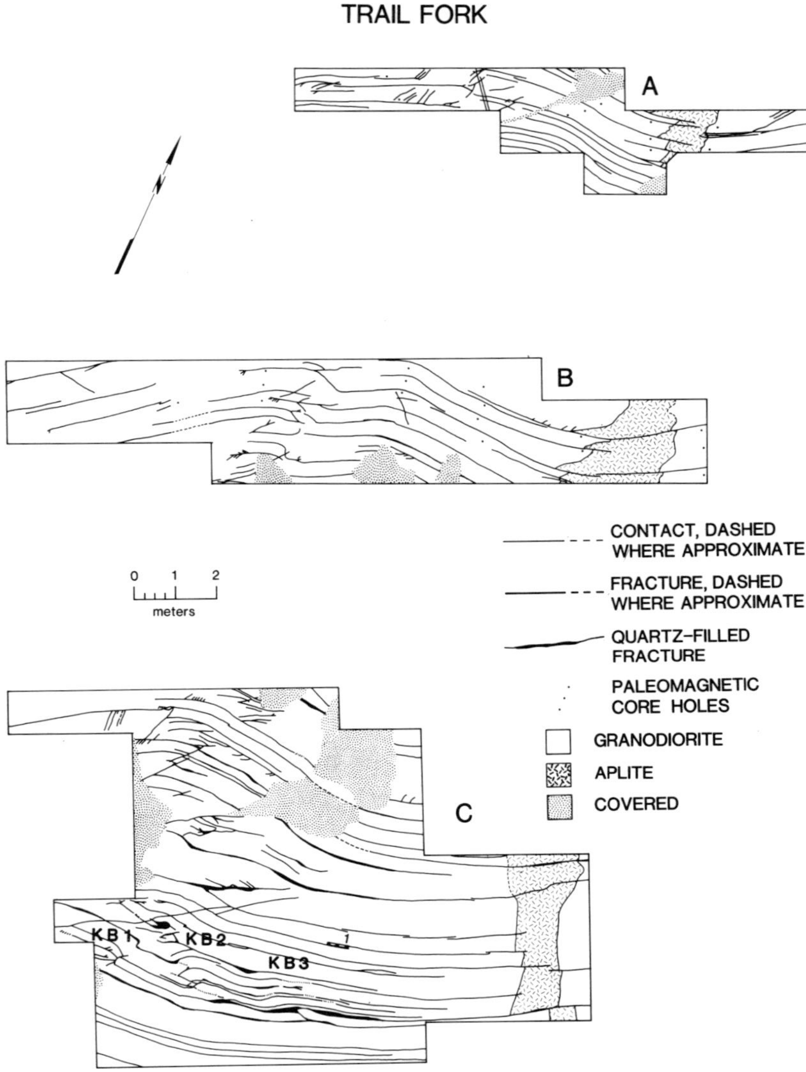

Figure 10
Detailed maps of selected parts of large kink band at Trail Fork, showing fault orientations and geometry, wide fillings between slabs, and small subparallel cracks at acute angle counterclockwise from trace of major fault set. For location see Figure 8.

The geometry of the larger kink band at Trail Fork is similar to that of the kink band at Kip Camp (Fig. 8 and Table 1). Detailed maps of segments of the Trail Fork kink band (Fig. 10) show that spacing between faults is smaller inside the band than outside it: the spacings between the faults range from 36 to 112 cm at different distances outside the band, and within the band they average 32 cm. The curvature of slabs in the west hinge is slightly tighter than in the east hinge. The difference in curvature becomes larger at the southern end of the band.

The southern termination of the large band at Trail Fork shows a local thickening or swell in the granodiorite slabs (Fig. 8 and 10C). HONEA and JOHNSON (1976) describe openings or puckers and crimps at kink-band boundaries in experiments with rubber strips. The displacement across the kink band in the vicinity of the swell decreases from the northern edge of the swell to the southern edge. Along the lower slabs in the swell (Fig. 10C) are several smaller kink bands. At the lower left corner (KB1 in Fig. 10C) is shown a tightly folded, narrow, right-lateral kink band, oriented N30E. The northern boundary of the band intersects a small left-lateral kink band, KB2, oriented N40W, forming a conjugate kink fold. To the right of KB2 is another left-lateral kink band, KB3. Both the left-lateral kink bands are oriented 130° to 140° CCW from the left-lateral faults. These are the only left-lateral kink bands observed at the mapped outcrops.

Another small, right-lateral kink band is directly west of the large kink band at Trail Fork (Fig. 10). The band is approximately 9 m long and 1.4 m wide.

The traces of the axial planes of the larger right-lateral kink bands at Kip Camp and Trail Fork are oriented approximately north–south which is 50° to 70° CCW from the trends of major left-lateral faults in the areas. The slabs within the band are gently folded, reoriented approximately 20° to 60° CW (clockwise) from the slabs outside the band. The geometry of the kink bands in granodiorite is similar to that of monoclinal kink bands in layered rocks and experimental materials (DONATH, 1968; RECHES and JOHNSON, 1976; GAY and WEISS, 1974).

Comparison of monoclinal kink bands and fault zones

The kink bands and fault zones in Bear Creek Valley are kinematically similar. They both are bounded by parallel planes, between which shearing deformation is concentrated. The differential displacement across the kink bands is right-lateral, however, and across the fault zones it is left-lateral.

Kink bands form in foliated rocks and minerals where the compression parallel to foliation, or to subdivided layers, is higher than that normal to foliation. The band is a zone within which foliation has been reoriented in response to the shear stress parallel to the trace of the axial plane (Fig. 11A) and across which the foliation outside the band has been differentially displaced. During folding the folia within the band slip in relation to one another in a sense opposite to the shear along the axial plane (Fig. 11A). For example, in Bear Creek Valley the shear along the axial planes is right-lateral and the slip between the slabs is left-lateral.

Fault zones in the granodiorite in Bear Creek Valley develop between two bounding, parallel fault surfaces. As slip occurs along the bounding faults, the rock between them develops parallel en-echelon tension cracks (Fig. 11B) that give the

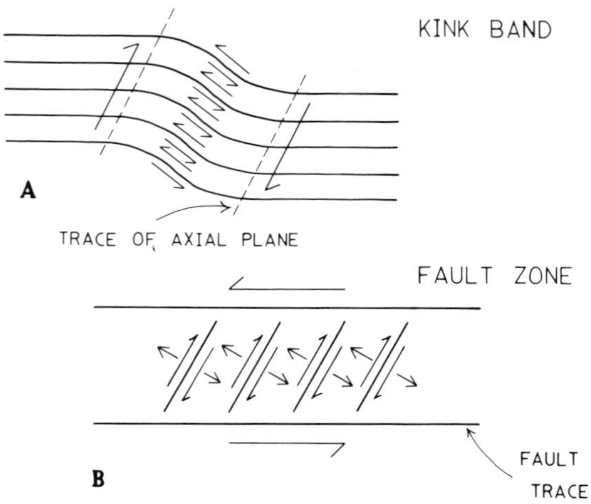

Figure 11
Comparison of monoclinal kink bands and fault zones. (A) Right-lateral shear parallel to trace of axial plane of kink band displaces preëxisting foliation in left-lateral sense. (B) Left-lateral shear parallel to two parallel fault planes causes subparallel extension cracks to open in echelon pattern.

fault zone an oblique foliation (SEGALL and POLLARD, 1983b). As left-lateral shear continues across the fault zone, the foliation within the zone may reorient by accommodation of right-lateral slip on folia. The foliation is initially parallel to the direction of maximum compression, but the shear stress along the folia increases with reorientation; the shearing will thus be unstable, as in kink bands (RECHES and JOHNSON, 1976).

Thus the kink bands and fault zones in the granodiorite are analogous structures, the fault zones accommodating left-lateral shear and the kink bands right-lateral shear. The bounding surfaces of the fault zones parallel preëxisting joints, whereas the boundaries of the kink bands transect the joints. The kink bands shear internally along preëxisting foliation, whereas cracks formed during slip along the bounding faults of fault zones become the foliation along which internal shear is accommodated. The shearing along foliation is unstable within both fault zones and kink bands.

Analysis of Minor Structures

Strain within kink bands

In this section macroscopic fiber strains and dilational strains are computed in the folded rocks, and the strains are then compared with crack widths and dilation of layers within the kink bands. Maximum fiber strains—the strains that result from pure bending of granodiorite slabs—can be calculated by means of elementary plate

Table 2

Average Maximum Fiber Strains (%)

Kip Camp, no. of observations		Trail Fork, no. of observations	
10, west hinge	7, east hinge	17, west hinge	15, east hinge
4.9 ± 2.5	2.9 ± 1.7	6.8 ± 4.0	3.7 ± 2.5

theory: the curvature is multiplied by half the thickness of the slab (JOHNSON, 1970). The maximum fiber strains due to bending in the kink bands mapped in Bear Creek Valley range between 1.0% and 17%. Table 2 shows the average values and standard deviation of maximum fiber strains for the two hinges of the granodiorite slabs in the kink bands at Kip Camp and Trail Fork.

Another strain recognized in kink bands is the dilational strain due to thickening or thinning of slabs within bands. The dilational strain can be calculated for an idealized band by a simple geometrical relationship that relates the angles between the layers, inside and outside the band, to traces of the axial planes (e.g., PATERSON and WEISS, 1966). With the angles as defined in Figure 12B, the equation for dilational strain d is

$$d = [(\sin \alpha/\sin \beta) - 1] \times 100 \qquad (1)$$

If α is different from β, dilation will occur within or between the slabs in the band. The graphs in Figure 12A are plots of Equation 1, with α held constant and β varied.

Dilational strains have been estimated for the kink bands at Kip Camp and Trail Fork by Equation 1, and plotted in Figure 12A. For most of the data β is less than 90°, and the dilational strains are positive (opening between layers), the strains ranging between −5% and 22%. Error bars, corresponding to errors in the measurements of α and β of ±5°, are shown for some data. They indicate that the strains computed are highly sensitive (±8%) to errors in the measurement of α and β. On the other hand, dilational strains calculated from the average change of thickness measured normal to slabs within the larger kink band at Trail Fork represent about 15% dilation, if compared with the thickness of slabs east of the band, and 19% dilation if compared with the thickness of slabs west of the band. Therefore, the strains and associated errors calculated from Equation 1 are reasonable estimates of dilation within the kink bands.

The fiber strains in the kink bands are accommodated partly by the development of short, filled cracks along the outer arcs of the folded layers, where fiber strains are extensional (upper left, Fig. 9). Many of the short cracks are wedge-shaped, wide at the base along the fault bounding a slab, thinning to a hairline crack within the slab. They do not extend across faults bounding the slabs, and their lengths are generally less than half the thickness of the slabs. The wedge-shaped cracks were observed

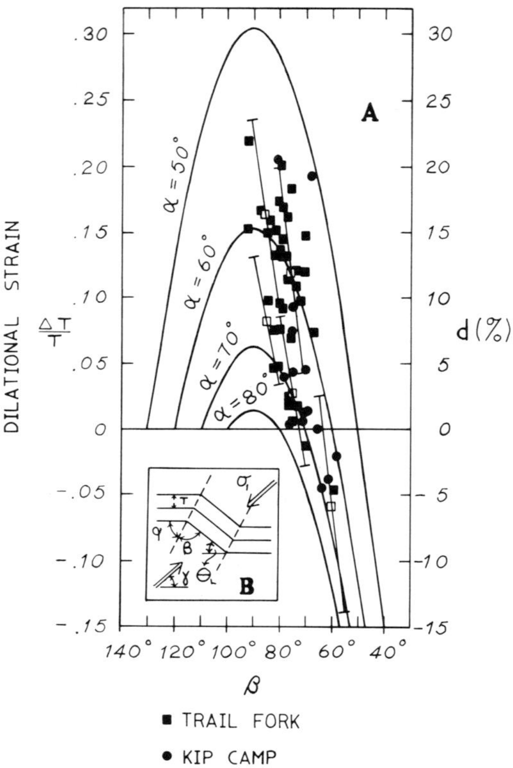

Figure 12

(A) Plots of percent dilational strain d due to thickening or thinning of layers within kink bands for range of β values with α constant; dilational strain estimated from kink bands at Kip Camp and Trail Fork plotted with error bars on some data points (open points). (B) Idealized kink band showing notation used for angles and thickness.

only in the folded granodiorite slabs, although not in every one. Figure 13A shows examples of the wedge-shaped cracks along one of the slabs; the wide fracture extending horizontally through the figure is the fault separating two slabs. Seven of the cracks are preferentially oriented 40° to 50° CCW from the fault trace.

At either side of an inflection point, where curvature reverses, short cracks open in slabs on opposites sides of a bounding fault (upper left, Fig. 9). Figure 13B shows an example of cracks developed near the inflection point along two slabs. The thick horizontal filling is the fault between the slabs, and the inflection point is at midlength along the fault. To the right of the inflection point, short wedge-shaped cracks are oriented 45° CCW from the fault and extend into the lower slab. To the left of the inflection point, the cracks are also oriented 45° CCW from the fault, but they extend into the upper slab.

The short wedge-shaped cracks can account for part of the extensional fiber strains in folded slabs. The strains due to the cracks range between 0.004% and

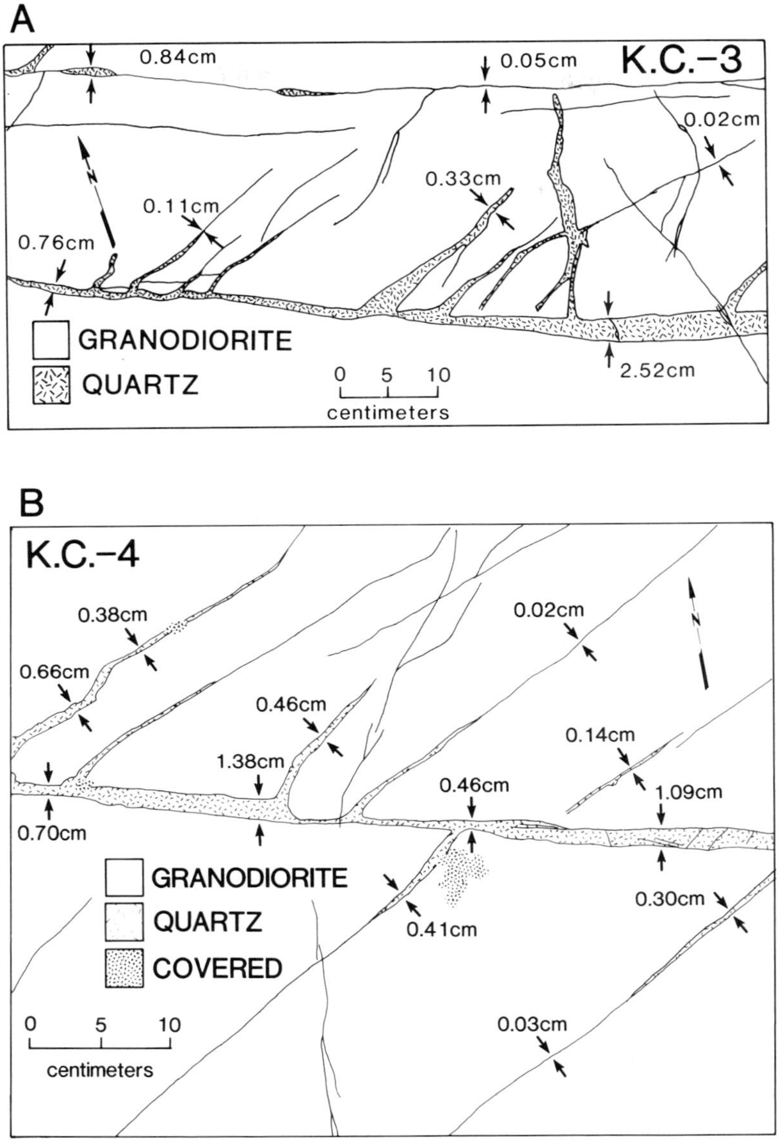

Figure 13

Quartz-filled wedge-shaped openings in granodiorite slabs, locations of figures indicated upper corners. A single line is a crack trace. (A) Wedges open upward into only one slab. (B) Wedges open into slab at top and bottom of figure, changing direction at inflection point where curvature reverses.

3.86%. Table 3 compares maximum fiber strains for particular folded slabs with strains due to the opening of the cracks in the same slabs, and shows that only a small part of fiber strain was accommodated by the opening of short cracks. Furthermore, short cracks are lacking in many slabs. It is most likely, therefore,

Table 3

*Comparison of Maximum Fiber Strains, E_f, with Strains due to Crack Openings, E_d, for Particular Slabs**

Kip Camp		Trail Fork			
West hinge		West hinge		East hinge	
E_f	E_d	E_f	E_d	E_f	E_d
1.8 ± 0.6	0.004 ± 0.002	7.0 ± 1.2	1.48 ± 0.06	3.5 ± 0.4	0.23 ± 0.01
8.3 ± 0.7	0.510 ± 0.014	9.4 ± 1.6	3.16 ± 0.09	5.4 ± 0.9	0.99 ± 0.02

* Strain in percent.

that grain-scale deformation accounts for much of the fiber strain due to bending.

Dilational strain within the kink bands is accommodated by local widening of faults and joints between granodiorite slabs. Figure 13A shows, for example, an increase in the thickness of the filling of a fault separating folded slabs. The filling thickens from left to right as the curvature increases. In general, the fillings of faults are thicker inside a kink band than outside.

The thick fillings within the faults and joints are predominantly lenticular, although several end bluntly against the granodiorite. Figure 9 shows, for example, several thick fillings in the upper left quadrant. Thick fillings at the Trail Fork outcrop (Fig. 10) occur within the swell at the southern end of the band.

The strains calculated due as to dilation (Fig. 12A) are generally much larger than the strains due to the fault and joint fillings. Table 4 compares dilational strains along the east and west hinge of the Trail Fork kink band, shown in Figure 10, A and B, with the strains due to the fillings. Less than 25% of dilational strain was accommodated by thickening of fillings, according to Table 4.

Thus the calculations of fiber strains and dilations show that mechanisms other

Table 4

*Comparison of Dilational Strains, D, with Strains Due to Fillings, E_{ext}, for Two Traverses**

Trail Fork			
East hinge		West hinge	
D	E_{ext}	D	E_{ext}
13.1 ± 8.0	2.98 ± 0.18	17.4 ± 8.0	1.78 ± 0.08
10.5 ± 8.0	0.31 ± 0.05	−0.92 ± 8.0	0.35 ± 0.05

* Strain in percent.

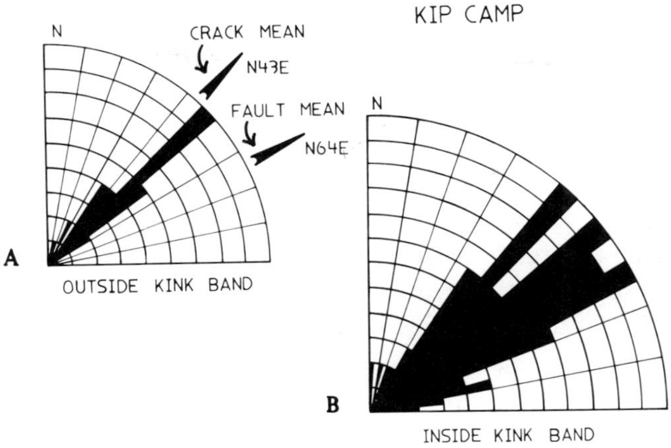

Figure 14
Orientations of crack traces associated with left-lateral faults at Kip Camp. Crack mean: average orientation of crack traces calculated from data shown. Fault mean: average orientation of fault traces outside kink band. Each semicircular segment represents one observation.

than opening of wedge-shaped cracks and opening of faults account for most of the fiber strain and dilation. These mechanisms probably operate at the grain scale.

Principal stress orientations

Orientations of extension cracks at an acute angle counterclockwise to the left-lateral faults and orientations of the slabs within the kink bands may be used to place bounds on the direction of regional maximum compression during faulting and folding in the Kip Camp and Trail Fork areas.

Cracks at terminations of isolated strike-slip faults grow out of the planes of the faults in the direction of maximum compression, as shown by Brace and Bombolakis (1963) and Segall and Pollard (1980). Extension cracks that form between tips of left-stepping left-lateral faults also grow subparallel to the regional compression direction (Segall and Pollard, 1980). Therefore, by measurements of the orientations of the cracks associated with the strike-slip faults, the direction of maximum compressive stress at the time the cracks formed can be determined.

We measured the orientations of the small cracks shown in the detailed map of the kink band at Kip Camp (Fig. 9), outside the band. The traces of the cracks have a mean orientation of N43E (Fig. 14A). The mean orientation of the traces of faults in that area is N64E, so the maximum compression direction at the time the cracks formed was about 20° CCW with respect to the faults outside the band at Kip Camp. Within the kink band at Kip Camp the orientations of cracks are more scattered (Fig. 14B), between about N40E and N70E.

The scatter of data within the kink band may be a result of reorientation of granodiorite slabs as the kink band formed; the cracks may have formed at different times as the slabs folded. Therefore, the cracks within the kink bands are poor indicators of maximum stress directions.

Orientations of maximum compression can also be calculated from the orientation of the granodiorite slabs within the kink bands. As shown by RECHES and JOHNSON (1976), the foliation within a kink band reorients until it reaches a locking angle. The locking angle is determined by the ratio of the maximum to the minimum effective compressive stress, the coefficient of friction between layers, φ, and the orientation of the maximum compression to the foliation outside the band, γ (Fig. 12B). The ultimate locking angle, θ_L, depends only on the coefficient of friction and the direction of maximum compression (COLLIER, 1978):

$$|\theta_L + \gamma| = 90^\circ - \varphi \tag{2}$$

Assuming the internal friction angle of the faults in the granodiorite to be $\varphi = 30^\circ$, and assuming that the slabs within the kink bands were at the ultimate locking angles, then $\theta_L = 45^\circ \pm 2^\circ$ at Kip Camp, and the orientation of maximum compression was $\gamma = 15^\circ \pm 2^\circ$, that is $13^\circ - 17^\circ$ CCW from the traces of the left-lateral faults outside the band. At Trail Fork angle $\theta_L = 40^\circ \pm 2^\circ$, so the orientation of maximum compression was $\gamma = 20^\circ \pm 2^\circ$ CCW from the traces of faults outside the band.

Thus, according to the analysis of the orientations of small cracks and kink bands, the orientation of the maximum compression during folding was 15° to 20° CCW with respect to the traces of the left-lateral faults outside the kink bands at Kip Camp and Trail Fork.

Interpretation

One interpretation for the structures in Mt. Abbot quadrangle is that the wide band, trending northwest along Bear Creek Valley (Fig. 1), is a large right-lateral kink band (Bear Creek kink band), analogous to the kink bands at Trail Fork and Kip Camp (Fig. 3 and 8). According to this interpretation, the joints within Bear Creek kink band originally trended N20E to N30E, roughly parallel to the lineaments that occur on either side of the band.

Assuming that the joints within Bear Creek Valley became reoriented, the direction of maximum compression computed for Bear Creek kink band is consistent with the direction of maximum compression deduced from the orientations of joints at Florence Lake. If the angle of internal friction along the faults and fault zones bounding the large slabs within Bear Creek kink band was $\varphi = 30^\circ$, and the slabs are at the ultimate locking angle, so that $\theta_L = 50^\circ$ to 60°, then Equation 2 indicates that the direction of maximum regional compression was 0° to 10° CCW from the direction of joint traces on either side of the band. Thus, it is most likely that the direction of maximum regional compression was N25E ± 5° during the time Bear Creek kink band formed. This direction is clearly similar to the direction, N20E, of maximum regional compression inferred from the trends of joint traces at Florence Lake.

The faults and kink bands that we studied in Bear Creek Valley are interpreted as minor structures on the limb of Bear Creek kink band. They may have formed as follows. After the formation of regional joints in the southern half of the Mt. Abbot quadrangle (Fig. 15A), granodiorite slabs along Bear Creek Valley began to buckle and reorient clockwise in response to maximum regional compression subparallel to the traces of the joints, in the direction N20–25E. By the time the joints (now transformed into faults) had become reoriented 15° to 20° CW from the maximum compression direction, or about N45E, slip along many of the preëxisting joints was producing small left-lateral strike-slip faults and right-lateral kink bands (Fig. 15B, inserts I and II). At fault tips, cracks opened in response to the shear along the faults and grew in the current direction of maximum regional compression, at an acute angle counterclockwise from the traces of the left-lateral faults (Fig. 15B, insert I). The cracks formed at all stages of reorientation, so their orientations are relatively scattered (Fig. 14). Small kink bands, such as those mapped at Kip Camp and Trail Fork, occurred locally where the granodiorite slabs were thin, that is, where preëxisting joints were closely spaced (Fig. 15B, insert II). Slip between the granodiorite slabs within the small kink bands occurred when the maximum compression was oriented 15° to 20° CCW with respect to trends of left-lateral faults outside the bands, and the small kink bands were locked before the reorientation within Bear Creek kink band was complete. Further reorientation of slabs in Bear Creek kink band caused en-echelon extension cracks to form between parallel, overlapping faults, nucleating fault zones, along which most of the subsequent slip was accommodated (SEGALL and POLLARD, 1983b); see Figure 15C, insert III. The kink band at Trail Fork (Fig. 8) was offset about 10 m by one of the fault zones, indicating major offset on the zone after the kink band had locked.

According to this interpretation of the structures in the area of Bear Creek Valley, only one tectonic event, with a single direction of maximum regional compression, is necessary to explain the development of joints, small left-lateral faults, small kink bands, and fault zones in the areas of Bear Creek Valley and Florence Lake. Right-lateral shear along the direction N27W ± 5°, parallel to the traces of the axial planes of Bear Creek kink band, reoriented granodiorite slabs and caused slip along preëxisting joints. Slip between layers folded into kink bands is commonly restricted to the limbs (e.g., RECHES and JOHNSON, 1976), and one would expect left-lateral slip between layers in the limb of a right-lateral kink band. Thus the left-lateral faults may well reflect slip on the limb of Bear Creek kink band, and the lack of shearing along the joints near Florence Lake would indicate that the rocks at Florence Lake were outside the limit of kink folding.

If the wide band in Bear Creek Valley is a kink band, there may have been a

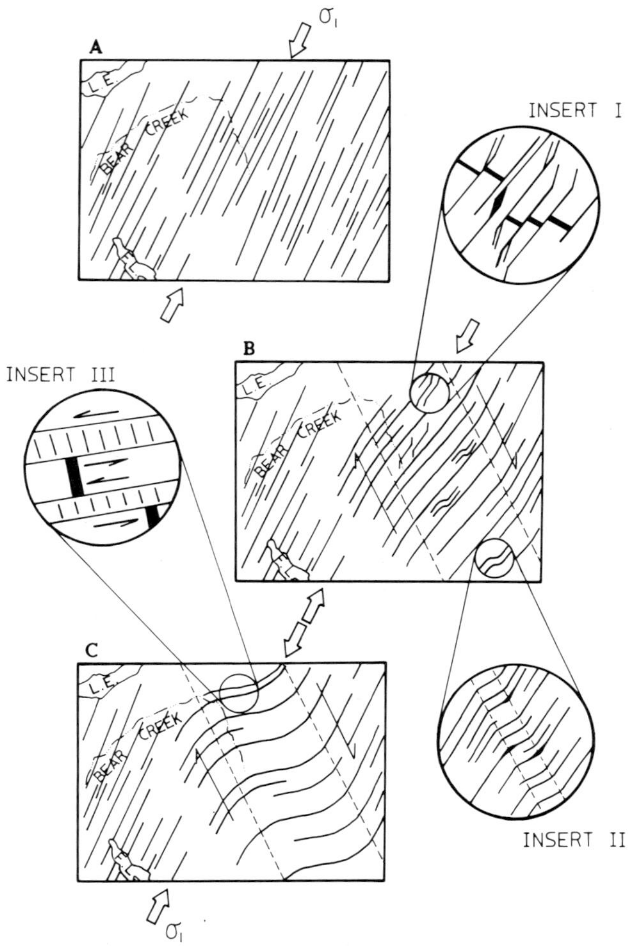

Figure 15
Outline of development of joints, faults, and kink bands in southern Mt. Abbot quadrangle. (A) Joints nucleate throughout quadrangle; trace of joints is parallel to maximum compression direction. (B) Joints, within wide band striking NW through quadrangle, begin to reorient in response to right-lateral shear parallel to boundaries of band, when reoriented 15° to 20° clockwise, left-lateral slip occurs on many of joints (insert I) and right-lateral kink bands nucleate along joints and faults (insert II). (C) Further reorientation with most of slip within band occurring along fault zones (insert III).

perturbation in the regional joint pattern in the vicinity of Bear Creek Valley that caused it to localize. Or the kink band may have localized as an accommodation to deep-seated right-lateral shear in the area. The trend of the Bear Creek kink band, about N27W, is subparallel to the trend of a pluton in the area of Bear Creek Valley (Fig. 2) and to major right-lateral strike-slip faults throughout California, such as the San Andreas fault zone. Field work in other parts of the Sierra Nevada (MAYO, 1941; MOORE, 1963; LOCKWOOD and MOORE, 1979) has shown that right-lateral faults

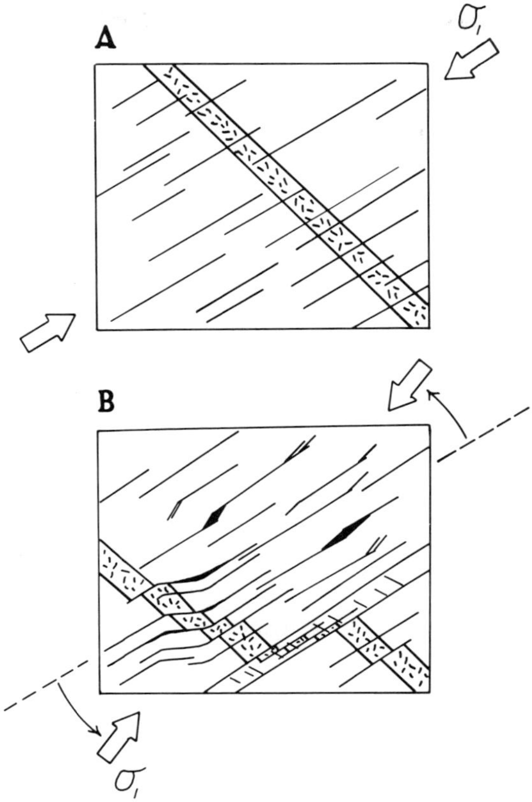

Figure 16
Mechanism proposed by SEGALL and POLLARD (1983b) for formation of faults, joints, and kink bands along Bear Creek Valley in southern Mt. Abbot quadrangle. (A) Joints nucleate with their traces parallel to maximum compression direction. (B) Maximum compression direction rotates 15° to 20° counterclockwise from traces of joints, thereby nucleating left-lateral faults and right-lateral kink bands.

are preferentially oriented northwest. Thus it is conceivable that the Bear Creek kink band and the minor structures at Trail Fork and Kip Camp reflect larger tectonic events.

Another interpretation, presented by SEGALL and POLLARD (1983b), is that the joints and faults formed with the orientations that are observed today. According to this interpretation, the orientation of the maximum compression direction at the time of jointing varied spatially across the quadrangle. Assuming that the joints were not reoriented, we obtain a range of regional compression directions. The traces of joints at Florence Lake trend about N20E, so one would infer maximum regional compression of N20E when the joints formed. The joints and faults in Bear Creek Valley have trends of N60–70E, so one would infer maximum regional compression of about N65E at the times the joints formed there (Fig. 16A). Orientations of small cracks associated with the faults and of the granodiorite slabs within kink bands indicate that the direction of maximum compression was 15° to 20° CCW with respect to the left-lateral faults outside the small kink bands. Thus one would infer a rotation of the maximum regional compression to about N45E in Bear Creek Valley (Fig. 16B) at the time the small fractures and kink bands formed there. This interpretation suggests that the joints represent a different tectonic event from the faults and kink bands, or that the stress field changed markedly through the quadrangle with time or position.

Acknowledgments

This research effort and the resulting conclusions were greatly facilitated by the research of Paul Segall in the Mt. Abbot quadrangle. Arvid Johnson contributed ideas and time towards this project, and has been a major influence in its development. Thanks also go to Jim Beaujon, who was invaluable as a field assistant, and Beccae Chaulot-Talmon, who helped with much of the drafting.

This research was funded by the U.S. Geological Survey, Sigma Xi, and the University of Cincinnati Geology Department.

References

Brace W. F. and Bombolakis, E. G. (1963), *A note on brittle crack growth in compression.* J. Geophys. Res. *68*, 3709–3713.

Collier, M. (1978), *Ultimate locking angles for conjugate and monoclinal kink bands.* Tectonophys. *48*, T1–T6.

Donath, F. A. (1968), 'The development of kink bands in brittle anisotropic rock,' in L. H. Larsen (ed.), *Igneous and Metamorphic Geology.* Geol. Soc. Amer. *Memoir 115*, 453–493.

Gay, N. C. and Weiss, L. E. (1974), *The relationship between principal stress directions and the geometry of kinks in foliated rocks.* Tectonophys. *21*, 287–300.

Honea, E. and Johnson, A. M. (1976), *Development of sinusoidal and kink folds in multilayers confined by rigid boundaries.* Tectonophys. *30*, 197–239.

Johnson, A. M., *Physical Processes in Geology.* Freeman, Cooper, San Francisco, Calif., 1970, p. 577.

Lockwood, J. P. and Lydon, P. A. (1975), *Geological Map of the Mount Abbot Quadrangle, Central Sierra Nevada, California.* U.S. Geol. Surv. Quad. Map GQ-1155, scale 1:62,500.

Lockwood, J. P. and Moore, G. (1979), *Regional Deformation of the Sierra Nevada, California, on Conjugate Microfault Sets.* J. Geophys. Res. *84*, 6041–6049.

Mayo, E. B. (1941), *Deformation in the Interval Mt. Lyell—Mt. Whitney, California.* Geol. Soc. Amer. Bull. *52*, 1001–1084.

Moore, G. (1963), *Geology of the Mt. Pinchot Quadrangle, Southern Sierra Nevada, California.* U.S. Geol. Surv. Bull. *1130*, p. 156.

Paterson, M. S. and Weiss, L. E. (1966), *Experimental deformation and folding in phyllite.* Geol. Soc. Amer. Bull. *77*, 343–374.

Reches, Z. and Johnson, A. M. (1976), *Asymmetric folding and monoclinal kinking.* Tectonophys. *35*, 295–334.

Segall, P. and Pollard, D. D. (1980), *Mechanics of discontinuous faults.* J. Geophys. Res. *85*, 4337–4350.

SEGALL, P. and POLLARD, D. D. (1983a), *Joint formation in granite rock of the Sierra Nevada.* Geol. Soc. Amer. Bull. *94*, 563–575.

SEGALL, P. and POLLARD, D. D. (1983b), *Nucleation and growth of strike slip faults in granite.* J. Geophys. Res. *88*, 555–568.

SEGALL, P. and POLLARD, D. D. (1984), *From joints and faults to photo-lineaments.* Proc. Fourth Internat. Conf. Basement Tectonics, 11–20.

(Received 1st September 1985, revised 20th March 1986, accepted 24th March 1986)

PAGEOPH, Vol. 124, No. 1/2, (1986)

0033-4553/86/020203-21$1.50 + 0.20/0

Structure and Deformational Character of Strike-Slip Fault Zones

QIDONG DENG,[1] DANING WU,[1] PEIZHEN ZHANG[2] and SHEFA CHEN[3]

Abstract—Strike-slip fault zones observed either in the field or in model experiments generally consist of several subparallel faults which make these zones complicated in geometry and kinematics. The geometry of a strike-slip fault or shear zone is dependent on arrangement (pinnate or en echelon), on step (left step or right step), and on the rank (smaller faults within larger faults) of the subparallel fault. The relations and interactions of these three factors create a variety of dynamic circumstances and tectonic settings within the strike-slip fault zones. These include pull-aparts in the release area between subparallel faults, push-ups in the jogs where the subparallel faults overlap, and pivotal movements, or rotation, of single faults along the whole fault zone. Each kind of tectonic setting is in itself characteristic, each setting consists of many subtypes, which are controlled chiefly by the geometric parameters of the subparallel faults. One of the most important phenomena revealed in the field work is two different kinds of evolution of strike-slip fault zones: one is the evolution of a zone with a tensile component, which is related to the growth of rock bridges, and the other, of one with a compressional component, which develops by the destruction of rock bridges. In this paper we discuss, on the basis of recent research on four strike-slip fault zones in China, the essential characteristics of strike-slip faults and the possible causes of the observed structural phenomena. Attention is focussed on the deformation, development, and distribution of horizontal displacements within strike-slip fault zones.

Key words: Pinnate, en-echelon, pull-apart, push-up, pivotal movement.

I. Introduction

Strike-slip faults often play an important role in the intraplate tectonics of continents. There are many strike-slip faults in China. Some of them acted as the key structures during Cenozoic deformation in Central Asia (MOLNAR and TAPPONNIER, 1975), and many of them still show intense activity. Earthquakes of magnitude greater than 7 are often related to them, directly or indirectly. Recently we studied several major strike-slip fault zones in China, such as the Fuyun fault in Xinjiang, the Haiyuan fault in Ninxia, the Xianshuihe fault in Sichuan, and a pull-apart area within a large strike-slip fault zone in northwestern Yunnan; see Figure 1. On the

[1] Institute of Geology, State Seismological Bureau, Beijing, China.

[2] Department of Earth, Atmospheric, and Planetary Sciences, Massachusetts Institute of Technology, Cambridge, Massachusetts 02139.

[3] Wuhan Geological College, Wuhan, China.

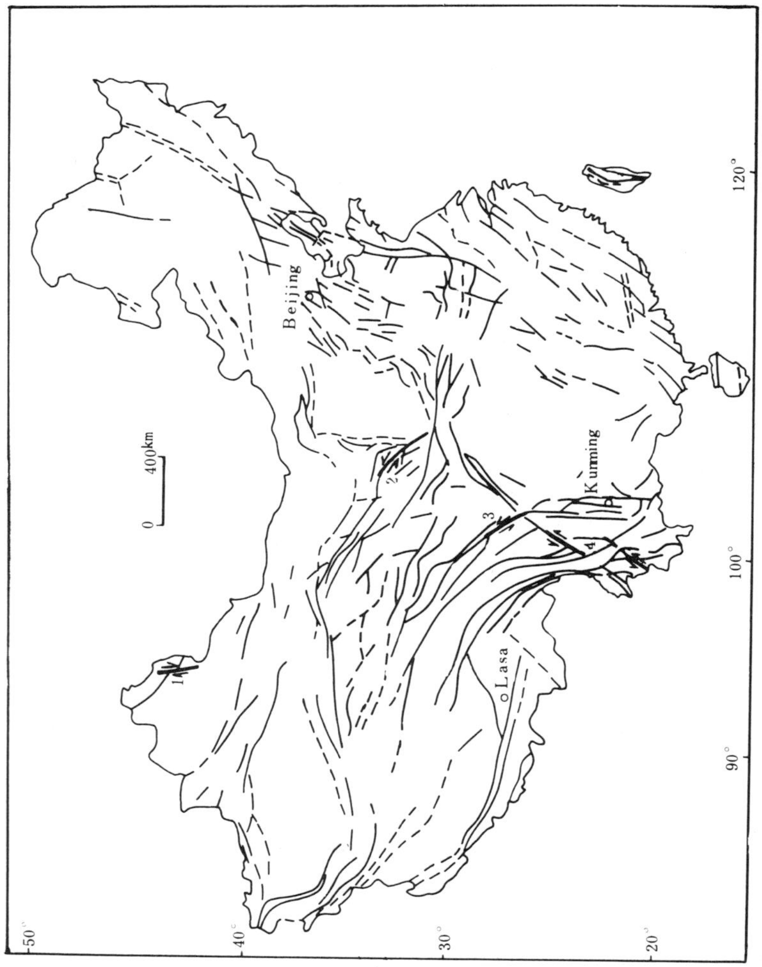

Figure 1
Tectonic setting of major strike-slip fault zones reported on: (1) Fuyun, (2) Haiyuan, (3) Xianshuihe, (4) pull-apart area in northwest Yunnan.

basis of the geology within and around these fault zones, we shall describe some of the characteristics that are of importance in understanding the deformation mechanism of this type of fault.

II. Geometry of strike-slip fault zones

As revealed both by Quaternary active faults and by surface rupture zones associated with or related to earthquake events, strike-slip faults often consist of a series of subparallel shear faults. DENG et al. (1966) and DENG and ZHANG (1984) showed that the arrangement of the subparallel faults is of two types, pinnate and en-echelon.

In the pinnate arrangement the angle between the orientation of the whole fault zone and that of the individual subparallel fault, which is called the pinnate angle, is less than 25° (Fig. 2a), and the subparallel faults are shear or shear-plus-tensile fractures. In the en-echelon arrangement the so-called en-echelon angle, which also is the angle between the orientation of the whole zone and the single subparallel fault, is usually greater than 25° (Fig. 2b), and the subparallel faults are tensional fractures. Figure 3 shows part of the surface rupture zone generated by the 1981 Daofu earthquake along the Xianshuihe fault; segments A, B, and C are composed of en-echelon extensional fractures, while segment D is composed of pinnate ones.

Within large strike-slip fault zones, subparallel faults are often arranged in a pinnate pattern, the en-echelon pattern being rare. For example, Figure 4 shows eight subparallel shear faults 1 to 4 km wide in the Haiyuan fault zone, and the pinnate angle is only 10° to 15°. The length of the subparallel faults varies from 70 to only 10 km (DENG and ZHANG, 1984). An analogous example appears in the Xianshuihe fault zone. When the strike-slip fault possesses a relatively large extensional component, the pinnate angle will be greater, as it in northwest Yunnan. We will discuss this later on.

Overlapping between two subparallel faults is a common feature in both the pinnate and en-echelon arrangements. The pattern of overlapping consists of right steps and left steps (SEGALL and POLLARD, 1980). Theoretical analysis and model experiments would show that right steps occur in left-lateral fault zones and left steps in right-lateral fault zones (Fig. 2), but in the field anomalous situations are common, left steps appearing in left-lateral fault zones and right in right. We call this anomalous pattern reverse overlapping (DENG et al., 1966, DENG and ZHANG, 1984). Within the left-lateral Haiyuan fault zone, for example, are seven subparallel shear faults (SF_2 to SF_8) reverse-overlapping each other, and only SF_1 and SF_2 overlap in the normal manner (Fig. 4). BOHAT (1982) also noticed this phenomenon. He suggested two mechanisms that can lead to the formation of reverse overlapping: the interacting of en-echelon or noncoplanar fractures under shear and extensional

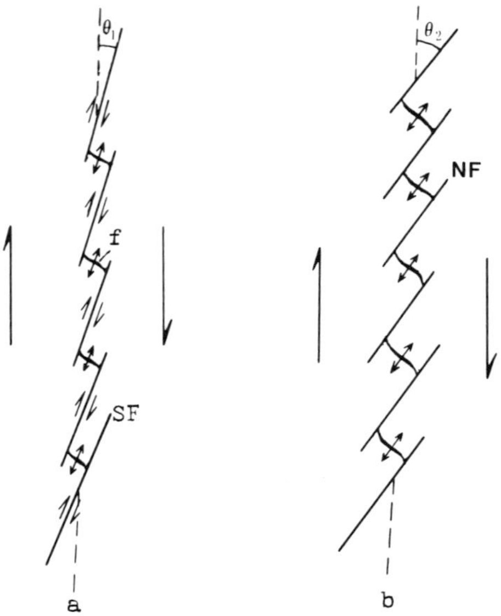

Figure 2
Arrangements of subparallel faults within strike-slip fault zone: (a) pinnate, $\theta_1 \leqslant 25°$; (b) en-echelon, $\theta_2 > 25°$. SF, subparallel shear fractures; NF, subparallel tensional fractures; f, compressional structure.

conditions and bifurcations connecting at the ends of fractures. DENG (1984) and CHEN and DENG (1985) proposed two other possible mechanisms: a reverse-overlapping fracture initiated from low-intensity points, and a pre-existing geometric

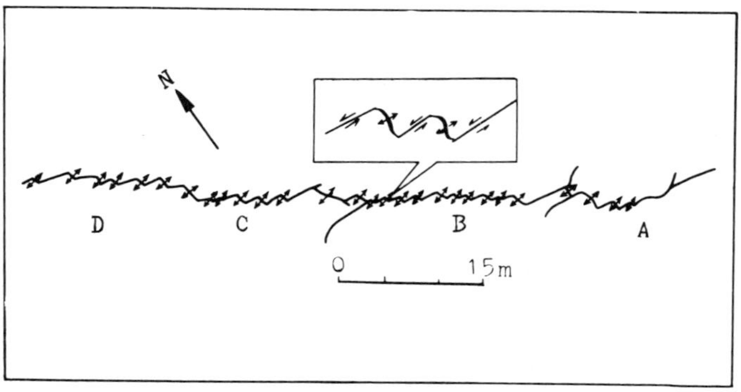

Figure 3
Surface rupture zone produced by Daofu earthquake (M_s = 6.9) in Sichuan, January 24, 1981. Modified from an original figure by the Sichuan Seismological Bureau. Segments A, B, C, and D together form pinnate pattern showing subparallel shear-fracture segments. Terminal tensional structures occur at ends of segments A and B. Segments A, B, and C each consist of low-rank fractures arranged in en-echelon pattern. Small fractures in segment D arranged in pinnate pattern. In overlapped area between ajdacent fractures mole tracks formed.

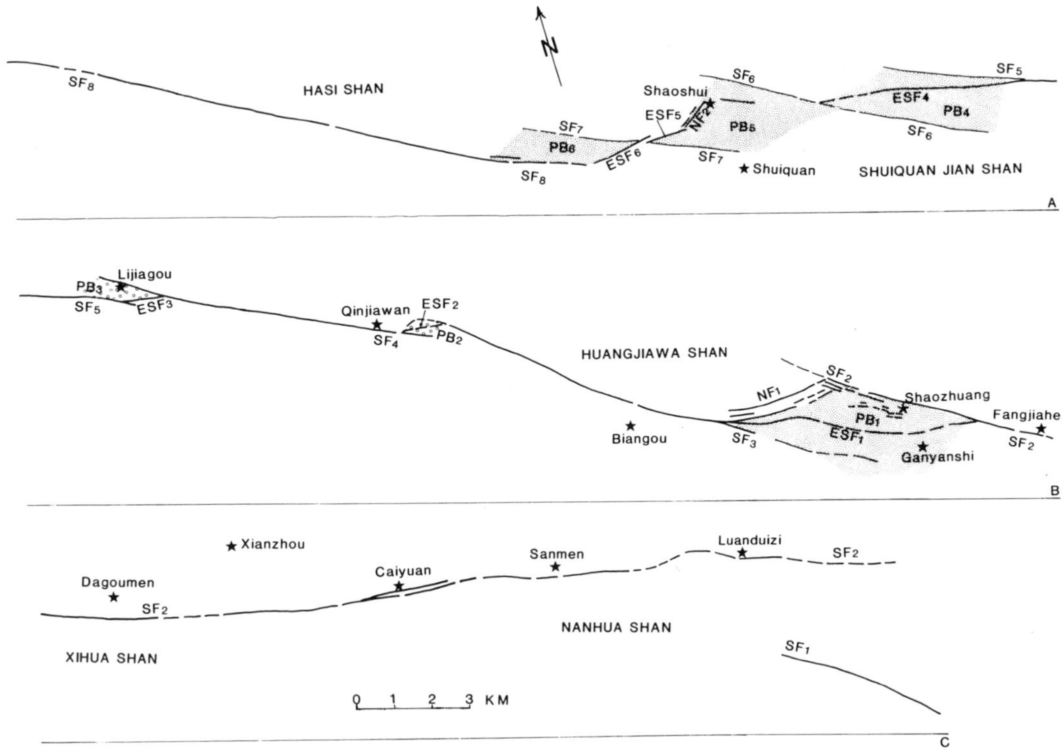

Figure 4

Tectonic sketch of Haiyuan fault zone and surface rupture zone of 1920 earthquake. A, B, C: western, middle, and eastern segments, respectively. Left end of segment B is continuation of right end of segment A, and left end of segment C is continuation of right end of segment B. Thick fault lines, ruptures made during 1920 earthquake; thin lines, no ruptures. Shaded areas, pull-apart basins (PB) formed in middle Pleistocene; small circles, pull-apart basins PB formed in early Pleistocene; SF, subparallel shear fault; NF, marginal normal fault; ESF, internal tensile shear fault.

pattern of faults with a change in the mechanical behavior of the fault zone.

In a strike-slip fault zone substructures are of different rank; that is, subparallel faults of a low rank (or 'low grade') will link together to form subparallel faults of a higher rank, whether in en-echelon or pinnate arrangement (KOIDE, 1983). Geological mapping and model experiments show that the lower-rank en-echelon faults intersect the general strike-slip fault zone at wider acute angles, while the higher-rank faults, which are composed of the former, intersect the zone at smaller angles. With an increase in the rank of subparallel faults, then, the angle is between their orientation and that of the strike-slip zone decreases; furthermore, the amount of overlapping between equal-rank subparallel faults increases. Figure 3 is an example of 'grade-integrated' subparallel faults: segments A, B, C, and D are the higher-rank secondary faults, each of which is made up of several lower-rank faults whose angles are greater than the segments' angles.

Extensional structures—bifurcations—sometimes develop at the ends of a strike-slip fault zone (ZHANG, 1982; DENG and ZHANG, 1984). Figure 5 shows an extensional structure at the southern end of a surface rupture zone generated by the 1931 Fuyun earthquake ($M_s = 8$). The zone extends 180 km, trending N20W. The termination consists of branching grabens extending more than 10 km. Although some horizontal offsets have been measured along these grabens, their magnitudes decrease toward the south; they amount to only a fifth of the average displacement along the main rupture zone. On the other hand, the boundary faults of the grabens form an obvious fault scarp, which implies that the grabens are shear in origin, with an extensional component. These grabens generally trend N15° to 20° E and are approximately parallel to the regional compression orientation. In the surface rupture Figure 3 also ($M_s = 6.9$) terminal extensional structures are found at the ends of segments A and B, intersecting the rupture zone at angles of 30° to 40°. These two examples have very different magnitudes but are very similar in geometry. Theoretical analysis, model experiments, and field observations all demonstrate that the terminal tensile fractures intersect the master strike-slip fault or shear rupture zone at angles of 30° to 40° (BRACE and BOMBOLAKISO, 1963; HOEK, 1965; LI, 1980; DENG and ZHANG, 1984). We believe this to be a common characteristic of strike-slip faults or shear rupture zones.

III. Interaction of subparallel faults and deformation of rock bridge areas

The geometry of strike-slip fault zones depends on the three factors described above: arrangement (pinnate and en echelon), overlapping pattern (left step and right step), and rank ('grade'). The combinations of these factors make for very complex tectonic patterns within a fault zone. A 'rock bridge'—the block between two overlapped subparallel faults—is an important unit within the fault zone. SEGALL and POLLARD (1980) and DENG and ZHANG (1984) pointed out that the interaction

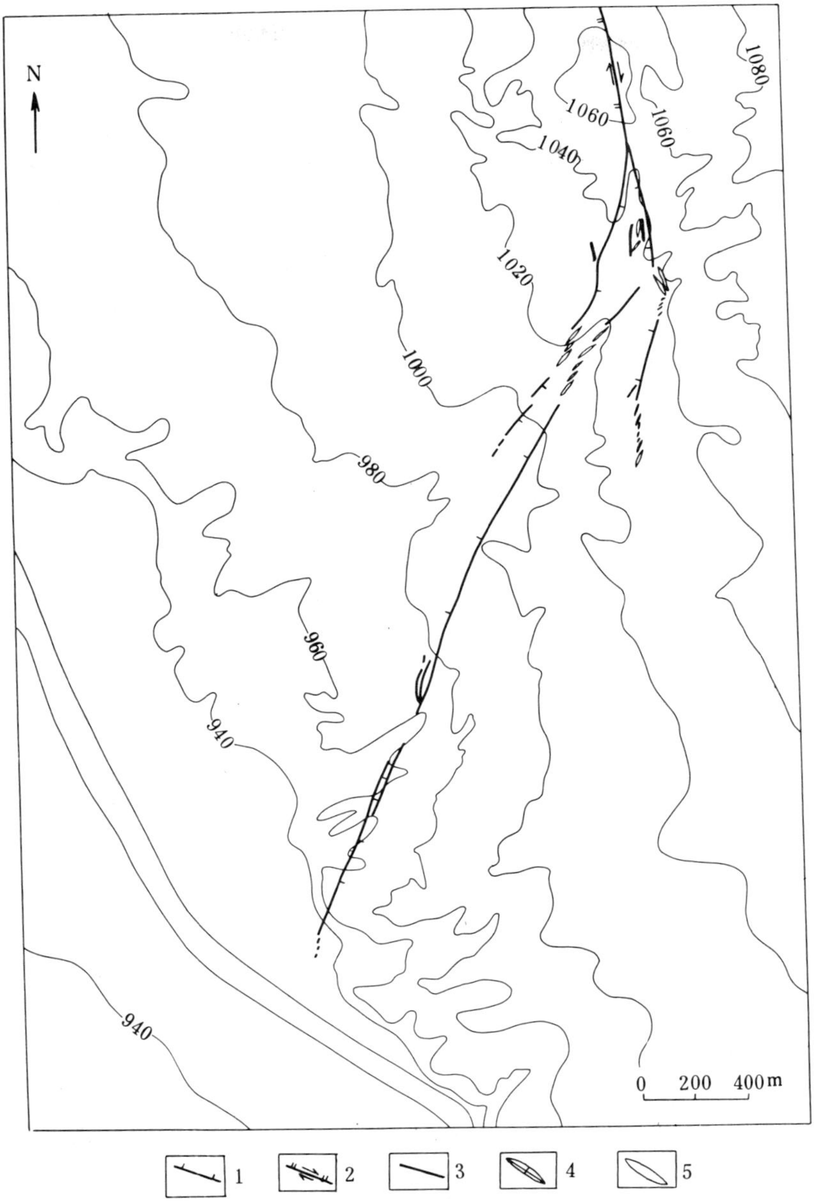

Figure 5
Terminal extensional structure in surface rupture zone of 1931 Fuyun earthquake, near Wulungu River. Legend: (1) normal fault, (2) reverse fault, (3) tensional rupture, (4) drumlin, (5) trough.

of several shear faults causes a superposition of the stress fields produced by the individual faults in the overlapped area. Rock bridges are located in such areas and therefore are subjected to more of accumulating strain and deformation than other areas within the fault zone. Geological observations and laboratory experiments show that the deformation in a rock bridge involves the pulling-apart and the pushing-up processes, which respectively produce two different types of geological structure.

1. Pulling apart

Pulling apart results from the antithetic movements of two overlapped subparallel faults with respect to the rock bridge. Extensional structures are formed, such as normal faults and grabens, which are often called pull-apart structures; they intersect the strike-slip faults at high angles. Figure 4 shows six pull-apart basins in the Haiyuan fault zone, each of them located in the rock bridge between two reverse-

Table 1

Characteristics of Pull-Apart Basins in Haiyuan Fault Zone and in Northwest Yunnan

Pull-apart basin	Shape*	Geometry of shear fault (km): Overlap	Separation	Depth	Ratios†: D/O	D/S	O/S	Number of subsidence centers	Sediment-thickness (m)
Haiyuan fault zone									
Salt Lake	Rho	5	5				1.0	1	176.15
Qingjiawan	Rho	0.9–1	1.1				0.82–0.91	1	very thin
Lijiagou	Rho	2.2	1.0				2.2		
Dayinshui	Rho	3.1	3.7	> 30	≫1		0.84	1	> 750.15
Shaoshui	Rho	3.2	3.8				0.84	1	> 335.15
Huangliang-tan	Rho	4.5	1.8				2.50	1	> 120.91
Northwestern Yunnan									
Pull-apart area, NW Yunnan	Rho	40	120	50–100	< 1		0.33	2	> 2000
Lijiao	S	3–5	10	> 20	> 1		0.3–0.5	1	> 500
Nujing	Rho	9	9	> 20	> 1		1.0	1	> 500
Bingju	Rec	9–10	7	> 20			1.28–1.3	2	1000–500
Qina	Rho	3	3	> 20	> 1		1.0	1	700
Nidu	Tri	8–10	15–20		> 1		0.5	1	> 300
Jianchuan-Hequin	Rec	38	8–20	> 20			2.1–1.9	2	500–700
Eryuan-Shonggui	Rec	30–35	18–20				1.7–1.75	2	300–100
Shaxi	Rho	3.2	3.5				0.91	1	> 300
Nandinghe fault zone (also NW Yunnan)									
Yunxian	Rho	3.3	3.5				0.94	1	
				> 50	> 1				
Gengma	Rho	12–15	18–20				0.67–0.75	1	

* Rho, rhombic; S, signoid; Rec, rectangular rectangular; Tri, triquetrous.

† D, depth; O, overlap; S, separation.

overlapped subparallel shear faults. The boundary normal faults of the basins intersect the subparallel shear faults at high angles. Table 1 presents the basic characteristics of these pull-apart basins. The Xianshuihe fault (Fig. 1) is a northwest-trending left-lateral strike-slip fault zone. In its western segment the subparallel shear fault overlapped in a left-step pattern, and the Daofu and Xialato pull-apart basins were formed. The Xianshuihe fault itself overlaps in a left-step manner with the Ganzhi fault, forming a pull-apart area in the rock bridge. This area, however, was dominated by northeast-trending normal faults, and only a small basin formed in its southwestern corner (WEN et al., 1985). It is worth noticing that the earthquake of 1967 ($M_s = 6.8$) occurred in just the pull-apart area; the isoseismal lines nearly parallel to the northeast-trending normal fault and the fault plane solution show a strike-slip faulting with a large normal dip-slip component (Table 2). In 1973 another earthquake occurred along the Xianshuihe fault; the surface rupture extends 90 km, and the maximum horizontal displacement is about 3.6 m. The seismogenic structures of these two earthquakes, then, appear to belong to two diffrent tectonic settings within the strike-slip fault zone. The first probably was produced by shear extension due to pulling apart, and the second, directly from shear faulting of the strike-slip fault.

RODGERS (1980) pointed out that deformation in the pull-apart area is controlled by the separation and overlapping distances of the subparallel faults, but his tectonic model experiment is for a homogeneous elastic material, and it differs somewhat from real conditions. Field geological observations indicate that the deformation in a pull-apart area is controlled by fault length and depth, besides separation and overlapping. According to the relative magnitudes of these parameters (Table 1), the deformation patterns in pull-apart areas are of four distinct types, as follows; see also Figure 6.

1. When the depth D of shear faults that control the pull-apart area is much greater than the separation S and the overlapping O (that is, $D \gg S, O$) and the ratio of separation to overlapping is equal to or greater than 1 (or $S/O \geqslant 1$), the pull-apart basin will be a rhombic graben, and there will be only one subsidence center, located usually in the middle of the basin. This is called the single-center type (Fig. 6a). The six pull-apart basins within the Haiyuan fault zone (Fig. 4 and Table 1) and the pull-apart basins in the middle segment of the Chenhai fault zone (Table 1) belong to this type.

2. When $D \gg O, S$ and $S/O < 1$, the basins will be narrow rectangular grabens and the subsidence centers will be located at both ends of the pull-apart area. This is called the double-center type (Fig. 6b). The Jianchuan-Hequin pull-apart area (in the southern segment of the Xiaojinhe strike-slip fault in northwestern Yunnan) is an example. Its length is about 38 km and its width about 18 km, so $S/O = 0.47$. The fault block basins are at the ends of the subparallel shear faults; normal faults and tilted blocks develop between the basins (Fig. 7). The Cariaco basin in the south Caribbean Sea also belongs to this type (SCHUBERT, 1982).

3. When $D < O, S$, according to the calculation by SEGALL and POLLARD (1980),

Table 2

*Fault Plane Solutions of Earthquakes in the Luhou-Gaanzi Fault Zone**

Event no.	Date	Lat. (N)	Long. (E)	Magn. (M_s)	Depth (km)	Nodal planes (deg)				P axis (deg)		T axis (deg)	
						Strike	Dip	Strike	Dip	A	P1	A	P1
1	Mar. 14, 1966	32°47′	97°45′	5.1	30	18	70	291	80	63	10	156.5	22
2	Aug. 30, 1967	31°27′	100°20′	6.8	8	245	45	38	48	236	76	141.5	1
3	Aug. 30, 1967	31°42′	100°20′	6.0	10	233	50	53	40	143	85	323	5
4	Feb. 6, 1973	31.5°	100.4°	7.9	17	25	88	295	89	250.2	1	340.4	1.8
5*	Feb. 8, 1973	31.4°	100.3°	6.3	35	21	60	201	30	111	75	291	15
6†	Sep. 9, 1973	31.5°	100.0°	5.8	32	31.5	73	211.5	17	302.5	62	122.5	28
7†	Jun. 16, 1982	31°50′	99°51′	6.0	17	5.7	82	275	86.5	50	12.5	141	3

* From Wen et al., 1985.

† The earthquakes occurred within the pull-apart area. Nos. 2 and 4 occurred in the segment of the subparallel shear faults. No. 4 produced a 90 km surface rupture zone along the subparallel shear fault.

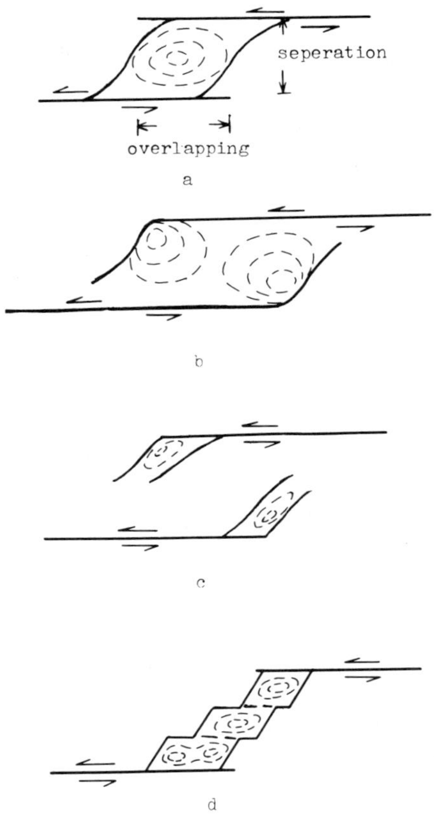

Figure 6

Deformation types in pull-apart area. In the following (see also text) D is depth, S is separation, O is overlapping. (a) Single center, $D \gg O, S$ and $S/O \geqslant 1$. (b) Double center, $D \gg O, S$ and $S/O < 1$. (c) Terminal extension type, $D < O, S$. (d) Combination.

the effect of the interaction between shear faults on the rock bridge will be very small. Extensional fractures or tensional basins will form at the ends of the shear faults, and there will be no pull-apart structure in the middle of the rock bridge (Fig. 6c). Figure 8 shows the pull-apart structures in northwest Yunnan. The northeast-trending Xiaojinhe fault and the Nandinghe fault form an en echelon and produce a large pull-apart area. The separation S of the area is about 120 km, the overlapping O is about 60 km, and the depth of the faults is probably no more than 50 to 100 km, so the tension fracture zone with a strike-slip component, composed of a graben system, formed at each end of the strike-slip fault zone. The interior of these two tensional fractures once again is cut by secondary shear faults and has created various pull-apart basins, including the Jianchuan-Hequin area mentioned before (Wu and Deng, 1985).

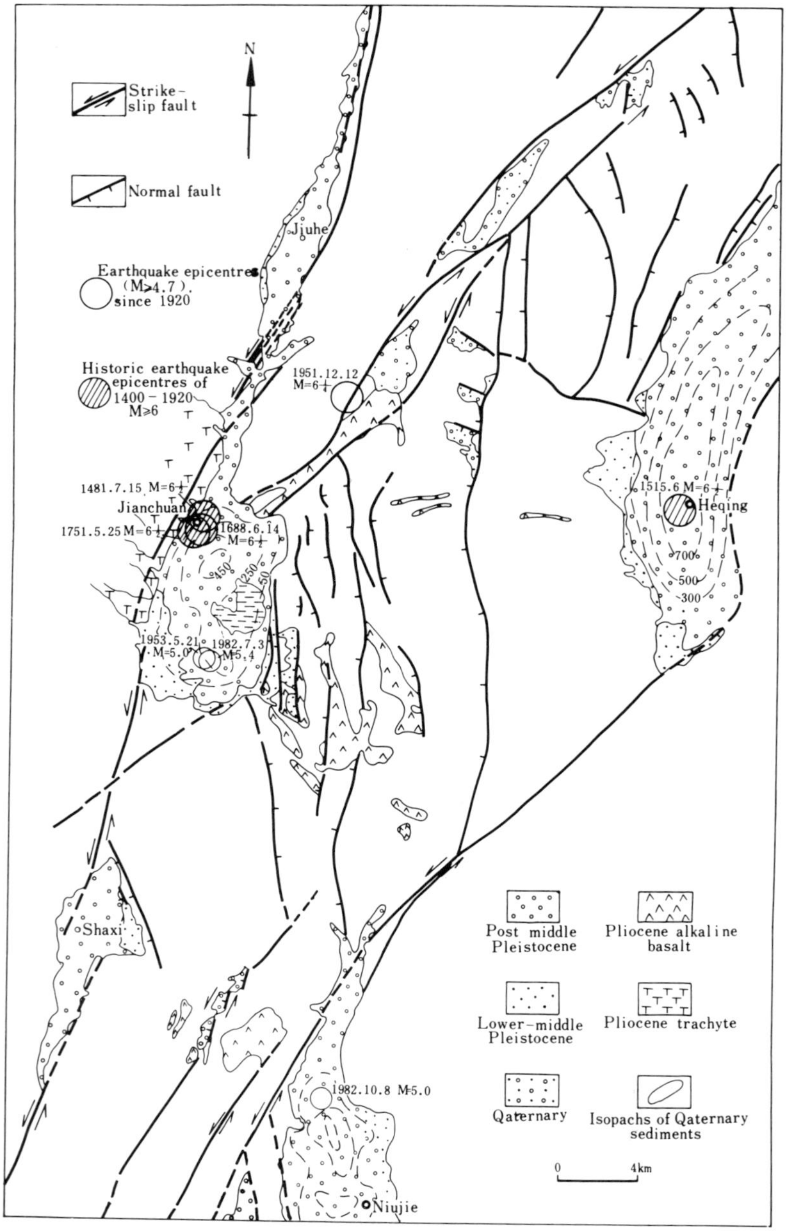

Figure 7
Tectonic sketch of Jianchuan-Hequin pull-apart area in northwest Yunnan.

4. If one segment of the strike-slip fault zone consists of a series of subparallel shear faults, the adjacent basins will become connected to each other as the dis-

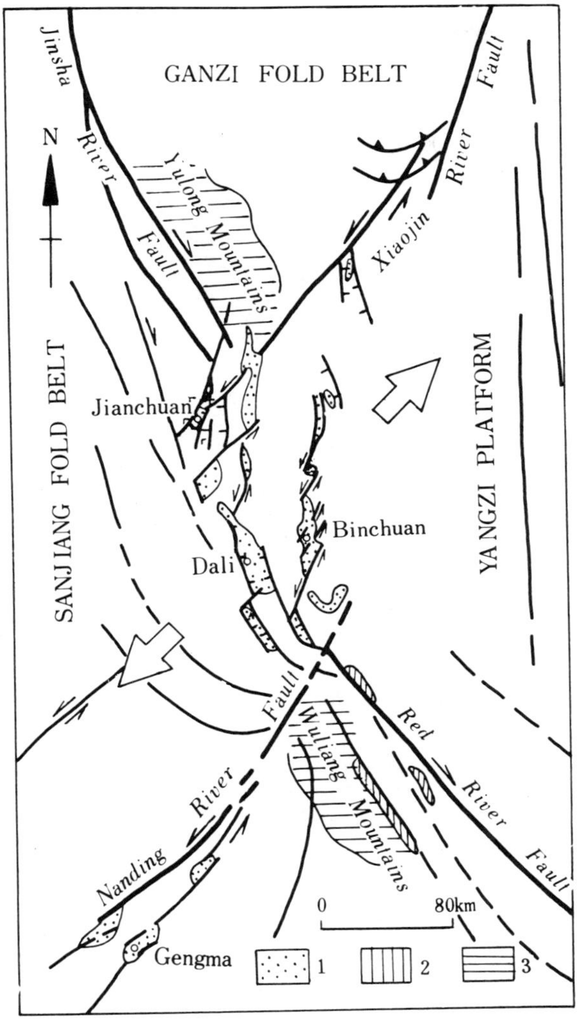

Figure 8
Tectonic sketch of pull-apart area in northwest Yunnan. Legend: (1) Quaternary basin, (2) late Tertiary basin, (3) uplifted area.

placement increases, creating one long, narrow pull-apart basin with several en-echelon subsident centers (Fig. 6d). In a statistical study of seventy pull-apart basins distributed world-wide AYDIN and NUR (1982) found that the ratio of length to width is a constant 3. The Bingchuan basin in the large pull-apart area of northwestern Yunnan belongs to this type (Fig. 9). It is controlled by four subparallel left-lateral strike-slip faults and forms three en-echelon subsident centers from north to south. Quaternary sediments 500 to 1000 m thick fill the basin. The length of the

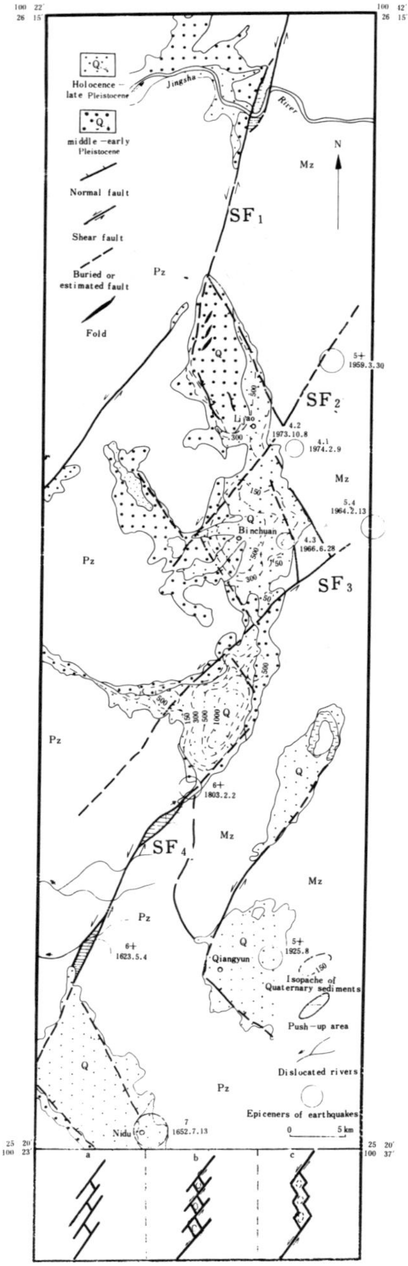

Figure 9
Tectonic sketch of Binchuan basin. Legend: a, b, c show sequence of development.

basis is about 40 km, the width is 11 to 13 km, and the ratio of length to width is about 3 (WU and DENG, 1985).

Figure 10
Mole tracks formed during Luhuo earthquake.

2. *Pushing up*

Pushing up occurs in normally overlapped areas between two subparallel shear faults. Because of synthetic movements of faults on both sides of a rock bridge, a compressional environment forms around the rock bridge. Compressive structures such as thrusts, folds, and uplifts commonly appear in such areas, and they intersect the strike-slip fault at high angles. The push-up structures in the Coyote Creek fault cited by SEGALL and POLLARD (1980) are typical. Mole tracks are typical push-up structures within earthquake surface rupture zones. For example, the 1973 Luhuo earthquake ($M_s = 7.9$) generated a 90 km surface rupture zone along the Xianshuihe fault; some mole tracks formed in the rock bridge between two adjacent pinnate fractures (Fig. 10). The southeastern segment of the Xianshuihe fault also contains several subparallel shear faults. Two of them were subjected to compression, so this area was uplifted intensively and formed a mountain, the Zheduoshan, that is 700 m above the land surrounding it.

IV. Destruction of rock bridges and evolution of strike-slip fault zones

During the history of a strike-slip faulting, discontinuous subparallel shear faults may become connected and form a relatively continuous fault zone. Rock mechanics experiments indicate that the joining is achieved by a breaking or rupturing of the rock bridge (WANG and ZHANG, 1984). Research on several strike-slip faults shows that the patterns of destruction in rock bridge areas are very complex.

Figure 4 and Table 1 refer to the Haiyuan fault and the pull-apart basins within it. Both the north and the south sides of the basins are controlled by subparallel shear faults with large dip-slip components. Normal faults developed at both ends of the basins. The thickness of sediments within the basins reaches 750 m. In the later stage of the basin's history a series of shear faults with normal dip-slip components (ESF) was generated in the interior of the basin. These new faults strike approximately parallel to the general orientation of the Haiyuan fault zone. They intersect the individual subparallel faults at angles of 10° to 15°. Because of them the subparallel faults are joined. They are very active structures as soon as they are formed. They control the subsidence centers and sedimentary terrains of the basins. We call this kind of fault the internal tensile shear fault of a pull-apart basin. The surface rupture of the 1920 Haiyuan earthquake ($M_s = 8.7$) was distributed along internal tensile shear faults when it crossed the pull-apart basins, creating large horizontal displacements (Fig. 4; 13). The pulling-apart parameters in the rock bridge between two overlapped subparallel faults are characterized by $D \gg S, O$ and $S/O \geqslant 1$. Rhombic grabens may be expected to form. In the further evolution of the Haiyuan fault the rock bridge is destroyed by the internal tensile shear fault, so that the subparallel faults will be joined and will form a continuous fault zone.

The Chenhai fault zone differs completely from the Haiyuan fault zone. There are six subparallel shear faults overlapping in a left-step pattern within 70 km along its middle segment. Five pull-apart basins were formed in a paternoster basin zone. Research on the history of this zone reveals that at first the individual pull-apart basins increased in area gradually and that later on adjacent basins joined together and formed one large basin zone (Fig. 9). The shear faults between two adjacent basins stopped controlling the later development of the basin as soon as the basin covered them. In Figure 9 the segments SF_2 and SF_3 are no longer active, and the basin is controlled by SF_1 and SF_4. The transverse fault system migrated outward gradually. The basin enlarged, and several rock bridge areas joined to form a larger pull-apart area. As mentioned above, the Chenhai fault, trending nearly north–south, has been acting as the terminal tensional structure of the large left-lateral Nandinghe strike-slip fault zone (Fig. 8). Because the angle between it and the Nandinghe fault is 30° or 35°, it possesses not only an extensional component but also a left-lateral strike-slip component (Wu and Deng, 1985).

The characteristics of these two typical strike-slip fault zones with pull-apart basins in their later stages of development suggest that, for a strike-slip fault zone with a compressional component normal to its general trend, the subparallel faults that bound the pull-apart basins will be linked through the formation of internal tensile shear faults, and the pull-apart basin will decrease in size until it disappears. This implies that shear strain is concentrated inside a narrower fault zone. For a strike-slip fault with a tensional component normal to its general trend, the adjacent rock bridges are enlarged and linked gradually, so as to form a large pull-apart basin with more than one subsidence center (Fig. 6d). Probably it would eventually evolve into a long and narrow basin zone or a tectonic valley like the Dead Sea Rift (Garfunkle, 1981).

V. Pivotal movement along strike-slip faults

It is well known that the strain field produced by shear motion on a strike-slip fault creates a quadripartite distribution. CHINNERY (1963) and LOO (1980) calculated the isopath maps for the stress fields and the strain fields, respectively, that are caused by strike-slip fault movement. The inhomogeneous stress state is accommodated variously, by means of deformation and displacement. In general, the compressional area is dominated by such structures as folds, thrusts, and uplifts, and the normal faults, basins commonly located in the extensional area. Because both tensional subsidence and compressional uplift occur in the same block of the fault, there is rotation about the vertical axis, a pivotal movement, in the fault plane.

The Fuyun fault in northwestern China is a right-lateral strike-slip fault with a length of more than 200 km. The 1931 earthquake with magnitude 8 occurred along this fault. Displacement since the middle Pleistocene has probably been 2 to 3 km, and the horizontal displacement during the 1931 earthquake reached a maximum of 14.6 m. The fault trends N20W and dips to the northeast at an angle of 60° to 80°. In the northern segment of the fault the motion is normal or strike-slip with a normal component. The eastern block dropped down along the fault, forming three basins, with the fault at their western boundaries (Fig. 11), while the western block was uplifted so that its regional planation surface is 500 m higher than that of the eastern block. In the south the fault motion is strike-slip with a reverse component. The western block of that segment is a Quaternary desert, which is 400 m lower than the mountains in the eastern block. Paleozoic rocks in the eastern block thrusted over the Quaternary gravels in the western block. It is clear that, except for the predominant horizontal displacement, the Fuyun fault has also a pivotal movement superimposed and that the axis of pivotal motion is subjected to intensive compression, which forms thrust faults.

Where the rock bridge area is relatively large in a strike-slip zone and the fault zone obviously ends, the pivotal motion may occur along individual subparallel faults, such as in the Haiyuan fault zone, shown in Figure 4. Almost every subparallel shear fault in that zone has pivotal movement to some extent. In the strike-slip faults with normal components, shear faults control the formation of the basin; strike-slip faults with reverse components are found outside the basins.

VI. Displacement distribution along the strike-slip fault

According to the results of both theoretical calculation and model experiments, horizontal displacements along a single strike-slip fault are normally distributed; the maximum displacement is located in the middle of the fault and decreases toward both ends. In the field the situation is much more complex. For many reasons it is impossible to measure displacement along a fault zone accurately. But the im-

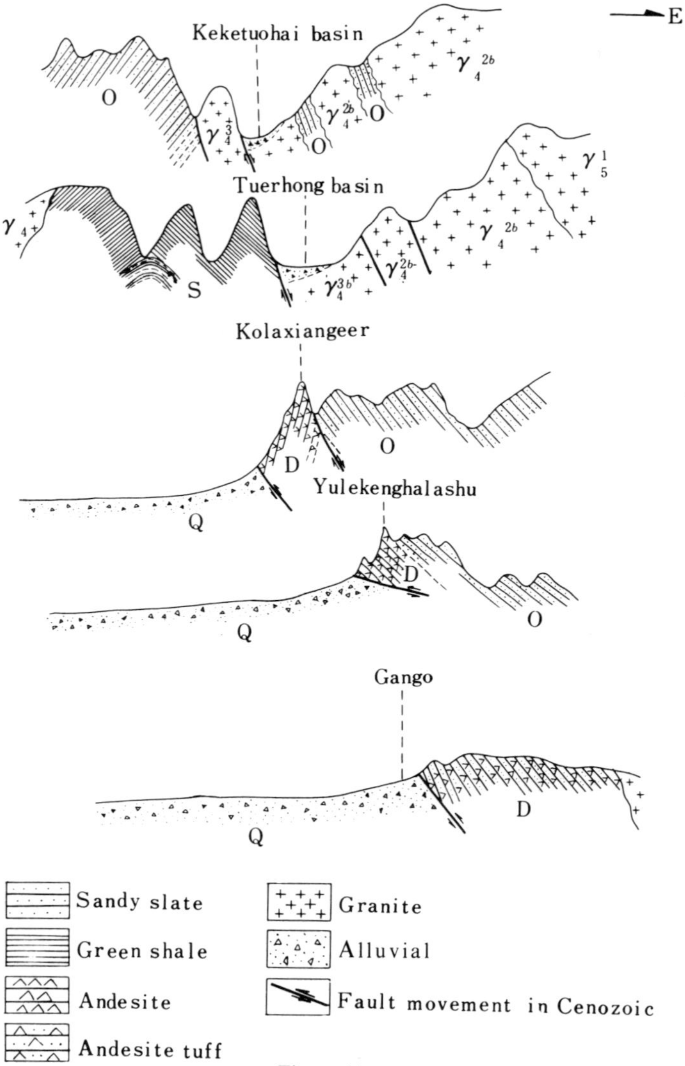

Figure 11

Synthetic cross section of Fuyun fault along the Keketuchai-Ertai strike-slip fault: O is Ordovician sandy slate, γ is granite, S is Silurian green shale, D is Devonian andesite and Andesite tuff, and Q is Quaternary sediments.

portance of understanding the displacement field along a strike-slip fault cannot be overemphasized. In the following we describe the measurements of displacement along two strike-slip earthquake surface-rupture zones in China.

The surface rupture of the Fuyun earthquake extends 180 km (ZHANG, 1982) with right-lateral strike-slip displacement. Along the surface rupture zone 171 measurements were obtained. The maximum offset is 14.6 m, located in the middle

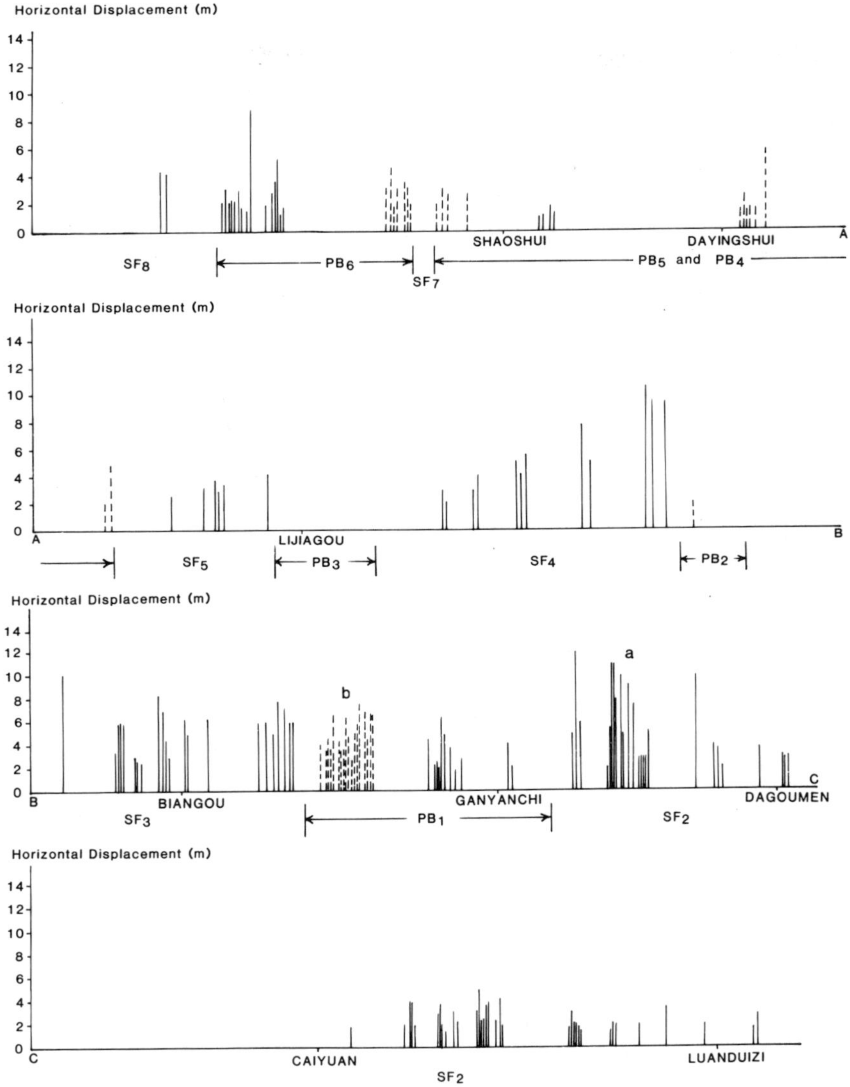

Figure 12
Horizontal displacement distribution along surface rupture of 1920 Haiyuan earthquake: (a) displacements along subparallel shear faults, (b) displacements in overlapping area.

segment of the rupture zone. The amount of offset attenuates toward both ends from the point of maximum displacement (Zhang, 1982; Deng and Zhang, 1984). The surface rupture zone of the Haiyuan earthquake extends 220 km. It is a left-lateral shear fault. The maximum of 10 to 11 m of displacement, among our 168 measurements, also is in the middle part of the zone and decreases toward both ends (Fig. 12).

Figure 12 shows that the displacement attenuate in an oscillatory pattern toward both ends. The local minimum displacement often corresponds to an overlapped rock bridge area, and the peak value corresponds to a subparallel shear fault. On the other hand, there are two types of offset distribution along individual subparallel faults. In the first the maximum displacement is located in the middle of the fault; in the second it is located near one of the ends. Subparallel faults of the Fuyun earthquake surface-rupture zone belong to the latter type. These observations imply that the shear component decreases in the rock bridge area while the extensional or compressional component increases. Thus the fracture processes of the individual subparallel faults may be different from that of the master fault zone and even from each other, giving rise to a complex deformation pattern within the strike-slip fault zone.

VII. Summary

Geological observations along four major strike-slip faults in China have revealed a complex variety of deformation patterns resulting from the interactions among subparallel faults within the master fault zones. Depending on the geometry, the sense of the stepping, and the rankings ('grades') of smaller subparallel faults within larger ones, distinct tectonic settings and dynamic circumstances are created in the regions between the overlapping subparallel faults, which can lead to diffferent histories of fault-zone evolution.

Acknowledgments

We thank our colleagues at the Ninxia and Xinjiang Seismological Bureau, with whom we worked on some of the strike-slip faults discussed in this paper. We thank the Sichuan Seismological Bureau for a wonderful tour along the Xianshuihe fault. We are especially grateful to Prof. C.-Y. Wang for his help and encouragement during the preparation and publishing of this paper. Finally, we thank Sheila Gardner for her kind help in typing this manuscript.

References

Aydin, A. and Nur, A. (1982), *Evolution of pull-apart basins and their scale independence.* Tectonics *1*, 91–105.

Bahat, D. (1982), *Extensional aspects of earthquake induced ruptures determined by an analysis of fracture bifurcation.* Tectonophysics, *83*, 163–183.

Brace, W. F. and Bombolakiso, E. G. (1963), *A note on brittle crake growth in compression.* J. Geophys. Res. *68*, 3709–3713.

CHEN, S. and DENG, Q. (1985), *The pull-apart basins within Haiyuan fault zone* (in Chinese). R. Crust. Mov. *1*, 38–45.

CHINNERY, M. A. (1963), *The stress changes that accompany strike-slip faulting.* Bull. Seis. Soc. Amer. *53*, 355–392.

DENG, Q. (1984), *The behavior of fault and basin type and their formation mechanism, Part II* (in Chinese). J. Earthquake Stud. *2*, 57–64.

DENG, Q., ZHONG, J. and MA, Z. (1966), *Characteristics of the shearing fracture zone and conditions of its formation* (in Chinese). Sci. Geol. Sin. *3*, 227–237.

DENG, Q. and ZHANG, P. (1984), *Research on the geometry of shear fracture zones.* J. Geophys. Res. *89*, B7, 5699–5710.

DU, Y., MA, J. and LI, J. (1986), *Interaction and stability of en echelon cracks* (in Chinese). J. Geophys. Sini., in press.

GARFUNKEL, Z. (1981), *Internal structure of the Dead Sea leaky transform (rift) in relation to plate kinetmatics*, Tectonophys. *80*, 81–108.

HOEK, E. and BIENIAWSKI, Z. T. (1965), *Brittle fracture propagation in rock under compression*, Int. J. Fract. Mech., *1*, 137–155.

KOIDE, H. (1983), *Seismotectonic problem of an echelon fault system and earthquake source mechanism* (in Japanese). J. Geol. Sci. *92*, 33–52.

LI, J. (1980), *Experimental study on the processes of deformation and failure of a typical rock mass and their relationship to both the seismic precursors and the foreshocks* (in Chinese). Seismo Geol. *2*, 11–18.

LOO, Z. (1980), *Stress deformation and tilt fields of the earthquake sources.* Acta Seismol. *2*, 169–185.

MOLNAR, P. and TAPPONNIER, P. (1975), *Cenozoic tectonics of Asia: Effect of a continental collision.* Science *189*, 419–425.

RODGERS, D. A. 'Analysis of pull-apart basin development produced by an echelon strike-slip faults,' in P. E. Ballance and H. G. Reading (eds.), *Sedimentation in Oblique-Slip Mobile Zones.* Assoc. Sed. Publ. p. 27–41, 1980.

SCHUBERT, C. (1982), *Origin of Cariaco basin, Southern Caribbean Sea.* Marine Geol. *47*, 345–360.

SEGALL, P. and D. D. POLLARD (1980), *Mechanics of discontinuous faults.* J. Geophys. Res. *85*, 4337–4350.

WANG, S. and ZHANG, L. (1984), *Shear fracture and strike-slip: A study on shock-generation mechanism of strong shallow earthquakes* (in Chinese). Seismol. Geol. *6*, 63–74.

WU, D. and DENG, Q. (1985), *Mechanism of graben formation in northwest Yunnan* (in Chinese). R. Crust. Mov. *1*, 118–132.

WEN, X., HUANG, S. and JIANG, Z., *Characteristics of neotectonics of Gangzi-Yushi fault zone of and its estimate of earthquake risk.* Seismol. Geol., in press.

ZHANG, P. (1982), *Deformation and fracture of Fuyun earthquake fault zone.* Ph.D. Thesis, Univ. Science Technol. China, Grad. School.

(Received 4th September 1985, revised 7th April 1986, accepted 17th April 1986)

PAGEOPH, Vol. 124, Nos. 1/2 (1986)

0033-4553/86/020225-44\$1.50+0.20/0

Fluid Infiltration into Fault Zones: Chemical, Isotopic, and Mechanical Effects

R. KERRICH[1]

Abstract—Fluid infiltration into fault zones and their deeper-level counterparts, brittle–ductile shear zones, is examined in diverse tectonic environments. In the 2.7 Ga Abitibi greenstone belt, major tectonic discontinuities, with lateral extents of hundreds of kilometres initiated as listric normal faults accommodating rift extension and acted as sites for komatiite extrusion and locally intense metasomatism. During reverse motion on the structures, accommodating shortening of the belt, these transcrustal faults were utilised as a conduit for the ascent of trondhjemitic magmas from the base of the crust and of alkaline magmas from the asthenosphere and for the discharge of thousands of cubic kilometres of hydrothermal fluids. Such fluids were characterised by $\delta^{18}O = +6 \pm 2$, $\delta D = -50 \pm 20$, $\delta^{13}C = -4 \pm 4$, and temperatures of 270 to 450 °C, probably derived from devolatilisation of crustal rocks undergoing prograde metamorphism. Hydrothermal fluids were more radiogenic ($^{87}Sr/^{86}Sr$ = 0.7010 to 0.7040) and possessed higher μ than did contemporaneous mantle, komatiites or tholeiites, and thus carried a contribution from older sialic basement. A provinciality of $^{87}Sr/^{86}Sr$ and $\delta^{13}C$ is evident, signifying that fault plumbing sampled lower crust which was heterogeneous at the scale of tens of kilometres. Mineralised faults possess enrichments of large ion lithophile (LIL), LIL elements, including K, Rb, Ba, Cs, B, and CO_2, and rare elements, such as Au, Ag, As, Sb, Se, Te, Bi, and W. Fluids were characterised by $X_{CO_2} \approx 0.1$, neutral to slightly acidic pH, low salinity $\leqslant 3$ wt-%, K/Na = 0.1, they carried minor CH_4, CO, and N_2, and they underwent transient effervescence of CO_2 during decompression. Clastic sediments occupy graben developed at fault flexures. The $^{40}Ar/^{39}Ar$ release spectra indicate that fault rocks experienced episodic disturbance on time scales of hundreds of millions of years.

At the Grenville front, translation was accommodated along two mylonite zones and an intervening boundary fault. The high-temperature (580 °C) and low-temperature (430 to 490 °C) mylonite zones, formed in the presence of deep-level crust-equilibrated fluids of metamorphic origin. Late brittle faults contain quartz veins precipitated from fluids with extemely negative $\delta^{18}O$ (−14 per mil) at 200 to 300 °C. The water may have been derived from downward penetration into fault zones of precipitation of low ^{18}O on a mountain range induced by continental collision, with uplift accommodated at deep levels by the mylonite zones coupled with rebound on the boundary faults.

Archean gneisses overlie Proterozoic sediments along thrust surfaces at Lagoa Real, Brazil; the gneisses are transected by brittle–ductile shear zones locally occupied by uranium deposits. Following deformation at 500 to 540 °C, in the presence of metamorphic fluids and under conditions of low water-to-rock ratio, shear zones underwent local intense oxidation and desilication. All minerals undergo a shift of −10 per mil, indicating discharge of meteoric-water-recharged formation brines in the underlying Proterozoic sediments up through the Archean gneisses, during overthrusting; ≈ 1000 km^3 of solutions passed through these structures. The shear zones and Proterozoic sediments are less radiogenic ($^{87}Sr/^{86}Sr$ = 0.720) than contemporaneous Archean gneisses (0.900), corroborating the transport of fluids and solutes through the structure from a large external reservoir.

Major crustal detachment faults of Tertiary age in the Picacho Cordilleran metamorphic core complex of Arizona show an upward transition from undeformed granitic basement through mylonitic to

[1] Department of Geology, University of Western Ontario, London, Ontario, Canada N6A 5B7.

brecciated and hydrothermally altered counterparts. The highest tectonic levels are allochthonous, oxidatively altered Miocene volcanics. This transition is accompanied by an increase of 12 per mil in $\delta^{18}O$, from +7 to +19, and a 400 °C decrease in temperature. Lower tectonic levels acted as aquifers for the expulsion of large volumes of higher-temperature reduced metamorphic fluids and/or evolved formation brines. The Miocene allochthon was influenced by a lower-temperature reservoir inducing oxidative potassic alteration; mixing occurred between cool downward-penetrating thermal waters and the hot, deeper aqueous reservoir.

In general, flow regimes in these fault and shear zones follow a sequence, from conditions of high temperature and pressure with locally derived fluids at low water-to-rock ratios, during initiation of the structures, to high fluxes of reduced formation or metamorphic fluids along conduits as the structures propagate and intersect hydrothermal reservoirs. Later in the tectonic evolution and at shallower crustal levels there was incursion of oxidising fluids from near-surface reservoirs into the faults. In general, magmatism, tectonics, and fluid motion are intimately related.

Key words: Fluid infiltration, geochemical transport, faults, isotopes, hydraulic fracturing.

1. Introduction

Fluid penetration through major structures—specifically faults, thrusts, and their deeper-level extensions, brittle–ductile shear zones—is a phenomenon well-established both by contemporary observations of seismic pumping (Sibson et al., 1975; Sibson, 1981, 1982) and by evidence of hydration or extensive hydrothermal metasomatism in ancient counterparts. On a global scale faults play a key role in the geochemical cycle, where volatiles from the hydrosphere are transferred to rocks as hydroxyl and carbonate minerals, subsequently being returned to the surface by release along faults during burial, compaction, and devolatilisation accompanying metamorphism. It has been pointed out, for instance, that faults at continental margins may act as one of the sites for return flow of the $\approx \times 1.5$ to 3×10^{21} gm of H_2O per Ma, released during the subduction of hydrated oceanic lithosphere (Fyfe and Kerrich, 1985; Wang, 1980).

The mechanical and chemical roles of fluids in fault and shear zones have been examined by a number of authors for eroded remnants of structures (Beach and Fyfe, 1972; Beach, 1976; Kerrich et al., 1977, 1980, 1984; Fyfe et al., 1978; Etheridge and Cooper, 1981; Fyfe and Kerrich, 1983, 1985) and for their seismically active modern counterparts (Irwin and Barnes, 1975; Sibson, 1982); there is, however, little quantitative information on the origin of the fluid reservoirs implicated, the chemical mass transport involved, the ambient temperature and pressure, physical and chemical properties, hydraulic conditions, or possible evolution of fluid regimes through a history of progressive incremental displacement along the structures.

This paper sets out to examine the variety of fluid regimes in fault zones, utilising information from structures in four areas, which encompass a variety of tectonic environments, styles of deformation, lithological groups, and ages. These are as follows.

1. Fault systems of the Abitibi greenstone belt ($\approx$ 2700 Ma).
2. Structures at the Grenville front ($\approx$ 1100 Ma).
3. Shear zones aged $\approx$ 820 Ma in Archean gneisses at Lagoa Real, Brazil.
4. Detachment faults in Cordilleran metamorphic core complexes, of Tertiary age.

The selection of these structural domains was based on existing mapping and documentation of the faults as major structural features tens to hundreds of kilometres in lateral extent. In addition, the presence of sporadically developed mineralisation provides the opportunity to evaluate fluid regimes associated with the faults.

Specific questions to be considered in the context of fluid transport through faults are the following.

1. Geochemical transport accompanying deformation and/or fluid infiltration.
2. Sources of volatile components, such as H_2O, CO_2, and H_2S.
3. The nature of rock reservoirs donating lithophile and other elements locally enriched in faults.
4. Ambient temperature–pressure and hydraulic conditions.
5. The history of incremental displacement and fluid activity.
6. The presence of highly sodic or alkaline magmas along some structural lineaments.

This paper draws on earlier articles concerning 'fluid participation in deep fault zones' (KERRICH et al., 1984; KERRICH, 1985), reports new data for previously described regions, and extends coverage of the subject with reference to additional tectonic situations. Faults almost invariably have poor surface exposure. In view of this, materials, for the study were obtained mostly from mines and their environs which were located on or near fault structures and thus afforded access to unweathered materials in open pits or underground.

2. *Abitibi greenstone belt faults*

2.1 Geological setting

The 2.7 Ga Abitibi greenstone belt, located in the Superior Province, is composed of an array of volcanic-plutonic complexes with intervening belts of volcaniclastic,

clastic, and chemical sedimentary rocks (GOODWIN and RIDLER, 1970). Prehnite-pumpellyite to lower greenschist facies of regional or hydrothermal metamorphism predominates, except in the periphery of batholiths, where amphibolite facies rocks may be present (JOLLY, 1978).

A prominent feature of the greenstone belt is a series of major east-west-trending tectonic discontinuities or fault zones, locally termed 'breaks', which extend for hundreds of kilometres (Fig. 1, a, b). Two of the dominant structures are the east-west-trending Destor-Porcupine and Kirkland-Lake–Cadillac-Bouzan fault zones; in the Kirkland Lake area these structures mark the north and south limbs of a regional east-plunging synclinorium; see Figure 1b (THOMSON, 1941; JENSEN, 1976; JENSEN and LANGFORD, 1983). The magnitude and direction of translation on these structures are not well quantified. The structures control the linear disposition of ultramafic rocks and locally delimit linear belts of clastic or chemical sedimentary rocks. The structures also acted as preferential sites for the emplacement of trondhjemitic and alkaline magmas and were locally domains of intense hydrothermal metasomatism, exemplified by rich lode gold deposits.

2.2 Metasomatism in fault zones: Calculations of chemical mass balance

Quantitative evaluation of chemical mass transport in metasomatised fault zones requires knowledge of the composition of rocks in the original and altered states together with the constraints on any possible volume changes. GRESENS (1967) suggested incorporating specific-gravity data into two-way mass-balance calculations, such that fixing either volume change or the behaviour of one component during a reaction provides a unique solution. For any metasomatic transformation of parent rock to altered product, volume factors f_v may be computed which correspond to the isochemical behaviour of individual components. Clustering of the volume factors of several components that are empirically known to be relatively immobile (such as TiO_2, Al_2O_3, Sc, Hf, and Zr) then provides a rational basis for estimating the volume change of the reaction as a whole. In the notation used by GRESENS (1967), $f_v = 1$ signifies constant-volume metasomatic alteration of reactants to products; $f_v = 0.8$ and $f_v = 2.0$, for example, correspond to 20 vol-% reduction and 100 vol-% increase, respectively. Furthermore, rigorous calculations of chemical mass balance require that the series of altered rocks, along with the unaltered precursor, belonged to an initially chemically uniform population. It is important to emphasise that the observed metasomatically altered products represent specified finite changes relative to the precursor, but that such changes may have been arrived at via a complex sequence of incremental chemical transfers.

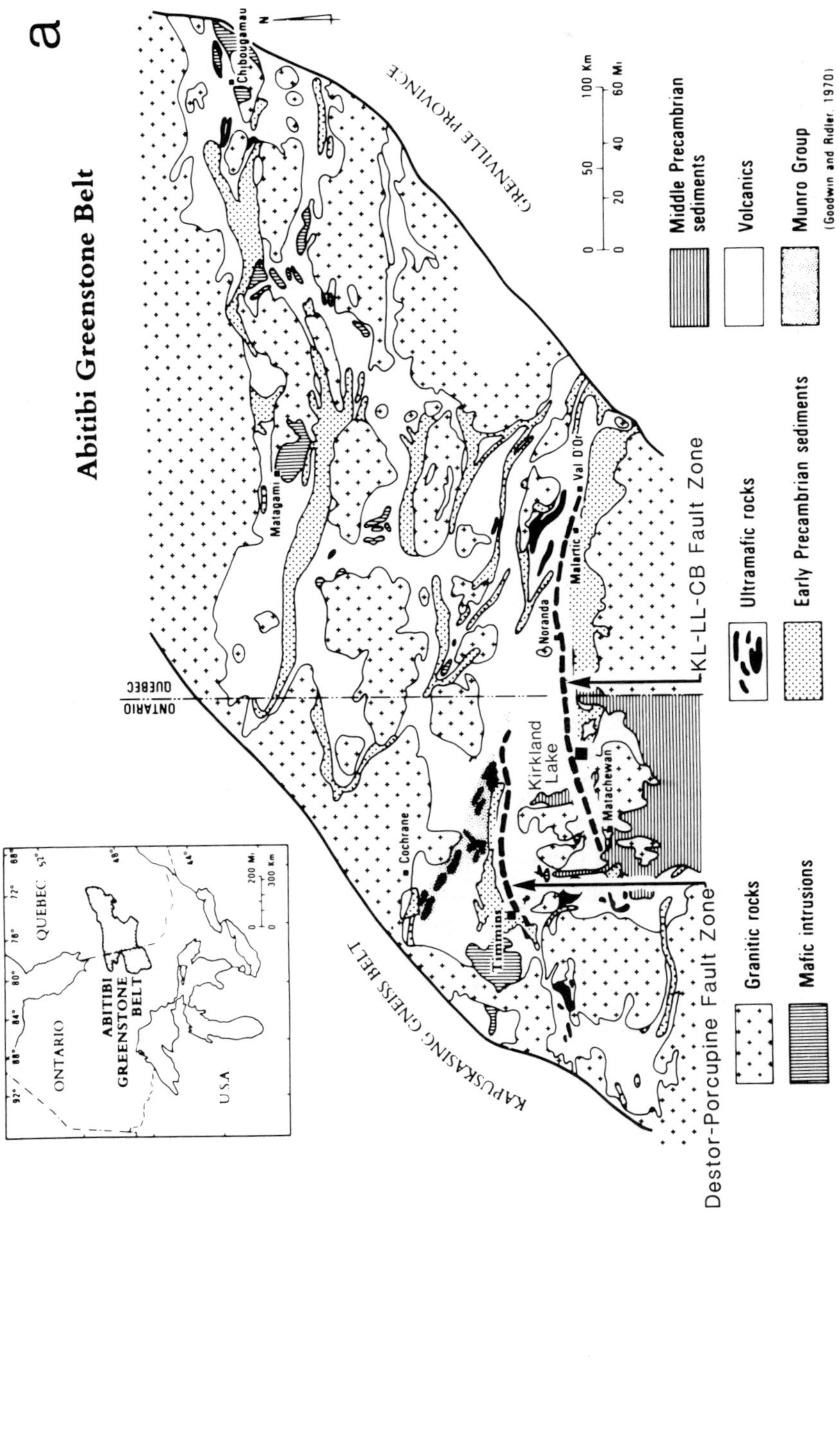
a
Abitibi Greenstone Belt
QUEBEC
ONTARIO
ABITIBI GREENSTONE BELT
U.S.A.
0 200 Mi
300 Km
Chibougamau
N
GRENVILLE PROVINCE
Matagami
Val D'Or
Malartic
Noranda
QUEBEC
ONTARIO
Kirkland Lake
Matachewan
Cochrane
Timmins
KAPUSKASING GNEISS BELT
0 50 100 Km
0 20 40 60 Mi
KL-LL-CB Fault Zone
Destor-Porcupine Fault Zone
Middle Precambrian sediments
Volcanics
Munro Group
(Goodwin and Ridler, 1970)
Ultramafic rocks
Early Precambrian sediments
Granitic rocks
Mafic intrusions

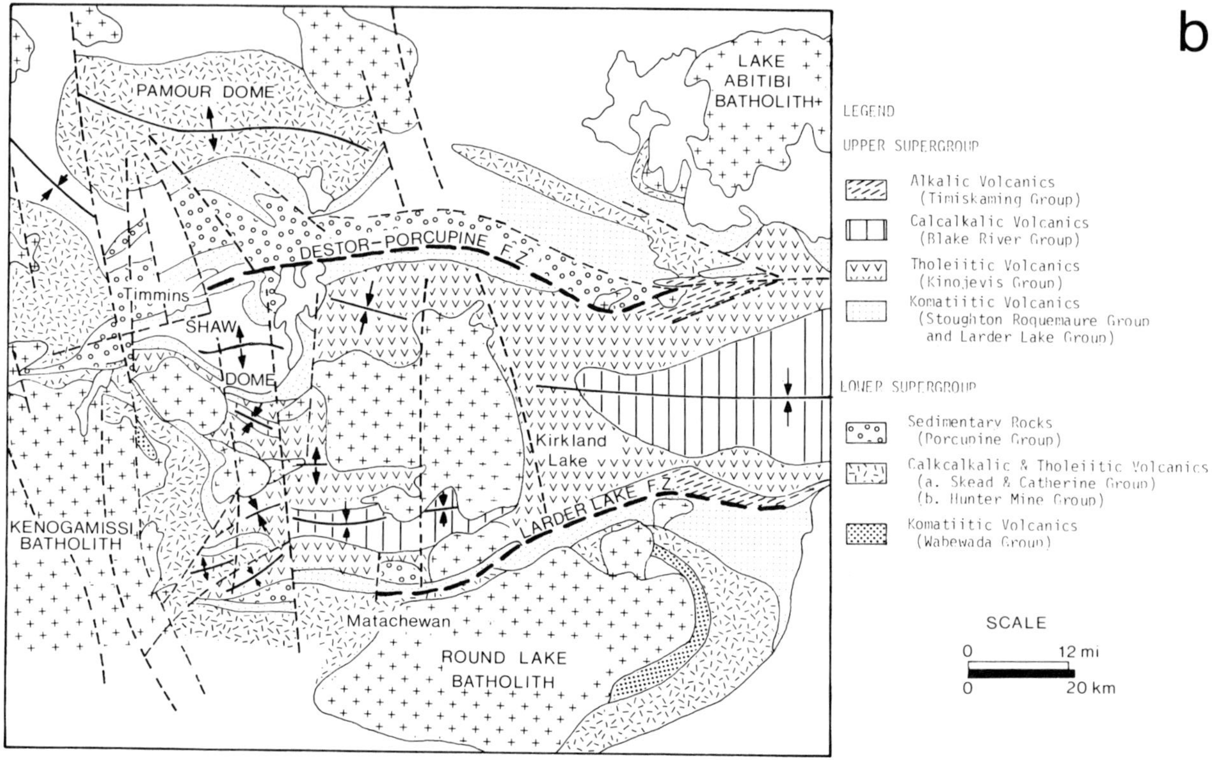

Figure 1
(A) Schematic geology of the Abitibi belt. After Goodwin and Ridler, 1970. KL-LL-LB refers to the Kirkland Lake–Larder Lake—Cadillac-Bouzan fault zone. (B) Geology of the Kirkland Lake area. After Jensen, 1978 a, b; 1980.

2.3 Hollinger Fault, Timmins: Patterns of alteration

Major structures in the Timmins district display at least three episodes of translation, which are associated with (a) magmatism, (b) fluid transport, and (c) sedimentation, respectively. Early extensional displacements on listric faults are associated with komatiitic flows and clastic sediments; spilitisation of the volcanics in the presence of marine water occurs at this stage, and may lead to coeval chemical sediments at volcanic centres or along faults, or both. Reverse motion is associated with trondhjemitic magmatism and major fluid discharge linked to gold deposits. At late stages during wrench faulting, domains of relaxation are the sites of molasse sedimentation in graben and alkalic magmatism, and also of the infiltration of meteoric waters. In this section we are concerned with the second 'main stage' of fluid discharge.

Intense metasomatic alteration along faults and their bounding wall rocks is marked by massive fixation of a select group of lithophile elements or components, including Si, CO_2, K, Rb, Ba, Cs, B. Such additions are reflected in the presence of veins typically composed of quartz, ferroan dolomite, muscovite, and sulphide minerals, collectively bounded by altered wall rocks. Numerous generations of veins, variably modified by intracrystalline deformation, attest to repeated cycles of volatile fluxing with intervening episodes of incremental ductile deformation (cf. KERRICH and ALLISON, 1978; KERRICH, 1983; KISHIDA, 1984). The magnitude of volatile flux through the structures, with its associated lithophile element fixation, is illustrated by the estimate that, for the known length of fault zones, about 8,000 km^3 of aqueous fluids were discharged along the faults, precipitating in the process 60×10^{15} gm of Si, 30×10^{15} gm of CO_2, and 6×10^{15} gm of K (KERRICH et al., 1986, in press).

The Hollinger fault, a splay of the Porcupine-Destor fault system, traverses a variety of different rock types where it outcrops in open pits at the surface. This fault is representative of the greenstone belt structures under consideration here, in respect of the variety of lithologies present, the tectonic style, and the range of metasomatic phenomena. Despite the combined effects of deformation and metasomatic alteration, host rocks of original trondhjemitic, basaltic, and ultramafic parentage can be identified on the basis of the ratios of the less mobile major and trace elements (KERRICH, 1983). Chemical transfers in the rocks of probable ultramafic origin are plotted in Figure 2, to illustrate the nature and magnitude of fault-related metasomatism.

Given that no perfectly fresh rocks are present, calculations of chemical mass balance have been conducted by matching the diagnostic 'immobile element' ratios of the three populations given above to analyses of fresh rocks from the Timmins area. An average of the analyses for three komatiite flows from Munro township (NESBITT and SUN, 1976, Table 1, 8–11) is utilised here as a hypothetical parental rock, based on compliance of Ti/Zr, Zr/Y, Al_2O_3/TiO_2, and Ni between this population of fresh rocks and the Hollinger altered counterparts. Considering the values of the above interelement ratios, coupled with the positive correlation of MgO to Ni and an inverse relationship of MgO to TiO_2 in fresh komatiites, the Hollinger ultramafic rocks probably had primary MgO contents in the range of 22 to 26 wt-%.

It is not possible with the present sampling to establish the extent of chemical modification of the ultramafic rocks arising from possible hydrothermal alteration shortly after extrusion, or the chemical effects of deformation prior to later hydrothermal alteration associated with the auriferous quartz-carbonate veins. Enhanced $\delta^{18}O$ values for silicate minerals likely signify isotopic exchange with marine, or evolved marine water at temperatures of $<200\,^{\circ}C$ ($\delta^{18}O$ for chlorite, +16 per mil).

Carbonate minerals, interlaminated with ultramafic tuff and of possible sedimentary origin, are cloudy, fine-grained, anhedral, and in the compositional range of high-magnesium dolomite, with no significant Mn substitution (KERRICH, 1983, Fig. 22). Where recrystallised at vein margins, the carbonates are compositionally ferroan-dolomite. Secondary carbonates texturally associated with quartz-chlorite-muscovite-gold veins in faults are clear, coarse-grained, anhedral, and within the magnesite-siderite solid-solution series. Overall, the ultramafic rocks have undergone a massive loss of iron, compared to their precursors; the observed trend of iron enrichment in secondary carbonates could partly be accounted for by transfer of iron from a silicate reservoir to a carbonate reservoir accompanying the hydrolysis of chlorite to muscovite in vein wall rocks. Both vein-related muscovite and chlorite contain ≈1% Cr_2O_3, reflecting the elevated chromium content of the ultramafic host rock.

The geochemical results for altered ultramafic rocks at Hollinger form two data arrays: a low K_2O series and a high one; see Figure 2. The former are remote from fault-hosted veins, have been leached to low volume factors (0.8 to 0.9), reflecting the extensive volumetric contraction accompanying the removal of major-element oxides, and possess low K_2O, Au, Rb, Ba, As, and Sb, whereas the latter group are adjacent to veins in faults and are characterised by massive additions of these elements. Taking the geological, textural, mineral, chemical, isotopic, and geochemical lines of evidence collectively, it is suggested that the ultramafic rocks at Hollinger underwent an early stage of seawater and/or metamorphic fluid-induced alteration under conditions of low T and P_{CO_2} in the submarine environment. This stage involved loss of Si, Fe, Mn, Mg, Na, and K, with gains of Sr plus volatiles, reflected in early laminated carbonates, during hydrolysis of primary silicate minerals to zeolites and clays. Subsequently, at deeper crustal levels under ambient conditions of higher T and P_{CO_2} during vein emplacement, the rocks experienced massive gains of CO_2, K, Rb, and Ba, and minor fixation of Fe, Ca, and Mn, as represented by the conversion of clays and zeolites, or their greenschist facies products, to secondary iron, magnesium-carbonates, muscovite, and chlorite (Fig. 2). Additional sodium may have been lost during hydrothermal alteration associated with vein emplacement, giving the low observed content of Na_2O.

Patterns of rare-earth elements for the series of low and high K_2O rocks, respectively have flat configurations characteristic of most Archean komatiites, but the

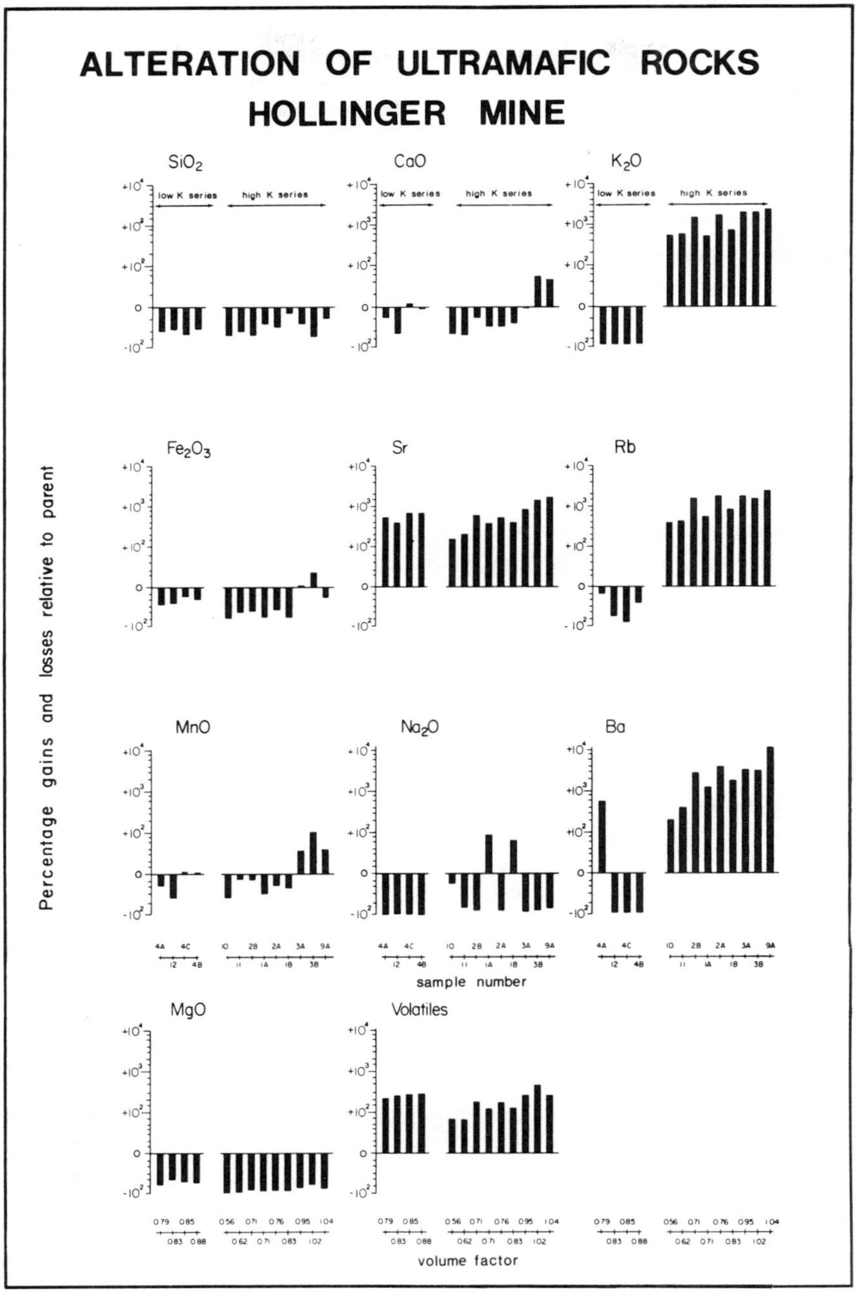

Figure 2

Gains or losses of major-element oxides and selected trace elements hydrothermally altered in rocks of primary ultramafic compositions (peridotitic komatiites), Hollinger Mine, number 12 vein, open pit. Gains and losses expressed as percentage changes relative to abundances in unaltered parent rock, and ordered according to increasing volume factor. After KERRICH, 1983.

high-K_2O mineralised rocks have systematically larger abundances by a factor of 1.5 to 2; see Figure 3, a, b). This cannot be fully accounted for by enrichment during volumetric contraction and therefore signifies either some absolute enrichment of the rare earth elements accompanying hydrothermal alteration (see also KERRICH and FRYER, 1979, LUDDEN, et al., 1984) or, alternatively, primary differences in the absolute abundances of rare-earth elements between neighbouring komatiitic units. Under low-temperature hydrothermal conditions the REE are relatively insoluble and accordingly behave isochemically. If appropriate ligands such as F or CO_3^{2-} are present, significant aqueous transport of the rare-earth elements is possible; such effects have been reported by KERRICH and FRYER (1979).

Many rocks have experienced volume losses of 20% during the successive episodes of hydrothermal alteration, with absolute cumulative depletions of SiO_2, Fe_2O_3, CaO, MgO, and Na_2O in excess of the total gains of volatiles and K_2O. Given the conservative behaviour of aluminium, these overall losses of chemical components have left the rocks relatively aluminious, and such hydrothermal leaching may promote flow focus within fault zones.

Fluids which induced the K–Si–CO_2 alteration in proximity to veins, and from which the auriferous veins were precipitated, were at temperatures of ≈ 400 °C, with fluid $\delta^{18}O$ of 7 ± 1 per mil, and at low redox potential; the latter property is indicated by the absence of Fe^{3+} silicates or oxides in the alteration mineral assemblage. Studies of primary liquid inclusions in vein quartz show filling temperatures of 360 °C (filling temperatures are a minimum estimate of the real temperature of hydrothermal mineral growth) and salinities of < 1.5 wt-% NaCl equivalent, and reveal the presence of significant liquid CO_2. Pressures are estimated at 2 to 3 kbar.

2.4 Source of volatile components in fault zones

The source, temperature, and magnitude of aqueous fluid reservoirs participating in major fault zones of the Abitibi greenstone belt have been evaluated from oxygen and hydrogen isotope data. Five distinct terrestrial fluid reservoirs have been identified on the basis of their oxygen and hydrogen isotope compositional ranges. These are marine, meteoric, magmatic, and metamorphic fluids, together with formation brines that generally contain a large component of 'evolved' meteoric recharge waters (TAYLOR, 1974); see Figure 4A. Oxygen isotope studies also permit estimates of ambient mineral and fluid temperatures at equilibrium, given that the fractionation of oxygen isotopes between coexisting silicates or metal oxides is a function of temperature alone (JAVOY, 1977; FRIEDMAN and O'NEIL, 1977). The problems of identifying fluid flow through rocks are particularly amenable to evaluation by oxygen isotope analyses, since isotopic exchange between rocks and an aqueous reservoir is reflected in shifts of $\delta^{18}O$ rock or mineral from an initial value by a magnitude that depends on the $\delta^{18}O$ fluid, ambient temperature, the fluid-to-rock ratio, and the extent of approach to equilibrium (TAYLOR, 1974, 1979).

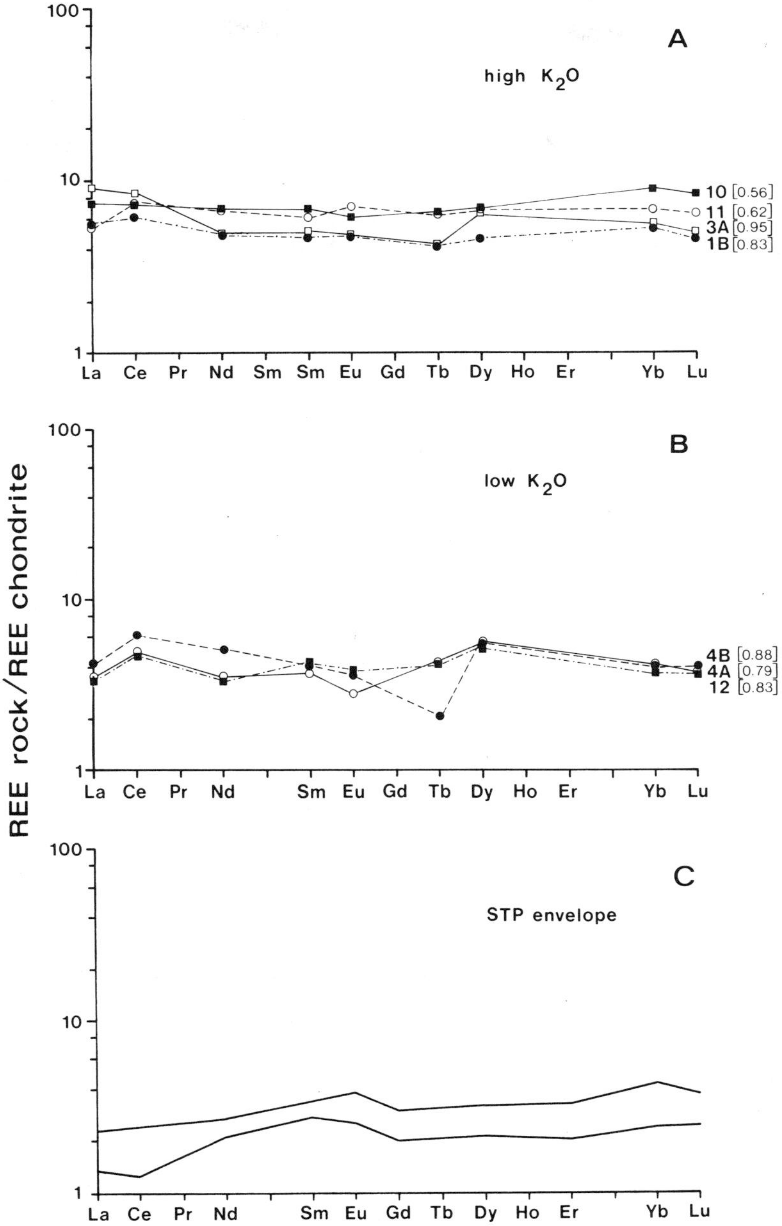

Figure 3

(A) Chondrite normalised rare-earth-element distributions for mineralised rocks of original ultramafic composition, number 12 vein, Hollinger Mine. Bold figures are sample numbers; figures in square brackets are deduced volume factors of alteration. (B) Rare-earth-element patterns for unmineralised counterparts, contiguous with those in A above. (C) Rare-earth-element envelope for relatively fresh komatiites (formerly termed 'peridotitic komatiites' and sometimes termed 'STP' for 'spinifex textured peridotites' (cf. NESBITT and SUN, 1976).

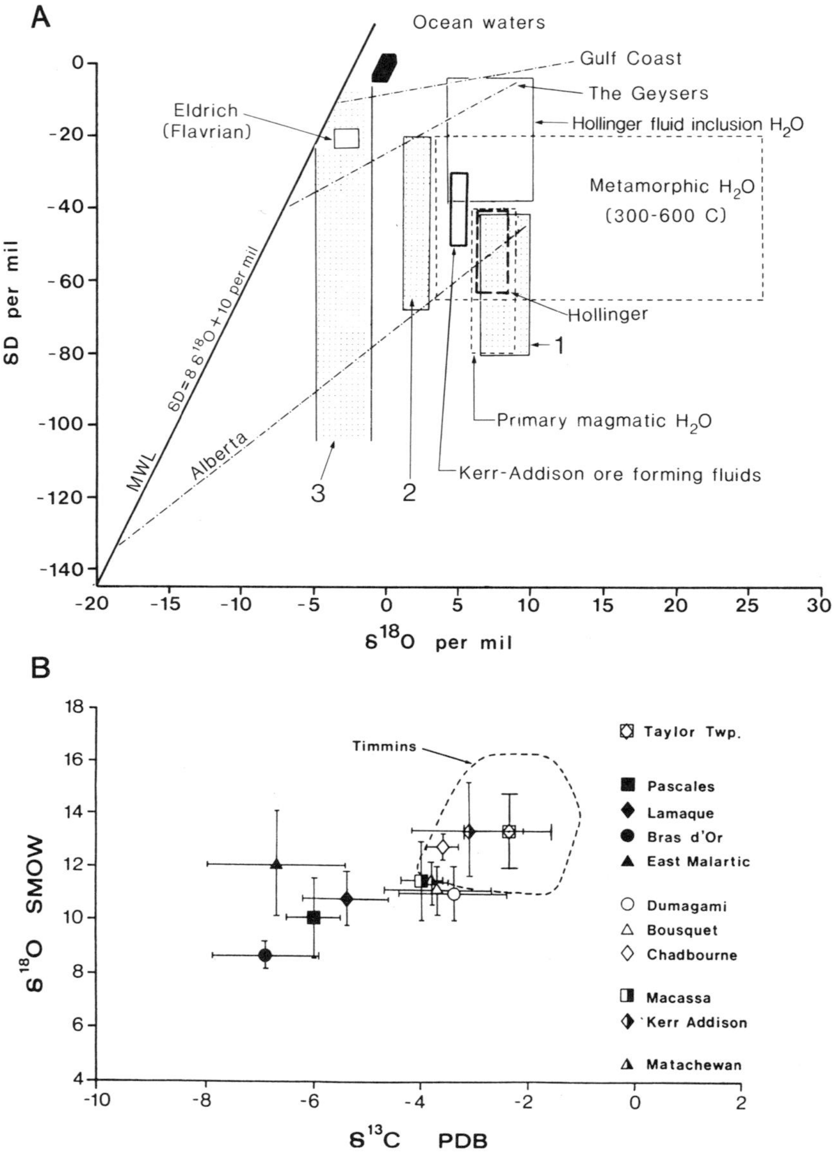

Figure 4

(A) Isotopic composition of natural terrestrial fluid reservoirs in δD versus $\delta^{18}O$ coordinate space. After TAYLOR, 1974. Superimposed are calculated fields for three vein stages at Macassa, one ore solution, two fluids associated with selenite deposition in faults, three late-stage quartz-magnetite stringers; after KERRICH and WATSON, 1984. Hollinger data from work reported, this paper; Kerr Addison data from KISHIDA and KERRICH, 1986; and Eldrich data from KENNEDY, 1985. Data for H_2O decrepitated from quartz veins of the Hollinger mine are from FYON et al., 1984. (B) Fields of different carbonates plotted on $\delta^{13}C$ versus $\delta^{18}O$ coördinate space; field for Timmins is after FYON et al., 1984.

A compilation of oxygen isotope data for hydrothermal silicates from mineralised fault zones is given in Table 1, and calculated fluid isotopic compositions are summarised in Figure 4A. A significant feature of the mineral results is that the $\delta^{18}O$ of hydrothermal quartz, muscovite, and chlorite lie within narrow intervals, signifying uniform hydrothermal conditions. For instance, at the Dome Mine, Timmins, located close to the Porcupine-Destor fault, all of the vein quartz from five different mineralised domains have a restricted range of $\delta^{18}O$ of 14 to 15.2; Δ quartz-muscovite and Δ quartz-chlorite are relatively constant at 3.5 to 4.0 and at 5.5 to 6.1 per mil respectively, corresponding to calculated isotopic temperatures of 270 to 450 °C and a fluid isotopic composition of 7 ± 1 per mil ($1\ \sigma$).

Considering the data for fourteen fault-hosted mineral deposits collectively distributed over distances of several hundred kilometres along the two main fault systems, the overall estimated range of ambient fluid temperatures is 270 to 450 °C, with a fluid isotopic composition given by $\delta^{18}O = 6 \pm 2$ per mil and $\delta D = -50 \pm 20$ per mil; see Figure 4A. The calculated fluid compositions are consistent with the range of most fluids implicated in metamorphism and overlap the magmatic field; see Figure 4A (cf. TAYLOR, 1974). Marine or meteoric waters may undergo isotopic evolution from primary signatures towards δD and $\delta^{18}O$ values regarded as characteristic of magmatic fluids, during high-temperature isotopic exchange with crustal rocks under conditions of low water-to-rock ratio. Fluids that have followed such an exchange trend, however, also tend to become saline, whereas the fluids implicated in fault metasomatism were of low salinity (see below). Accordingly, participation of evolved marine or meteoric waters at this stage of fault-related fluid activity and metasomatism probably may be ruled out. The inferred metamorphic fluids, with a possible magmatic component, are considered to be generated during progressive accumulation and burial of the volcanic-sedimentary sequences and, hence, broadly contemporaneous with volcanic-plutonic-tectonic activity.

The observed narrow interval of 11 to 15 per mil for $\delta^{18}O$ in vein quartz is independent of wall-rock lithology. Quartz in wall rocks is shifted from initial values towards that of vein quartz in proximity to zones of intense metasomatism. These relations signify both an approach to isotopic equilibrium and fluid-dominated conditions along fault structures. A corollary of these results is that lateral diffusion of such chemical components as CO_2, K, and Au from wall rocks into fault zones (cf. BOYLE, 1961, 1979) is incommensurate with the isotopic data.

The ubiquitous presence of hydrothermal carbonates, along with quartz, in mineralised fault zones provides evidence that the fluid reservoir carried significant CO_2, raising the question of the source of the carbon. The oxygen and carbon isotope compositions of hydrothermal carbonates, principally from fault-hosted lode gold deposits, are summarized in Figure 4B. For the majority of data, carbonates occupy a restricted field, where $\delta^{18}O$ spans 10 to 17, and $\delta^{13}C$ -8 to -0.5, per mil.

In general, quartz-carbonate pairs are out of oxygen isotope equilibrium,

Table 1

*Oxygen Isotope Composition of Minerals Separated from Gold-bearing Veins and their Host Rocks in Mines from the Yellowknife Timmins, and Val d'Or Districts, together with Calculated Isotopic Temperatures and $\delta^{18}O$ Fluid**

		$\delta^{18}O$ vein quartz	$\delta^{18}O$ mineral	$\delta^{18}O$ host quartz	Δ quartz-mineral	Temp. (°C)	$\delta^{18}O$ fluid
Yellowknife	Gold-bearing quartz-	12.48		12.06			
District, Campbell	carbonate veins, 2300 in	12.53	6.58 c		5.95	440	8.2
shear	sericite-chlorite schist	11.50	5.97 c		5.53	480	7.9
		11.59	7.03 m		4.56	320	7.5
	4900	11.43		11.29			
	5100	10.80					
Con shear	Gold-bearing quartz-	12.25					
	carbonate veins, 250	11.50	5.97 c		5.53	480	7.9
	in sericite-chlorite schist 1550	12.06		12.13			
Giant shear,	Gold-bearing quartz-	12.28					
east zone	carbonate veins in sericite-	12.77		12.57			
H.G. zone	chlorite schist	12.01					
open pit							
Western	Gold-bearing quartz veins	13.03		13.70			
granodiorite	emplaced into shear zone						
	within graniorite						
Ptarmigan	Gold-gearing quartz veins in	12.60	6.67 c		5.93	440	8.2
	metagreywacke	13.16		13.91			
Surface	Gold-bearing quartz-	13.51	6.70 c		6.81	370	7.6
	carbonate veins in metabasalts	13.56	7.45 c		6.11	420	8.9
Dome Mine,	Banded quartz-muscovite	14.74	10.98 m		3.76		
Timmins	veins in ultramafic schist	14.41	11.28 m	14.66	3.13		
		14.37	10.78 m		3.59	430	9.8
		14.52	10.54 m		3.98	380	8.8
		14.63		15.17			
	Banded quartz-tourmaline	15.07					
	veins	15.24					
	Veins in quartz-feldspar	14.97		15.31			
	porphyry	13.93		13.87			
		14.19	8.05 c	14.26	6.14	420	9.6
	Quartz-tourmaline veins in	15.09	9.21 c		5.88	440	10.8
	dacite	15.17	9.38 c		5.79	450	11.1
	Quartz veins in Timiskaming	14.65		14.10			
	slates	14.52		15.77			
		14.76	11.26 m		3.50	440	10.5
	Quartz-carbonate veins in	14.53	8.99 c	14.84	5.54	480	10.9
	metabasic schists	15.20					
	Quartz-carbonate veins parallel to auriferous stratiform carbonate	14.87	9.31 c	15.32	5.56	480	11.6

Table 1 *continued*

		$\delta^{18}O$ vein quartz	$\delta^{18}O$ mineral	$\delta^{18}O$ host quartz	Δ quartz-mineral	Temp. (°C)	$\delta^{18}O$ fluid
Aumor Mine, Timmins	Quartz veins parallel to auriferous stratiform carbonate	14.65 15.01					
MacIntyre Mine	Quartz-carbonate veins 1526 stope	15.02 14.96	10.62 m	14.86	4.40	340	8.2
	Quartz-carbonate veins, 4675-5 vein	14.81 14.74	10.80 m	14.90	4.01	380	9.2
Paymaster	Quartz-carbonate-chlorite veins; mine dump	14.87 14.89	9.17 c		5.70	460	10.9
	Silicified Au-bearing porphyry	14.80 14.62					
Hollinger	Quartz-carbonate veins in altered ultramafic tuffs	13.5	7.7		5.8	440	9.1
Chadbourne	Quartz-albite-carbonate veins in fractured basalt and andesite	12.0 to 13.5	6.1 to 6.9		5.9 6.8	380 to 450	8.1
			8.1 to 8.9		3.6 to 4.0		
Kerr Addison Mine	Quartz-muscovite-carbonate veins in carbonate sediment	14.48	10.54 m		3.94	390	9.0
	'Flow ore,' albitic-pyrite tuffs and flows	14.4	12.0 ab 6.9 c		4.2	350	8
	'Flow ore', muscovite pyrite tuffs and flows	14.2	10.1 m				
Lamaque	Quartz-tourmaline-carbonate veins Main granodiorite	12.84 12.71 12.79	6.54		6.30	410	7.8
Pascales	Quartz-tourmaline-carbonate veins	11.86 11.58	5.71 c		6.15	420	8.1
Bras d'Or	Quartz-tourmaline-carbonate veins	11.61 12.70	5.61 c		6.00	430	8.2
East Malartic	Albite-orthoclase-quartz-tuffaceous chert	13.4 to 21.7	10.4 to 12.3 (ab)			200	0.1
			14.1 to 16.7 (or)			150	0.1

* Abbreviations: c, chlorite; m, muscovite; ab, albite; or, orthoclase. Isotopic temperatures and fluid $\delta^{18}O$ calculated from the mineral-water fractionation equations reported by CLAYTON et al., (1972) for quartz, O'NEIL and TAYLOR (1967) for feldspars, O'NEIL *et al.* (1969) for muscovite, and WENNER and TAYLOR (1971) for chlorite.

probably because of the susceptibility of carbonates to retrograde oxygen exchange. The carbon isotope compositions define a narrow field, compared to most other hydrothermal systems.

Carbon of -8 to -0.5 per mil could be derived from a number of different carbon reservoirs, given that the average $\delta^{13}C$ of igneous, sedimentary, and metamorphic rocks is about -5 and that of juvenile carbon is -5 ± 2 (OHMOTO and RYE, 1979). KERRICH (1983) and GOLDING and WILSON (1983) have argued on the basis of carbon isotope data and geological considerations that the hydrothermal CO_2 in Archean structurally hosted gold deposits is derived from a mixture of juvenile carbon and marine carbonate with variable proportions of reduced carbon. FYON et al. (1984) propose a mantle origin for the CO_2. A contribution of carbon from the oxidation or hydrolysis of hydrocarbons low in ^{13}C in interflow sediments at source is likely and would account for the most depleted values (Fig. 4B).

A key feature is the provinciality of $\delta^{13}C$ carbonate values (Fig. 4B). Redox-dependent fractionations accompanying the precipitation of carbonates account for, at most, a shift of $+2$ per mil, and some of the most depleted regions are in rocks essentially devoid of carbonaceous material. The observed provinciality is interpreted in terms of lateral heterogeneities in the lower crust supplying CO_2 to structures, represented by variations in the relative proportions of oxidised to reduced carbon, and implies that the faults are tapping very large crustal volumes.

Evidence from fluid inclusions concerning the properties of hydrothermal solutions involved in the fault-zone mineralisation is sparse, owing to pervasive transgranular fracturing and intracrystalline deformation of vein minerals. Some of the fluid-inclusion data is compiled in KERRICH (1983) and COLVINE (1984). Such fluids were of low salinity ($\leqslant 3$ wt-% NaCl equivalent), contained minor quantities of reduced gas species such as CH_4, CO, and COS, and minor N_2, possessed relatively high K/Na (0.1), and were characterised by variable CO_2 contents, probably signifying transient effervescence of CO_2 during decompression.

In the Timmins, Matheson, Kirkland Lake, and Val d'Or regions of the fault systems sulphates occur close to or within the structures. The sulphates, typically anhydrite or selenite, coexist with albite, magnesium-chlorite, calcite, and hematite. On the basis of stable-isotope data and mineralogical considerations these domains have been interpreted as sites for incursion of Archean marine water into submarine volcanic rocks along zones of high permeability, such as faults (KERRICH and HODDER, 1982; KERRICH and FYFE, 1983); ambient temperatures were 150 to 200 °C, and $\delta^{18}O$ fluid was 0 ± 1.5 per mil (Fig. 4A).

A third fluid regime is evident from stable-isotope data in plutonic rocks of the Kirkland Lake and Noranda districts. In these districts quartz-magnetite and quartz-chlorite veins were precipitated at temperatures of 120 to 250 °C from meteoric water of low ^{18}O (Fig. 4A).

Thus, cooling of the volcanic-plutonic complexes was initially dominated by ocean water, in the submarine environment, with a transition to devolatilisation of large crustal volumes and, finally, to meteoric-water infiltration as the Archean volcanic-plutonic terrain emerged above sea level. This scheme is corroborated by the presence of late fluviatile clastic sediments.

2.5 *Age relationships*

A tightly constrained chronological framework exists for volcanic-plutonic activity in the Timmins and Kirkland Lake areas of the Abitibi greenstone belt, based on the classic uranium-lead isotopic age determinations conducted by KROGH and co-workers (Table 2). PERCIVAL and KROGH (1983) summarise the three-stage development of the Abitibi belt as a volcanic-sedimentary stage spanning 2750 to 2696 Ma, a metamorphic-tectonic event at 2700 to 2685 Ma, and late plutonism over the interval 2685 to 2665 Ma. The timing, however, of fluid transport through the major structures and the source of solute components have remained controversial issues.

Existing age determinations on mineralisation in such structures as the Kirkland Lake and Larder Lake fault zone are principally Stacey-Kramers model lead isotopic ages on galenas and leached coexisting pyrite, which define ages of 2656 to 2700 Ma (FRANKLIN et al., 1983). Rb–Sr isochron and incremental $^{40}Ar/^{39}Ar$ spectra yield closure times of 2550 to 2700 Ma; see Figure 5 (KERRICH et al., 1984; KERRICH, 1985). The least disturbed isotopic systems are in good agreement with model lead ages, whereas the younger ones reflect resetting during continued displacement along the major structures into the Proterozoic. For instance, the Kirkland Lake fault zone locally displaces Proterozoic Cobalt Group sediments, which are in the order of 2250 Ma.

According to these results in conjunction with field relations, marine water in-

Table 2

Age Constraints, Abitibi Belt

	Age (Ma)	
Timmins (NUNES and PYKE, 1981)		
Lower supergroup (top)	2725 ± 2	
Upper supergroup (top)	2703 ± 2	
Kidd Creek felsic host	2717 ± 4	
Kirkland Lake (NUNES and JENSEN, 1980)		
Hunter Mine group	2710 ± 2	
Blake River group	2703 ± 2	
Timiskaming		
Noranda		
Lake Dufault granodiorite	2701	
Abitibi batholiths (KROGH et al., 1982)	2675 – 2685	
Matachewan dykes (GATES and HURLEY, 1973)	2690 ± 93	$^{87}Sr/^{86}Sr = 0.700 \pm 0.001$

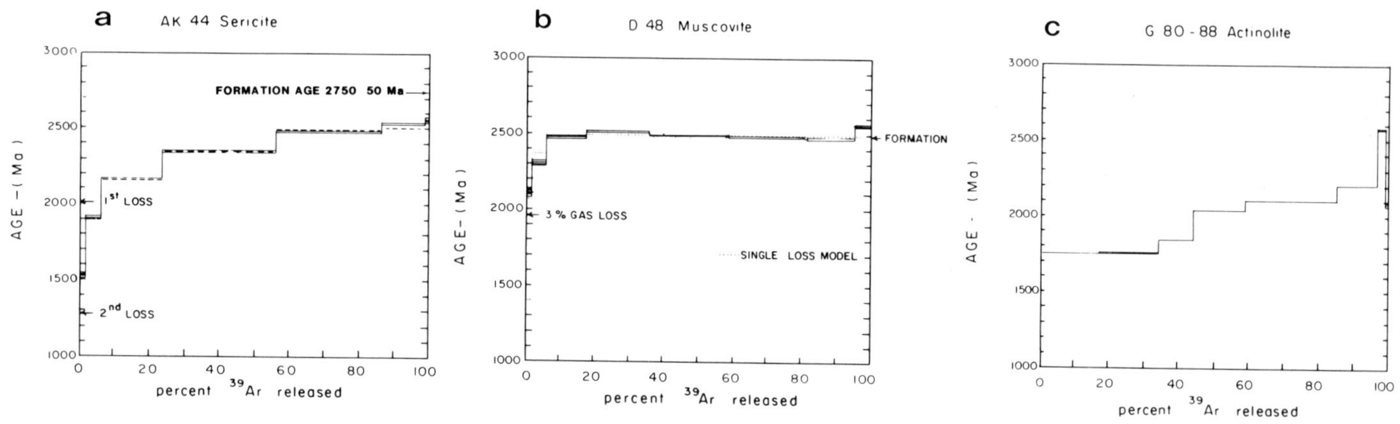

Figure 5

Incremental $^{40}Ar/^{39}Ar$ spectra from fault-related hydrothermal minerals. (a) Kerr Addison muscovite. (b) Dumagami muscovite. (c) Macassa actinolite.

cursion occurred along faults during early extension, but the dominant hydrothermal discharge in Abitibi belt fault zones is interpreted as having occurred during an episode of major fracturing and ductile deformation that accompanied emplacement of granitic batholiths into the supracrustal sequences; hydrothermal mineralisation was thus an integral part of the thermal and mechanical energy of the greenstone belt.

2.6 *Source of hydrothermal solutes*

The origin of fluid reservoirs that infiltrated structural lineaments has been considered above on the basis of light-stable isotope evidence. The source rock reservoir, however, for hydrothermally transported solutes, such as Si, CO_2, K, Rb, Li, Cs, Sr, and rare metals, including Au, Ag, As, Sb, B, Se, Te, Pb, Bi, and W, now present in mineralised faults, is a more subtle problem. The hypothesis of lateral secretion from wall rocks as the origin of mineralisation in fault zones formerly received wide acceptance (cf. BOYLE, 1961, 1979). Alternative hypotheses appeal variously with respect to granitic rocks, sediments, or komatiites as potential donor reservoirs. Herein the question of solute source is explored by means of radioisotope tracers, namely $^{87}Sr/^{86}Sr$, $^{207}Pb/^{204}Pb$, and $^{206}Pb/^{204}Pb$, which may be diagnostic of the specific rock reservoir from which the solutes were derived.

Contemporaneous $^{87}Sr/^{86}Sr$ ratios of the hydrothermally mineralised faults have been estimated from analyses of tourmaline, scheelite, and piemontite, all of which possess $^{87}Rb/^{86}Sr \leqslant 0.004$ and therefore have a negligibly small radiogenic-strontium evolution over the time elapsed since formation (Table 3). Moreover, these minerals, unlike ferrodolomites in the fault zones, appear to have behaved as isotopically closed systems.

If the estimated range of 0.7011 to 0.7040 for initial $^{87}Sr/^{86}Sr$ ratios of the source rock reservoir for hydrothermal solutes has validity, then this reservoir was more radiogenic than the upper mantle at 2690 Ma (0.700 ± 0.001), the Matachewan diabase dyke swarm which transects all major lithological units and structures (2690 ± 50 Ma, 0.701 ± 0.001), or the contemporaneous Abitibi belt volcanics of mafic to ultramafic composition (0.7000 to 0.7012); see Tables 2 and 3 and Figure 6 (FAURE, 1977; GATES and HURLEY, 1973; HART and BROOKS, 1974). An inferred radiogenic character for the source may implicate contributions from sialic basement to the greenstone belt, and/or felsic volcanic rocks of the Lower Supergroup, as well as volatiles released from the mafic to ultramafic volcanic sequence. FRANKLIN et al. (1983) have shown, on the basis of lead isotopic work, that in general fault-hosted gold mineralisation in the Abitibi greenstone belt possesses a characteristically more radiogenic signature and higher μ ($\mu = {}^{238}U/^{204}Pb$) than contemporaneous mafic volcanic rocks.

Thus the deduced Sr and measured lead isotopic characters of the fault mineralisation are more evolved than their contiguous mafic volcanic bounding media, collectively implicating an external, deeper-level, more radiogenic source of these

Table 3

Rb–Sr Isochron and $^{40}Ar/^{39}Ar$ Closure Ages for Mineralised Faults, and Age-corrected Mineral $^{87}Sr/^{86}Sr$

District, Mine	Age (Ma)	Method	^{87}Sr/^{86}Sr, initial	^{87}Sr/^{86}Sr, age-corrected	^{87}Sr/^{86}Sr mineral, age-corrected *
Timmins					
Hollinger	2440 ± 26	Rb–Sr	0.70262 ± 0.0014	0.7015	
	2570 ± 30	^{40}Ar/^{39}Ar (muscovite)			
Parnour					0.7010 t
Paymaster					0.7012 to 0.7015 t
Aunor					0.7014 to 0.7020 t
Dome					0.7015 t, s
Kirkland Lake					
Macassa	2575 ± 50	^{40}Ar/^{39}Ar (actinolite)			0.7012 to 0.7011 a, s
Kerr Addison	2510 ± 50	Rb–Sr	0.70249 ± 0.0009	0.7010	
	2550 ± 40	^{40}Ar/^{39}Ar (muscovite)			
Bousquet					
Dumagami	2420 ± 19	Rb–Sr	0.70406 ± 0.0001	0.7020	
	2550 ± 30	^{40}Ar/^{39}Ar (muscovite)			
Bousquet					0.7017 to 0.7019 t
Val d'Or					
Lamaque					0.7031 ± 0.7041 t
Pascales					0.7007 to 0.7014 t
Bras d'Or					0.7010 to 0.7022 t

* Abbreviations: t, tourmaline; s, scheelite; a, actinolite.

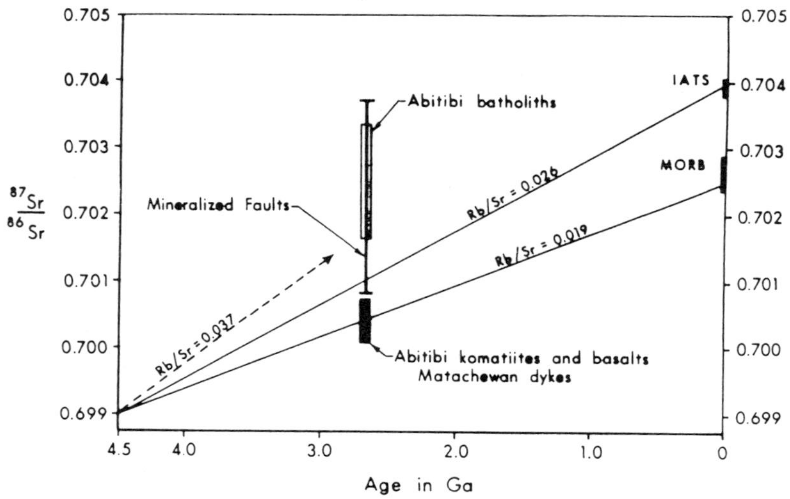

Figure 6
Contemporaneous $^{87}Sr/^{86}Sr$ of mineralised structures, compared with mantle, komatiites, and granodiorite batholiths, for the Abitibi greenstone belt at 2.7 Ga.

components. As a corollory, the lateral-secretion mechanism is ruled out. This inferred deeper source is plausibly the same as that donating Si, CO_2, K, Rb, Li, Cs, Au, Ag, As, Sb, B, etc. Long-range communication to a lower crustal source is corroborated by the presence of the major structures themselves, and also independent estimates of transport distances based on enrichments of rare elements in the structures (KERRICH and FRYER, 1979; KERRICH, 1983). Granulite facies rocks are typically depleted in LIL elements relative to the upper crust. The hydration and LIL-element enrichments observed in the faults may represent the discharged products of coeval devolatisation and depletion of LIL elements of the lower crust.

A distinct provinciality of $^{87}Sr/^{86}Sr$ initial ratios exists (Table 3) and, as in the case of regional variations in $\delta^{13}C$, this is interpreted in terms of large-scale lateral heterogeneity of the lower crust supplying volatiles and solutes to the structures.

2.7 *Sodic and alkaline igneous rocks*

A characteristic feature of the major fault zones, in addition to intense hydrothermal metasomatism, is the sporadically distributed colinear arrays of subvolcanic stocks, along with their extrusive equivalents, of highly sodic or alkaline composition (Fig. 1B). Prominent examples include trondhjemites of the Malartic, Matachewan, and Timmins areas, and alkaline rocks at Kirkland Lake (Fig. 1 A and B).

The sodic rocks are dominated by quartz and oligoclase or albite; they conform in most respects to the definition of a trondhjemite based on major-element oxides, according to BARKER (1979): SiO_2 at $>68\%$, usually $<75\%$, and Al_2O_3 typically at $>15\%$ at 70% SiO_2 and $<14\%$ at 75% SiO_2, and Na_2O at 4.0–5.5% (etc.). For instance, a population of six trondhjemites at Canadian Arrow possesses SiO_2 at $65.9 \pm 1.81\ 1\sigma$, Al_2O_3 at $16.7 \pm 0.22\ 1\sigma$, and Na_2O at $8.32 \pm 1.16\ 1\sigma$ (MCNEIL and KERRICH, 1985). Most such trondhjemites have elevated primary Na_2O contents, but this is incremented during pervasive spilitic alteration in the presence of marine water, under conditions of low temperature, as evidenced by reversed quartz-feldspar fractionations ($\delta^{18}O$ for quartz, 9 to 11, and $\delta^{18}O$ for albite, 10 to 16; FRYER et al., 1979; KERRICH and FRYER, 1979; KERRICH, 1983).

Groups of alkaline igneous intrusions and trachytic tuffs are abundant at Kirkland Lake; formerly termed 'syenites', these rocks are in fact potassic basalts and intrusive equivalents. A general consensus exists that alkaline mafic rocks are the products of low degrees of partial melting of asthenosphere, or 'enriched' asthenosphere, and such fusion products may have a preferred spatial association with linear fracture zones, such as rifts (cf. FITTON, 1985). Along with the alkaline magmas are lamprophyre dykes and domains of fenitisation, highly enriched in large-ion lithophile elements. Representative chondrite normalised diagrams are depicted in Figure 7, (A) to (D); a characteristic feature of the alkaline rocks is the pronounced chondrite-normalised troughs at Ta–Nb, P, and Ti; see Figure 7, (C) and (D). Magmas of trondhjemitic composition are generally regarded as having been formed by a partial

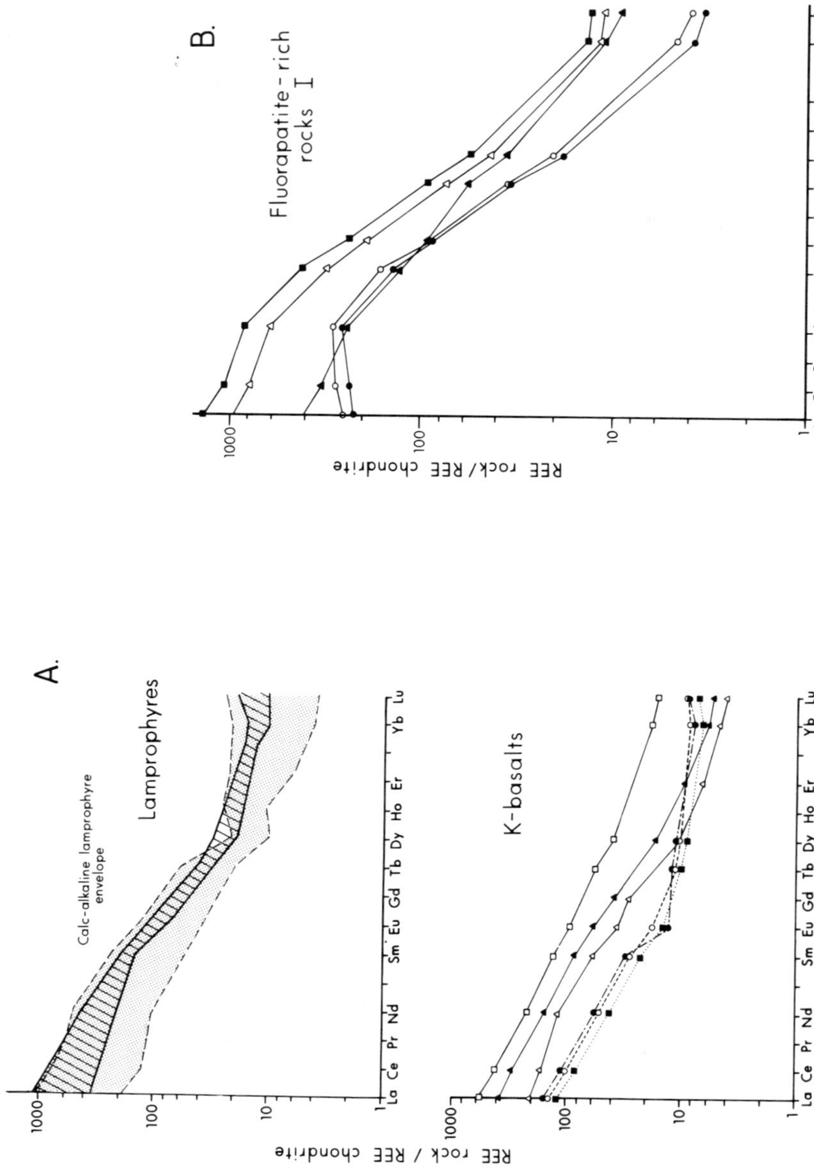
A.
Lamprophyres
Calc-alkaline lamprophyre envelope
K-basalts
B.
Fluorapatite-rich rocks I
REE rock / REE chondrite
1000
100
10
1
La Ce Pr Nd Sm Eu Gd Tb Dy Ho Er Yb Lu

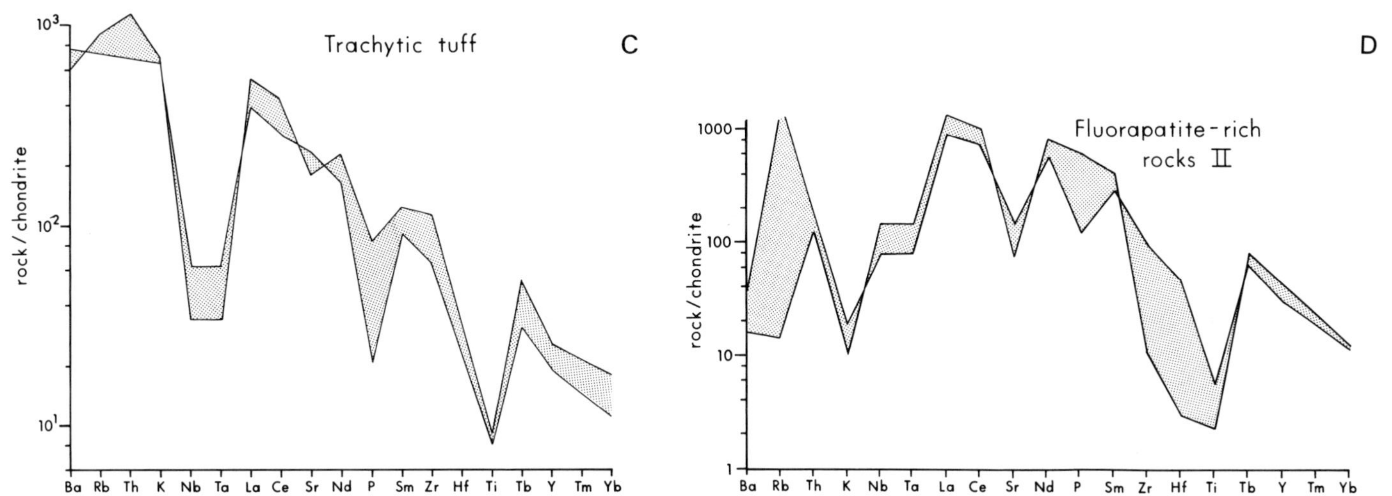

Figure 7

Chondrite normalised element distributions for specified suites of alkaline magmas. (A) Lamprophyres diagram, after McNeil and Kerrich, 1986; alkali-basalts diagram, after Kerrich and Watson, 1984. (B) Fluorapatite-bearing fenitised trondhjemites, from King, 1983; trachytic tuffs (Kirkland Lake) and fluorapatite-rich fenites, after Kerrich and Watson, 1984, and King, 1983, respectively.

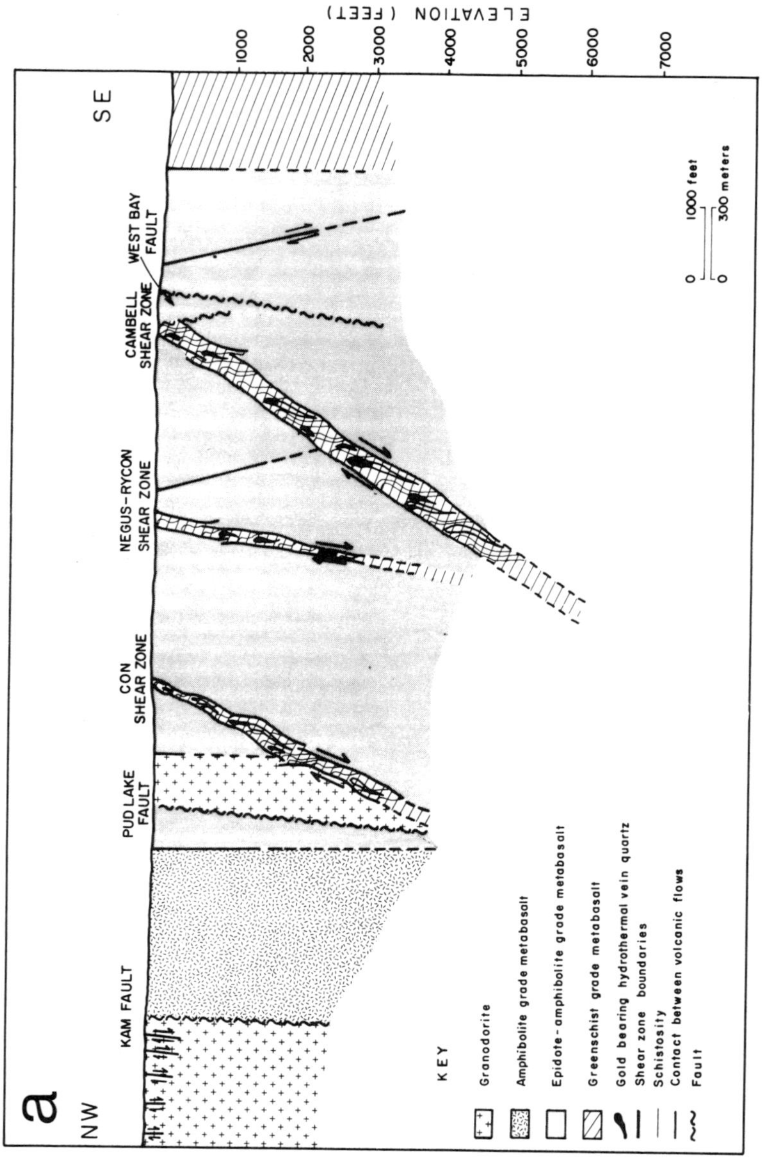
a
NW
SE
KAM FAULT
PUD LAKE FAULT
CON SHEAR ZONE
NEGUS-RYCON SHEAR ZONE
CAMBELL SHEAR ZONE
WEST BAY FAULT
ELEVATION (FEET)
1000
2000
3000
4000
5000
6000
7000
KEY
Granodorite
Amphibolite grade metabasalt
Epidote-amphibolite grade metabasalt
Greenschist grade metabasalt
Gold bearing hydrothermal vein quartz
Shear zone boundaries
Schistosity
Contact between volcanic flows
Fault
0 1000 feet
0 300 meters

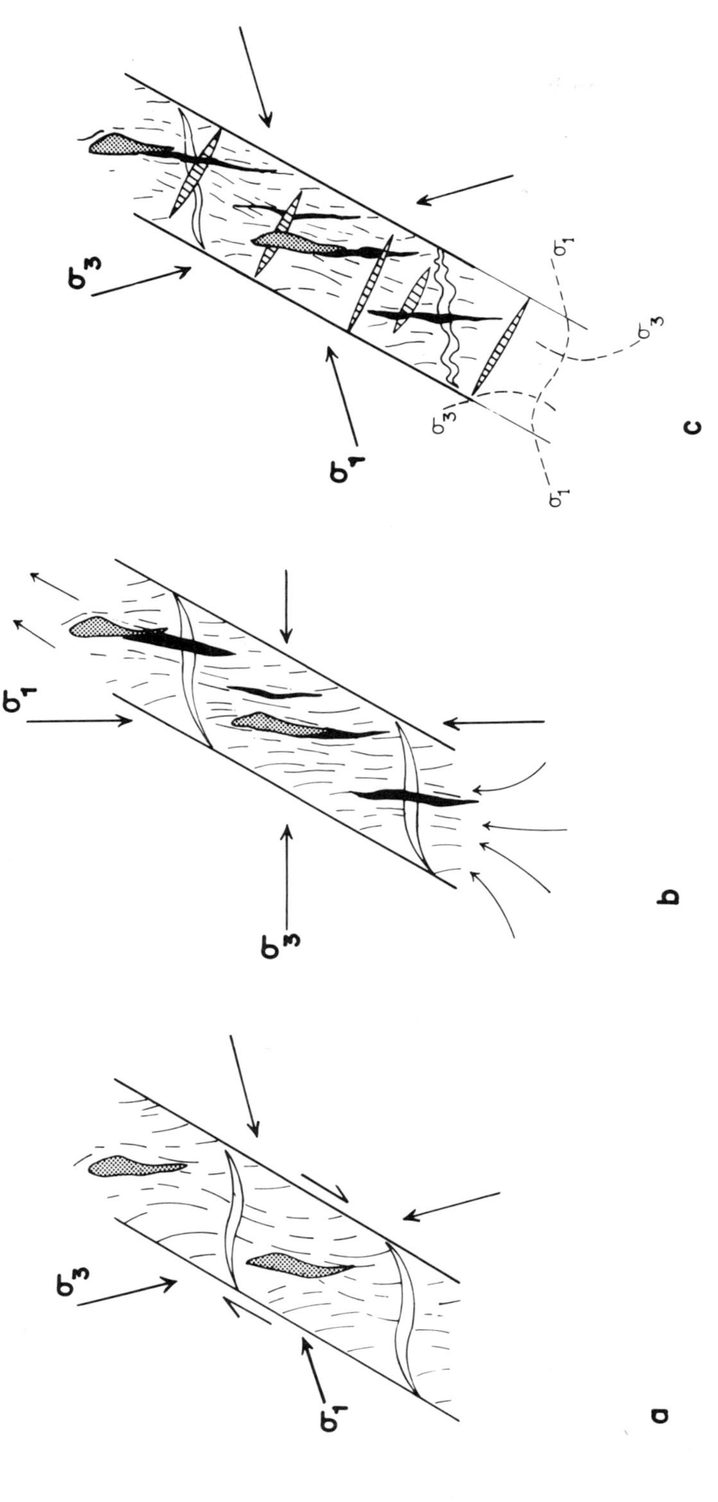

Figure 8

(a) Schematic geological cross section of Yellowknife greenstone belt, illustrating disposition of principal rock types and geometry of brittle–ductile shear zones. (b) Possible switching of stress regimes accompanying hydraulic fracturing and generation of the three sets of vein arrays. After KERRICH and ALLISON, 1978.

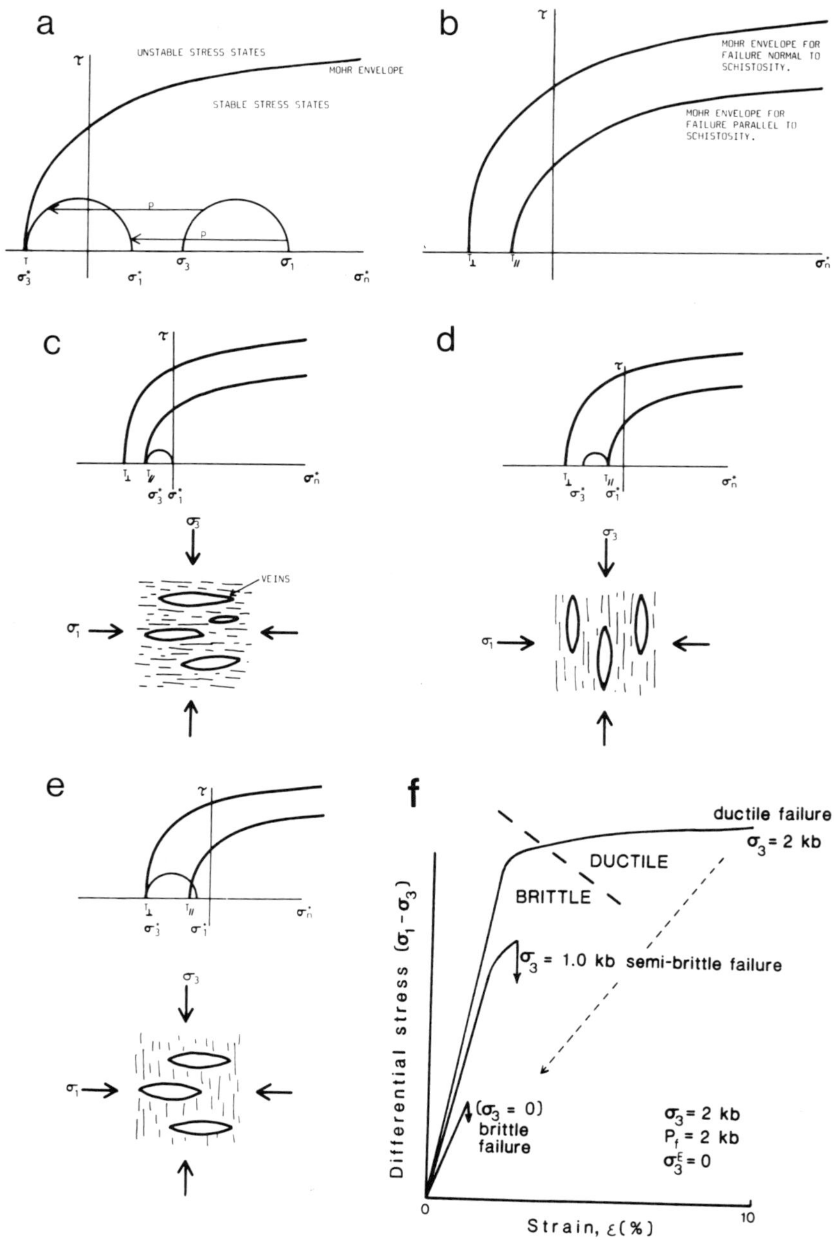

Figure 9

Mohr diagrams illustrating inferred stress conditions for formation by hydraulic fracturing of en-echelon veins of first stage; effective normal stresses $\sigma_n = \sigma_n - p$ versus shear stresses τ. (a) Mohr's stress semicircle at left represents external stress condition; semicircle at right, external stresses modified by fluid pressure p at failure. Mohr envelope is that of Griffith criterion in tensile regime and of Navier-Coulomb criterion in compressive region of diagram; T is tensile strength of undeformed metabasalt. Note that all diagrams are symmetrical across the $\tau = 0$ line. (b) Failure parallel and normal to schistosity: $T_\perp$ and $T_\parallel$ are melting of mafic granulites at the base of the greenstone belt sequence (COLLERSON and FRYER, 1978; BARKER et al., 1979).

The spatial association of trondhjemites and alkaline magmas with major linear structures is interpreted in terms of translithospheric fractures, which tap magmas of sodic composition generated at the base of the crust, and deep asthenospheric partial melts of alkaline composition. Such magmatic activity is broadly coeval with venting of hydrothermal fluids, the latter probably from a crustal reservoir; thus magmatism, tectonic activity, and fluid motion are all linked to a thermal anomaly.

2.8 Hydraulic conditions during vein propagation

One of the fundamental questions pertaining to mineralised Archean faults is the hydraulic regime of the fluid system. Are the geothermal fluids undergoing thermally driven convective circulation under approximately hydrostatic conditions and in equilibrium with surface waters or, alternatively, do the extensive vein stockworks form by hydraulic fracturing in the presence of geopressurised aqueous reservoirs? This question has been addressed by KERRICH and ALLISON (1978) and ALLISON and KERRICH (1979) with reference to Yellowknife, and by ROBERT and BROWN (1984) for the Sigma Mine, Val d'Or, where orientations of the principal stresses and ambient fluid pressures during vein propagation may be deduced, given the well-constrained geometry of shear zones and the vein arrays they host.

A three-stage sequence of veining has been recognised in major brittle–ductile shear zones of the Yellowknife greenstone belt (KERRICH and ALLISON, 1978; ALLISON and KERRICH, 1979), as follows.

1. Formation of the major shear zones, with concomitant development of the early en-echelon barren veins.
2. Development of the massive to banded gold-quartz veins approximately parallel to the schistosity, in the shear zones.
3. Formation of late 'ladder' vein systems, which are also locally auriferous, at high angles to the schistosity within the shear zones.

---

uniaxial tensile strengths of schist. (c) Hydraulic fracturing parallel to schistosity when it is oriented normal to the σ_3 direction. This is one limiting case; the other is shown in Figure 8. (d) Hydraulic fracturing across a plane perpendicular to schistosity when σ_1 direction is normal to this plane; this will occur only if $\sigma_1 - \sigma_1 \leqslant T_\perp - T_\parallel$. (e) Inferred stress conditions for development of late 'ladder' veins: differential stress exceeds difference in strengths of schist parallel and perpendicular to schistosity. After ALLISON and KERRICH, 1979. (f) Generalised rheological behaviour of rock, illustrated in differential stress, $\sigma_1 - \sigma_3$, versus strain, coördinate space. Note transition from brittle to ductile behaviour at elevated values of confining stress σ_3; the transition can be induced by transient increases of fluid pressure which act to diminish effective confining stress, $\sigma_3 - P_{\text{fluid}}$.

This sequence is illustrated in Figure 8b. The origin of each vein system will now be considered in turn.

The veins at Yellowknife were formed by the infilling of tensile fractures. Close to the Earth's surface tensile fractures may be formed, but at depth the increase of confining pressure σ_3 induces propagation of shear fractures. Under conditions of high fluid pressure P_f, however, tensile failure by hydraulic fracturing may occur in the crust at depths in excess of 30 km. The principles involved are discussed by FYFE et al. (1978). The stress conditions under which hydraulic fracturing can occur may be represented by

$$\sigma_3 - P_f \leqslant T$$
$$\sigma_1 - \sigma_3 \leqslant -4T$$

where σ_1 and σ_3 are the maximum and minimum principal stresses respectively and T is the tensile strength of the rock. The first inequality represents the condition where the fluid pressure exceeds the confining pressure by an amount equal to or greater than the tensile strength of the rock. The second inequality is the condition for suppression of the shear failure mode. These stress conditions may be portrayed on a Mohr diagram (Fig. 9a).

The shear zones are characterised by an intense schistosity, the mechanical properties of which are highly anisotropic. In particular, the tensile strength in a direction normal to the schistosity, $T_\perp$, will be low compared with that parallel to the schistosity, $T_\parallel$. On a Mohr diagram the envelope for failure parallel to the schistosity will appear at lower stresses than the envelope for failure normal to the schistosity (Fig. 9b). As the pore-fluid pressure increases, however, and the effective stresses are reduced until σ_3 intersects the Mohr envelope, failure will occur only if the schistosity is oriented normal to the direction of σ_3 (Fig. 9c). If the schistosity is oriented orthogonal to the σ_1 direction, then failure along the schistosity will occur when the value of σ_1 intersects the Mohr envelope, as long as σ_3 is less than the tensile strength of the rock normal to the schistosity (Fig. 9d). It is thus possible to achieve hydraulic fracturing within the shear zone parallel to the schistosity with the same external stress regime that caused the initial development of the shear zones, and the switch in orientation of stresses, postulated by KERRICH and ALLISON (1978) and illustrated in Figure 8b, is plausible, but not necessary. The stress conditions that must be satisfied for these veins to develop are the following.

$$\sigma_3 - P_f \leqslant T_\parallel)$$
$$\sigma_1 - \sigma_3 < -(T_\perp - T_\parallel)$$
$$\sigma_1 - \sigma_3 \leqslant -4T_\parallel$$

The latest system of fractures occurs as an array of 'ladder' veins oriented approximately normal to the schistosity and dipping at shallow angles towards the east. These veins cross-cut those of the second stage, which now have a pinch and swell structure as a result of progressive incremental deformation in the shear zones.

The stress conditions for this late stage of hydraulic fracturing may be formulated as

$$\sigma_3 - P_f \leqslant T$$
$$\sigma_1 - \sigma_3 < -4T$$
$$\sigma_1 - \sigma_3 < -(T_\perp - T_\parallel)$$

and are portrayed on the Mohr diagram as in Figure 9e.

The geometry of the early veins and the schistosity within the shear zones accords well with the concept of shear-zone development in rocks showing material behaviour that is transitional between brittle and ductile.

The geochemical evidence indicates local redistribution of material via migration in a fluid phase and the absence of any significant mineralisation during the initial development of the ductile shear-zone structures. The second-stage veins have a conspicuous banding and often contain slivers of schist oriented parallel to the vein margins. Later intracrystalline deformation has largely obliterated the original quartz microstructure, but the banded nature of the veins and the volumes of hydrothermal fluids involved suggest that the veins were formed by repeated hydraulic fracturing and infilling by quartz precipitated from the hydrothermal fluids. One may envisage a repeated cycle of buildup of a fluid reservoir accompanied by elevation of fluid pressure to a point sufficient to induce hydraulic fracture, leading to discharge of the geopressurised reservoir, with a resultant volume increase and a consequent reduction of fluid pressure in the source region. Repeated episodes of hydraulic fracturing under conditions of high P_f have cycled the effective confining stress ($\sigma_3 - P_f$) such that the rocks have in turn cycled through the brittle ($\sigma_3 - P_f$, low) to ductile ($\sigma_3 - P_f$, high) transition (Fig. 9f).

Mechanical anisotropy of schists within the shear zones will cause refraction of the stress, and hydraulic fracturing will tend therefore to form parallel or normal to the fabric as in the second and third stages of vein development (Fig. 8 and 9).

Development of the second-stage veins requires fluid pressure to exceed the maximum principal stress by an amount equal to the tensile strength. This will occur if the fluid pressure has a value close to the geostatic pressure for a given depth. Hydraulic fracturing will permit upward movement of this fluid, the pressure of which is equivalent to the geostatic pressure at greater depth minus the hydraulic gradient, and which, we suggest, is capable of exceeding the maximum principal stress at higher levels (Fig. 8). The hydraulic fractures will therefore propagate from depth towards the surface.

These structural considerations apply equally to many other examples of the fault-hosted mineralisation discussed above, which also appear to have formed under conditions of hydraulic fracturing induced by geopressurised hydrothermal reservoirs.

3. *Grenville front*

The Grenville front, near Coniston, Ontario, is defined by the Grenville front boundary fault and associated mylonite zones. These structures collectively form the demarcation between low-grade metasediments of the southern province and high-grade Grenville gneisses (La Tour, 1981). Translation on the fault zone is thought to be of Grenville age ($\approx$1100 Ma), although the magnitude and direction of displacement are not precisely known. Bordering these structures immediately to the north is a major fault, the Wanapitae.

La Tour (1981a) has distinguished two discrete and distinct mylonite zones at the Grenville front, designated MZI and MZII respectively. Metamorphic reactions associated with MZII are prograde in nature. On the basis of garnet-biotite geothermometry the ambient temperature of deformation is estimated at 600 °C, and oxygen isotope fractionations (Δquartz-muscovite is 2.6 to 2.8 per mil) yields temperatures of 580 to 640 °C, in close accord with that given above (Kerrich et al., 1984). The $\delta^{18}O$ of quartz in mylonitic schists is uniform at 11.0 to 11.8; syntectonic quartz veins have $\delta^{18}O$ values within the same range as quartz in the host rock, signifying a growth of veins in isotopic equilibrium with schists, probably from the same fluid. Fluids present during syntectonic veining and deformation have a calculated $\delta^{18}O$ of 9.6 ± 0.6.

Metamorphic reactions in mylonite zone MZI are exclusively retrograde in nature, and garnet amphibole thermometry indicates that deformation occurred at temperatures below 540 °C (La Tour, 1981). Oxygen isotope data for MZI yield temperature estimates of 420 to 490 °C and fluid isotopic compositions of 7.0 ± 0.6 per mil. Fluids present during deformation along MZII and MZI were probably of metamorphic origin, indigenous to the immediate host rocks.

Rock units transected by the Wanapitae fault have undergone extensive brittle fracturing, accompanied by local quartz-albite-carbonate veining. Oxygen isotope data on cataclastically deformed metadiabase yields temperature estimates of 280 ± 40 °C, and a fluid $\delta^{18}O$ of $+3.2$. Fluid involvement in the Wanapitae fault was at relatively low temperatures, possibly from a reservoir of formation brines (Kerrich et al., 1984).

An array of quartz-hematite veins ($\delta^{18}O$ for quartz of -0.8 to -1.3) occupies fracture arrays superimposed on the mylonitic fabric of MZI. Quartz crystallised at temperatures of 200 to 350 °C in the presence of fluids of low ^{18}O with an isotopic composition of -8 (350 °C) to -14 (200 °C) per mil. Continental meteoric water is the only terrestrial fluid reservoir with negative $\delta^{18}O$ (Taylor, 1974). The fluid $\delta^{18}O$ values calculated above are commensurate with that of precipitation on high mountain ranges at mid to high latitudes, such as the Sierra Nevada, Rocky, or Andean mountains (cf. Taylor, 1974). This episode of fracturing is interpreted as having) resulted from the infiltration of waters with low ^{18}O derived from precipitation on a high mountain range, down a brittle fault zone. The mountain range could have been induced by continental collision, with deformation accompanying crustal thickening accommodated at deep levels by translation on the high-temperature mylonite zone MZII, followed by isostatic rebound on the low-temperature zone MZI and the associated Grenville front boundary fault (LA TOUR, 1981; KERRICH et al., 1984). This scheme could account for the tectonic juxtaposition of high-grade gneisses that had been deformed in a ductile mode in the presence of metamorphic fluids at elevated temperatures and low-grade metasediments that had experienced brittle fracturing during the incursion of an ^{18}O-depleted surface reservoir.

4. Shear-zone-hosted uranium in overthrust Archean gneisses, Bahia, Brazil

Ductile shear zones transecting high-grade Archean basement gneisses over a band 40 by 100 km near Lagoa Real, State of Bahia, Brazil, are host to extensive uranium mineralisation. The mineralisation is associated with retrograde metamorphism and extensive sodium metasomatism of host felsic gneissic basement rocks (LOBATO et al., 1983, a, b, c). Preliminary oxygen isotope studies of the uranium deposits together with their host gneisses revealed that the uraninite was precipitated from hydrothermal solutions of meteoric origin ($\delta^{18}O \leqslant -2$ per mil) at temperatures of ≈ 550 °C; see Figure 10A (LOBATO et al., 1983, a, b, c).

A model in which amphibolite and granulite facies basement gneisses were transported over Proterozoic greenschist facies metasediments of the Espinhaço sequence along a thrust surface has been proposed by those authors (LOBATO et al., 1983, a, b, c). Such a mechanism involving tectonic emplacement of portions of hot basement over colder, wet rocks would induce dehydration and overpressure in meteoric-water-charged aquifers, causing the high fluid pressures necessary to maintain thrust motion (cf. FYFE et al., 1978).

Calculated or measured $\delta^{18}O$ and δD values of hydrothermal solutions implicated in the mineralisation can provide information on the original fluid reservoir. Identifying the rock reservoir(s) that donated metals to any specified deposit, however, is a more subtle and less tractable problem. The question of uranium provenance for the shear-zone-hosted uraninite deposits has been addressed by means of Rb/Sr isotopic tracer studies. Coenrichment in strontium and uranium during hydrothermal alteration can be deduced on the basis of mineralogical and textural relations and chemical mass balance in the shear zones, but this does not necessarily imply that uranium and strontium have been leached from a common rock reservoir. Despite this, strontium isotope tracer studies can constrain the possible rock reservoirs with which the hydrothermal solutions equilibrated.

Four possible strontium donor reservoirs may be envisaged: Archean gneisses, Proterozoic rocks, seawater, and alkalic intrusive rocks. The Archean gneisses have a Rb/Sr isochron age of 2.86 Ga (CORDANI, 1982) and an elevated Rb/Sr ratio and

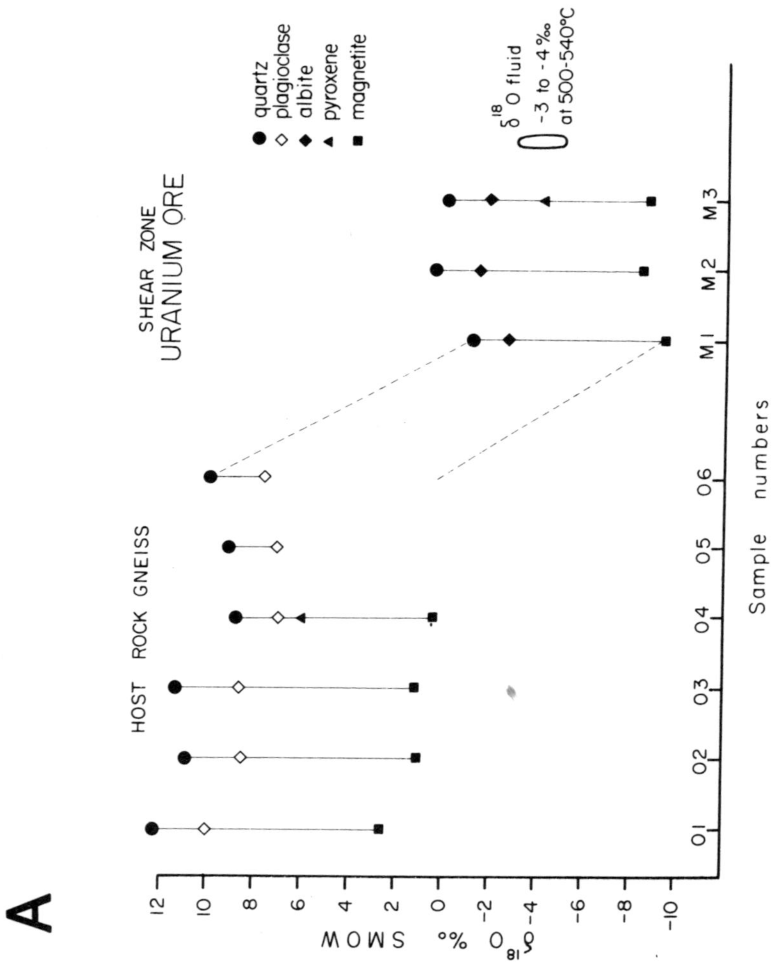
A
HOST ROCK GNEISS
SHEAR ZONE
URANIUM ORE
quartz
plagioclase
albite
pyroxene
magnetite
δ18O fluid
-3 to -4‰
at 500-540°C
δ18O ‰ SMOW
12
10
8
6
4
2
0
-2
-4
-6
-8
-10
01
02
03
04
05
06
M1
M2
M3
Sample numbers

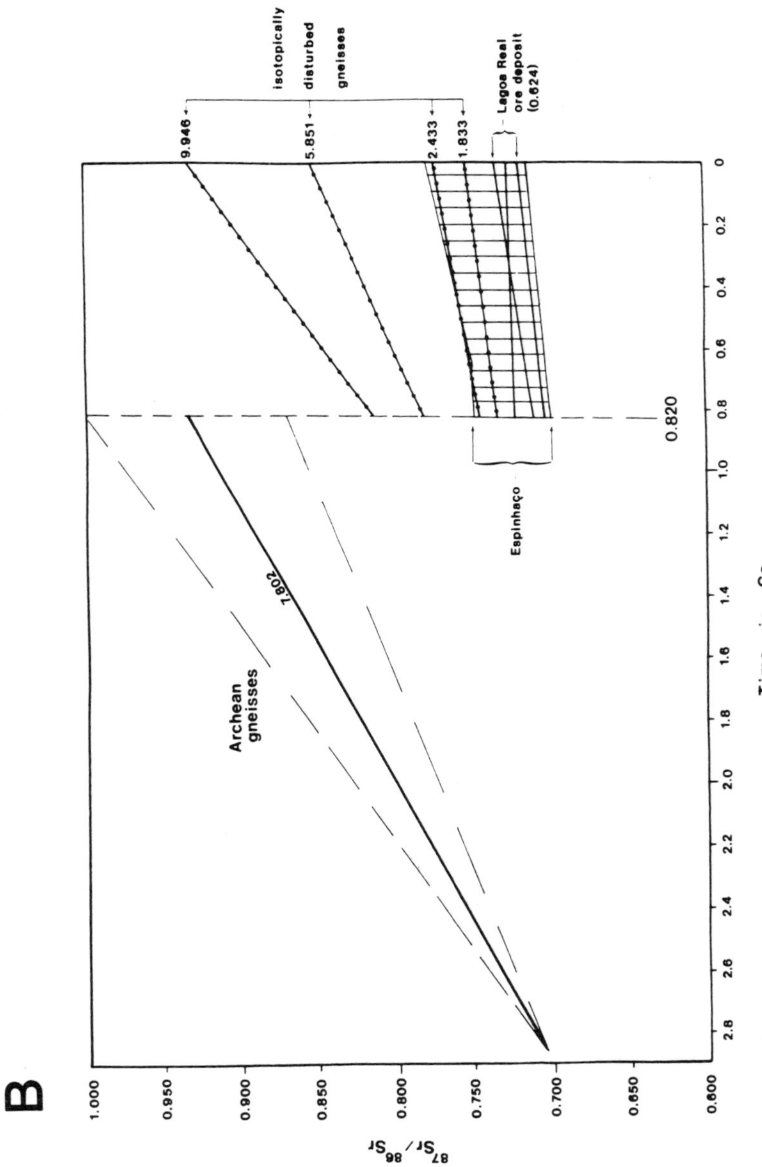

Figure 10

(A) The $\delta^{18}O$ of specified minerals in shear zones and wall rocks, Lagoa Real, Brazil. After LOBATO et al., 1983, b. (B) The $^{87}Sr/^{86}Sr$ of specified rock reservoirs at Lagoa Real, Brazil. After KERRICH et al., 1986. See text for explanation.

therefore have evolved along a steep $^{87}Sr/^{86}Sr$ trajectory through time. The Espinhaço sequence is substantially younger, with a Rb/Sr isochron age of 1.0 Ga for volcanic rocks of the Boquira Formation. Hydrothermal alteration of deformed gneisses accompanying uraninite deposition resulted in an almost total loss of rubidium with concomitant gains of strontium (LOBATO et al., 1983c). Thus the Rb/Sr ratios in the deposit and its altered host rock are <0.01, such that little ^{87}Sr has evolved.

Establishing the source of strontium in the uranium deposit requires a knowledge of the mean $^{87}Sr/^{86}Sr$ ratio of each of the possible donor reservoirs at the time of uranium introduction. The best current estimate of the age of uranium mineralisation is 820 Ma, according to $^{207}Pb/^{206}Pb$ age determination (STEIN et al., 1980). An average value for the $^{87}Sr/^{86}Sr$ ratio of the gneisses at 820 Ma can be estimated from isochron data (CORDANI, 1982). From the average $^{87}Rb/^{86}Sr$ (7.8), the initial $^{87}Sr/^{86}Sr$ ratio at the time of last strontium isotope homogenisation (0.705 at 2.86 Ga), and the elapsed time to mineralisation, a mean $^{87}Sr/^{86}Sr$ of 0.935 at 820 Ma is obtained. This assumes that the samples used in constructing the isochron are representative of the gneisses as a whole. An array for the strontium isotope evolution of the gneisses is portrayed in Figure 10B, calculated from the mean $^{87}Rb/^{86}Sr$ and Rb/Sr $\pm$ 2 of the mean (KERRICH et al., 1986).

A second approach to estimating the $^{87}Sr/^{86}Sr$ of the gneisses at 820 Ma is by means of a suite of samples collected from the vicinity of orebodies but external to the shear zones. Back-extrapolation of analytically determined $^{87}Sr/^{86}Sr$ ratios, with the use of measured $^{87}Rb/^{86}Sr$, gives a mean of 0.770 at 820 Ma. This second estimate is significantly lower than that calculated from the isochron data and is interpreted as reflecting a lowering of the $^{87}Sr/^{86}Sr$ ratio of the gneisses by an infiltration of hydrothermal fluids with relatively low $^{87}Sr/^{86}Sr$ during uranium mineralisation; see Figure 10B (KERRICH et al., 1986).

At 820 Ma the Proterozoic Espinhaço had a calculated range in $^{87}Sr/^{86}Sr$ of 0.700 to 0.740 and a mean of 0.732. Though slightly higher, this ratio is close to that of the uranium deposit (Fig. 10B). Contemporaneous marine water was 0.708 (VEIZER and COMPSTON, 1974). Measured $^{87}Sr/^{86}Sr$ ratios of the ore deposit are uniform, in keeping with their low Rb/Sr ratios. The deposit is projected to have been characterised by an $^{87}Sr/^{86}Sr$ of 0.719 at 820 Ma. Strontium evolution arrays are depicted for all of the rock reservoirs considered in Figure 10B. From these data it is clear that at 820 Ma the Proterozoic Espinhaço had an $^{87}Sr/^{86}Sr$ ratio close to that of the uranium deposit.

Other possible rock reservoirs cannot be ruled out as strontium donors, and mixtures of strontium from seawater and the gneisses remains an alternative explanation. If this is taken in conjunction, however, with constraints imposed by field relations, mineralogy, and $^{18}O/^{16}O$ data, the conclusion is that strontium in the orebody was originally indigenous to the Proterozoic Espinhaço, having been transported to the site of deposition by high-temperature hydrothermal fluids originating as formation brines in the Espinhaço and discharged through shear zones in response to the thrust mechanism, as discussed by LOBATO et al. (1983, a, b). Estimates of the fluid volume discharged are ≈ 1000 km^3 on the basis of considerations of uranium solubility.

5. *Metamorphic core-complex detachment faults*

Metamorphic core complexes constitute an unique structural entity in the tectonic architecture of the North American Cordillera. They are broadly domal physiographic features, characterised geologically by a basal core of intensely deformed metamorphic and/or plutonic rocks, grading upwards through mylonitic equivalents to low-angle planar breccia zones delimiting a detachment fault. This 'infrastructure', or lower plate, is overlain by a 'suprastructure', or upper plate, composed of listrically faulted, allochthonous, and generally unmetamorphosed Tertiary rocks (for an overview see (CONEY, 1980). Twenty-five such core complexes have been identified to date, discontinuously arrayed along the axis of the North American Cordillera (CONEY, 1980).

Chloritic breccias, characterised by pervasive transgranular fracturing, propylitic alteration, and hydraulic fractures occupied by veins, superimposed on the mylonitic rocks, are a prominent feature of metamorphic core complexes. Breccia zones are coplanar with detachment faults, which are also the sites for local Cu–Au mineralisation, with associated vein networks bearing hematite and manganese oxide. These features raise questions as to the nature of the fluid regimes responsible for metasomatic alteration and fracturing along the plane of detachment.

The Picacho Mountains are located near Phoenix in southern Arizona and constitute the southerly extension of the array of metamorphic core complexes aligned along the North American Cordillera (CRITTENDEN et al., 1980). Geological descriptions of the area are given by several authors, most recently by REHRIG and KEITH (1984). The geological relations of the Picacho area resemble those of a typical metamorphic core complex; see Figure 11 (cf. CRITTENDEN et al., 1980). The key geological components are three tectonic plates, or thrust blocks, separated by two intervening detachment thrusts. Undeformed to mylonitic granite of the lower plate is demarked by a lower detachment thrust, underlain by chloritic breccia. Variably altered and fractured granitic rocks of the middle plate are bounded by an upper detachment zone, upon which allochthonous volcanic and sedimentary rocks of Miocene age lie (Fig. 11). The volcanics have experienced intensive low-temperature oxidative alteration, such that the dominant secondary minerals are K-feldspar, calcite, hematite, and manganese oxides.

At the Picacho tectonic section the salient isotopic features are an upwards trend of whole-rock δ^{18}O from 7.6 to 18.9 in conjunction with a decrease of estimated ambient temperature from 550 to 150 °C (Fig. 11, A, B). In undeformed Oracle

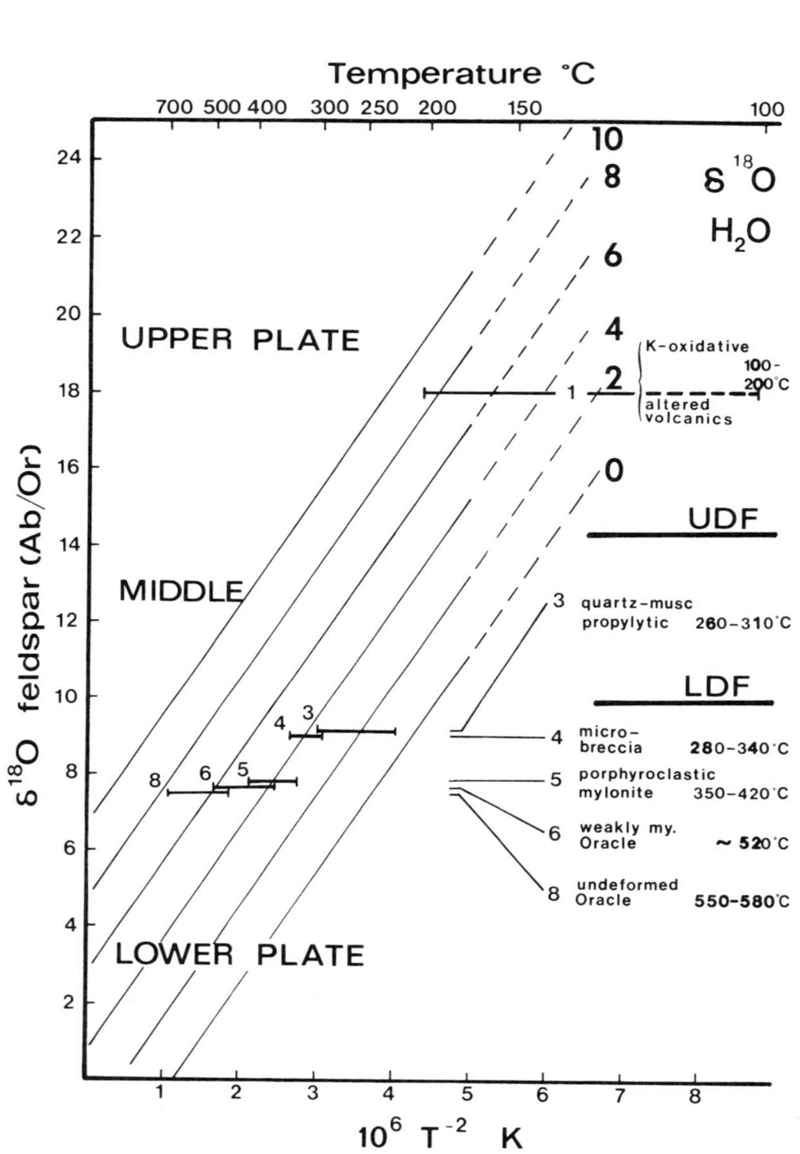

granite of the lower-plate quartz ($\delta^{18}O$, 9.1 to 10.8 per mil), feldspar (7.5 to 8.2) and hornblende (4.9 to 5.6) possess near-magmatic isotopic fractionations. The calculated isotopic temperatures are 500 to 600 °C. A calculated $\delta^{18}O$ of H_2O in equilibrium with feldspars of about +7 is appropriate for magmatic fluids of granites with normal $\delta^{18}O$ (Fig. 11A).

Mylonitic counterparts of the Oracle granite within the lower plate have a small

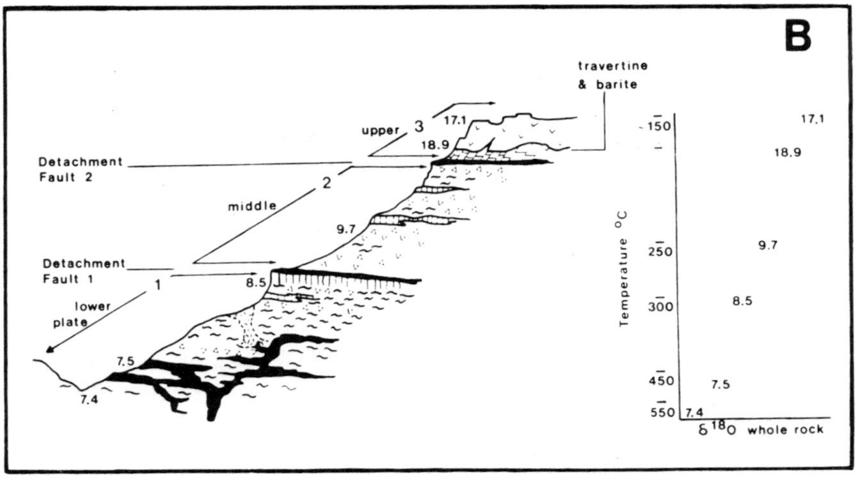

Figure 11

(A) The ^{18}O of feldspar versus temperature, estimated on basis of mineralogical criteria and oxygen isotope fractionations for Picacho section. Family of lines represents $\delta^{18}O$ of water in equilibrium with albite or K-feldspar of specified $\delta^{18}O$ at given temperatures. (B) Tectonic cross section with measured $\delta^{18}O$ of whole rock. Note increase of $\delta^{18}O$ up section. After KERRICH and REHRIG, 1986.

but systematic ^{18}O shift in quartz relative to their undeformed precursors; accordingly, the quartz-feldspar fractionations are larger. The calculated isotopic temperatures are 350 to 500 °C; these are regarded as a dynamic and metamorphic overprint of the granitic protolith and pertain to the ambient thermal conditions of mylonite formation. Whole-rock oxygen isotope compositions of the mylonitic rocks are identical with undeformed Oracle granite, within the limits of analytical uncertainty. Hence the mylonites probably developed under conditions of low water-to-rock ratio in the presence of fluids derived from a breakdown of hydrous minerals in the granitic precursor. The $\delta^{18}O$ of these fluids is calculated at 3 to 6 per mil, and they probably are predominantly of metamorphic or magmatic origin or evolved formation brines, or all three reservoirs.

Brecciated mylonite, which underlies the lower detachment surface, has a quartz $\delta^{18}O$ of 11.6 and thus has experienced a positive ^{18}O shift of +2.3 per mil relative to quartz in the underlying ductile mylonitic precursors of the breccia. Fractionations amongst quartz, feldspar, and chlorite correspond to isotopic temperatures of 300 to 350 °C. The positive shift in quartz is accompanied by a whole-rock shift of +1 per mil relative to mylonitic and pristine Oracle granite; these relations require that the breccia has experienced fluid fluxing, in order to account for the measured isotopic excursion. Hydrothermal fluids in equilibrium with the chlorite breccia have a calculated $\delta^{18}O$ of 4 ± 1. Thus the breccia marks a structural-hydrological domain representing the boundary from a lower zone of low water-to-rock ratio to an upper zone of high fluid fluxes.

Veined and hydrothermally altered rocks of the middle plate have elevated quartz $\delta^{18}O$ (12 per mil) and whole-rock values (9.7 per mil) relative to the underlying plate and thus define a progressive trend of ^{18}O enrichment up the structural section. The temperatures of alteration are estimated to be 220 to 330 °C in the presence of a fluid $\delta^{18}O$ of 2 to 5 per mil.

The temperatures of hydrothermal alteration in volcanic rocks of the upper plate cannot readily be estimated. Assuming a span of 100 to 200 °C, fluids of -3 to $+8$ per mil are implicated in the oxidative event that produced K-feldspar, illite, hematite, carbonate and manganese oxides.

In summary, upwards in the tectonic section temperatures progressively diminish and the whole-rock $\delta^{18}O$ increases dramatically. These relations are interpreted as signifying dual fluid regimes: expulsion of high temperature, crust-equilibrated and reduced fluids of metamorphic, magmatic, or formation origin from lower structural levels, through the middle plate, accompanied by cooling, in contrast to cool, oxidising thermal waters into the upper plate. Thus the tectonic section records an upward transition from high to low ambient temperature, low to elevated water-to-rock ratio, ductile to brittle deformation, and a fluctuating interface of deep aqueous reservoirs at lithostatic pressure, with shallow surface ones under hydrostatic conditions.

6. *Summary and conclusions*

In general, fluid regimes in the fault and shear zones studied follow a sequence, from conditions of high temperature and pressure with locally derived fluids at low water-to-rock ratios, during initiation of the structures, to high fluid fluxes along discrete conduits, as the structures propagate and develop large-scale permeability and penetrate hydrothermal reservoirs. At this second stage the discharging fluids were at high temperature and pressure and were reduced, owing to their transport upwards from deeper crustal levels; such fluids were predominantly of metamorphic origin, their characteristic high $\delta^{18}O$ reflecting equilibrium of the reservoir with crustal rocks at elevated temperatures in the source region.

Later in the tectonic evolution, as deeper crustal levels were exposed by erosion, such that conditions of lower ambient temperature and pressure pertained, an incursion of near-surface waters into faults occurred; these fluids were either formation brines in aquifers intersected by the fault or they were from direct downward penetration of surface fluids into the structures. Such fluids were oxidised and of lower $\delta^{18}O$, the former fact reflecting near-surface environments and the latter being commensurate with a component of meteoric continental surface water of lower $\delta^{18}O$ and/or metamorphic fluids of shallower level, where the $\delta^{18}O$ is buffered to smaller values by larger rock-water fractionations at the lower ambient temperatures. The high-temperature high ^{18}O and low-temperature fluid regimes of faults may proceed coevally, the former at deep crustal levels and the latter at shallow ones, juxtaposition of the two taking place by subsequent vertical displacements on the structure. The flow of both deep-level (metamorphic) and shallow-level (meteoric) fluids has been implicated in contemporary seismic activity on the San Andreas fault by IRWIN and BARNES (1975) and SIBSON (1981, 1982), respectively.

At higher crustal levels a significant infiltration of meteoric water along major thrust faults has been identified. LEE et al. (1984) have shown that the δD and $\delta^{18}O$ and the apparent K-Ar ages of Mesozoic and Tertiary plutons in Nevada have been disturbed in proximity to the Snake Range décollement. In Oligocene plutons (37 Ma) deformed cataclastically, the $\delta^{18}O$, the δD, and the K-Ar ages are as low as -2.5 per mil, -155 per mil, and 18 Ma respectively, owing to deformation combined with incursion of isotopically light thermal waters. For undeformed Jurassic plutons the $\delta^{18}O$ values are undisturbed, but both the δD and the K-Ar ages have been perturbed over distances of tens of metres below the décollement.

The Bitterroot Lobe, Sapphire Block detachment zone of western Montana is defined by a sequence of mylonitic granites 500 m thick, locally overprinted by chloritic breccias characterised by intense transgranular fracturing. KERRICH and HYNDMAN (1986) have shown that the ductile mylonites were formed at 550 ± 50 °C in the presence of magmatic or high-temperature metamorphic fluids. Chloritic breccias are depleted in ^{18}O by about 10 per mil relative to mylonitic precursors and appear to have formed under conditions of lower temperature (250 to 370 °C) and effective confining stress, acting as an aquifer for the infiltration of surface meteoric waters of low ^{18}O ($\delta^{18}O$ of -7 to -12 per mil).

At Lagoa Real, discharge of the fluids with low ^{18}O, of surface-water origin, probably took place at a depth of ≈ 15 km in the crust and at a relatively high temperature of ≈ 550 °C. This situation appears to have occurred during overriding of a sedimentary sequence with meteoric-water-recharged aquifers by high-grade gneisses during overthrusting. Thrusting of basement rocks over low-grade sediments is a typical feature of most mountain belts, and it has been suggested as an important component of Appalachian structure, as judged from high-resolution seismic profiling (COOK et al., 1980). Major fluid discharge must accompany such thrusting.

The role of metamorphic fluids in shear zones at deep crustal levels has been emphasised by BEACH and FYFE (1972), BEACH (1976), and KERRICH and FYFE (1981). ETHERIDGE et al. (1984) present evidence for high pore-fluid pressures during prograde regional metamorphism, with resultant low effective confining stress, leading to enhanced permeabilities and, in turn, to the local dominance of advective flow. All of these studies, including the present one, emphasise that large-scale fluid flow may occur at deep crustal levels under appropriate conditions. Recently this has received decisive corroboration in the encountering of high-temperature fluids at depths of ≈ 12 km in deep continental drilling. On the other hand, some shear zones appear to have been essentially chemically and hydrologically closed systems. This is the case in the Mieville shear zone transecting Hercynian granodiorite of the Aiguilles Ranges in Switzerland, where no significant excursions of $\delta^{18}O$, oxidation state, or composition occur from the protolith across the structure (KERRICH et al., 1980).

In addition to the mechanical effects of transient high fluid pressures in faults (KERRICH and ALLISON, 1978; SIBSON, 1981), a modification of ambient chemical conditions by fluids streaming through faults may play a role in silicate deformation. As discussed above, the oxidation state of iron in some structures indicates significant reduction by fluids, probably because of elevated partial pressures of hydrogen, and it is possible that such high P_{H_2} may influence the cracking behaviour of minerals in the fault zone. For instance, hydrogen embrittlement is a well-known phenomenon in metals and ceramics and is thought to be due to the chemisorption of hydrogen at fracture surfaces, which promotes nucleation of dislocations at crack tips (LYNCH, 1979; MICHALSKE and FREIMAN, 1982). HOBBS (1984) notes that, at an intracrystalline level, the incorporation of a hydrogen defect that is capable of acting as an acceptor in silicates leads to a strong dependence of point-defect chemistry on the fugacities of both water and oxygen.

Strontium isotope data have been employed in a number of studies of hydrothermal systems in an effort to address the solute source problem. Shear zones in the Willamaya Complex, Broken Hill, Australia, are plotted above the 1665 Ma isochron of the Potosi Gneiss host. ETHERIDGE and COOPER (1981) attribute the introduction of radiogenic strontium, coupled with SiO_2, K_2O, and Ba leaching, to large-scale fluid flow through the shear zone, where fluids ($^{87}Sr/^{86}Sr = 0.794$) had interacted with the Willamaya Complex metamorphics.

These, along with many other studies, demonstrate that where geological boundary conditions are appropriate, combined light-stable isotope and strontium isotope tracer studies may resolve the question of fluid and solute source reservoirs for domains of hydrothermally metasomatised faults.

Acknowledgments

Appreciation is expressed to C.-Y. Wang for the invitation to submit an article for the special issue of *The Journal of Pure and Applied Geophysics* on the internal structure of faults. D. Myers, D. S. Rogers, E. Van Hees, G. Nemcsok, S. Comline, and P. Bedard are thanked for permission to visit the Con, Dome, Aunor and Hollinger, Macassa, Darius, and Lamaque mines, respectively, and for supplying additional material. I am grateful to B. McLarty, L. M. Willmore, and J. Forth for analytical assistance. Mrs. G. McIntyre and Mrs. J. Miller are kindly thanked for conducting the typing and L. M. Willmore for drafting the diagrams. The Canadian Institute of Mining and Metallurgy is thanked for permission to reproduce some of the figures from Special Volume 27. I acknowledge with special gratitude numerous discussions with W. S. Fyfe and N. J. Price and the benefit of their incisive thoughts on 'fluids in the crust'. L. Lobato collected the rocks from Lagoa Real.

This research was supported by the National Research Council of Canada and

Ontario Geological Survey Research Funds. The light-stable isotope laboratory at the University of Western Ontario has been kept operational, thanks to the engineering skills of R. Jackman, R. Daddar, R. Kager, and R. Shirran.

REFERENCES

ALLISON, I. and KERRICH, R. (1979), 'History of deformation and fluid transport in shear zones at Yellowknife', in R. D. Morton (ed.). Proc. Gold Workshop, Yellowknife, N.W.T.

BARKER, F., 'Trondhjemites: Definition, environment and hypothesis of origin', in F. Barker (ed.), *Trodhjemites, Dacites, and Related Rocks.* Elsevier, New York, 1979.

BARKER, F., ARTH, J. G. and MILLARD, H. T., 'Archean trondhjemites of the southwestern Big Horn Mountains, Wyoming', in F. Barker (ed.), *Trondhjemites, Dacites, and Related Rocks.* Elsevier, Amsterdam, 1979, p. 401–414.

BEACH, A. (1976), *The interrelations of fluid transport, deformation, geochemistry and heat flow in early Proterozoic shear zones in the Lewissian complex.* Phil. Trans. Roy. Soc. London, Ser. A280, 569–604.

BEACH, A. and FYFE, W. S. (1972), *Fluid transport and shear zones at Scourie, Sutherland: Evidence for overthrusting?.* Contrib. Mineral. Petrol. *36*, 75–180.

BOYLE, R. W. (1961), *The geology, geochemistry, and origin of the gold deposits of the Yellowknife district.* Geol. Surv. Canada, Mem. *310*, 193.

BOYLE, R. W. (1977), *Discussion of: Iron reduction around gold-quartz veins, Yellowknife District, N.W.T.* Canada Econ. Geol.

BOYLE, R. W. (1979), *The geochemistry of gold and its deposits.* Geol. Surv. Canada Bull. *280*, 548 pp.

CLAYTON, R. N., O'NEIL, J. R. and MAYEDA, T. K. (1972), *Oxygen isotope exchange between quartz and water.* Geophys. Res. *77*, 3057—3067.

COLLERSON, K. D. and FRYER, B. J. (1978), *The role of fluids in the formation and subsequent development of early continental crust.* Contrib. Mineral. Petrol. *67*, 151–167.

COLVINE, A. C., ANDREWS, A. J., CHERRY, M. E., DUROCHER, M. E., FYON, A. J., LAVINGE, M. J., MACDONALD, A. J., MARMOT, S., POULSEN, K. H., SPRINGER, J. S. and TROOP, D. C. (1984), *An integrated model for the origin of Archean gold deposits.* Ontario Geol. Surv. Open File Rept. 5524, 98 p.

CONEY, P. J. (1980), *Cordilleran metamorphic core complexes: An overview.* Geol. Soc. Amer. Mem. *153*, 7–34.

COOK, F., BROWN, L. and OLIVER, J. (1980), *The Southern Appalachians and the growth of continents.* Sci. Am. *243*, 156–168.

CORDANI, U. G., DE (1982), *Interpretacao geocronologica de regiao de Lagoa Real, Ba.* Relatorio Integrado, unpublished ed.

CRITTENDEN, M. D., JR., CONEY, P. J. and DAVIS, G. H. (1980), *Cordilleran metamorphic core complexes.* Geol. Soc. Amer. Mem. *153*.

ETHERIDGE, M. A. and COOPER, J. A. (1981), *Rb/Sr isotopic and geochemical evolution of a recrystallized shear (mylonite) zone at Broken Hill.* Contrib. Mineral. Petrol. *78*, 74–84.

ETHERIDGE, M. A., COX, S. F., WALL, V. J. and VERNON, R. H. (1984), *High fluid pressures during regional metamorphism and deformation: Implications for mass transport and deformation mechanisms.* J. Geophys. Res. *89*, No. B6, 4344–4358.

FAURE, G., *Principles of Isotope Geology.* Wiley, Toronto, 1977, 464 p.

FITTON, J. G. (1985), *Basic volcanism associated with intraplate linear features.* Trans. Roy. Soc. London, Ser. A., in press.

FRANKLIN, J. M., ROSCOE, S. M., LOVERIDGE, W. D. and SANGSTER, D. F. (1983), *Lead isotope studies in Superior and Southern provinces.* Geol. Surv. Canada Bull. *351*, 60 p.

FRIEDMAN, I. and O'NEIL, J. R. (1977), 'Compilation of stable isotope fractionation factors of geochemical interest', in *Data of Geochemistry.* U.S. Geol. Surv. Prof. Paper 440.

FRYER, B. J., KERRICH, R., HUTCHINSON, R. W., PEIRCE, M. G. and ROGERS, D. S. (1979), *Archean precious metal hydrothermal systems, Dom Mine, Abitibi greenstone belt. I.* Canada J. Earth Sci. *16*, 421–439.

FYFE, W. S. and KERRICH, R. (1983), *Fluid release and mineralization associated with thrusting.* Fourth Int. Symp. Water-Rock Interaction, Misasa, Japan, Int. Assoc. Geochem. Cosmochem., 150–151.

FYFE, W. S. and KERRICH, R. (1985), *Fluids and thrusting.* Chem. Geology *49*, 353–362.

FYFE, W. S., PRICE, N. J. and THOMPSON, A. B., *Fluids in the Earth's Crust.* Elsevier, Amsterdam, 1978, 383 p.

FYON, J. A., SCHWARCZ, H. P. and CROCKET, J. H. (1984), *Carbonatization and gold mineralization in the Timmins area, Abitibi greenstone belt: Genetic links with Archean mantle CO_2-degassing and lower crust granulitization.* Geol. Assoc. Canada Prog. with Abstr. *9*, 65.

GATES, T. M. and HURLEY, P. M. (1973), *Evaluation of Rb–Sr dating methods applied to Matachewan, Abitibi, Mackenzie, and Sudbury dike swarms in Canada.* Canad. J. Earth Sci. *10*, 900–919.

GOLDING, S. D. and WILSON, A. F. (1983), *Geochemical and stable isotope studies of the No. 4 lode, Kalgoorlie, Western Australia.* Econ. Geol. *78*, 438–450.

GOODWIN, A. M. and RIDLER, R. H. (1970), *The Abitibi orogenic belt.* Canada Geol. Surv. Paper 70-40, 1–31.

GRESENS, R. L. (1967), *Composition – volume relations of metasomatism.* Chem. Geol. *2*, 47–65.

HART, S. R. and BROOKS, C. (1974), *Clinopyroxene-matrix partitioning of K, Rb, Cs, Sr, and Ba.* Geochim. Cosmochim. Acta *38*, 1799.

HOBBS, B. E. (1984), *Point defect chemistry of minerals under a hydrothermal environment.* J. Geophys. Res. *89*, No. B6, 4026–4038.

IRWIN, W. P. and BARNES, I. (1975), *Effects of geologic structure and metamorphic fluids on seismic behaviour of the San Andreas fault system in central and northern California.* Geology *4*, 713–716.

JAVOY, M. (1977), *Stable isotopes and geothermometry.* J. Geol. Soc. London *133*, 609–636.

JENSEN, L. S. (1976), *Regional stratigraphy and structure of the Timmins-Kirkland Lake area, District of Timiskaming, Ontario.* Dept. Mines Misc. Paper 67, 183 p.

JENSEN, L. S. (1978a), *Archean komatiitic, tholeiitic, calc-alkalic, and alkalic volcanic sequences in the Kirkland Lake area.* In Toronto '78 Field Trip Guidebook, Geol. Assoc. Canada, 237–259.

JENSEN, L. S. (1978b), *Larder Lake synoptic mapping project, District of Cochrane and Timiskaming.* Ont. Geol. Surv. Misc. Paper 94, 64–69.

JENSEN, L. S. (1980), *Gold mineralization in the Kirkland Lake–Larder Lake areas* Ont. Geol. Surv. Misc. Paper 97, 59–65.

JENSEN, L. S. and LANGFORD, F. F. (1983), *Geology and petrogenesis of the Archean Abitibi Belt in the Kirkland Lake area, Ontario.* Ont. Geol. Surv. Open File Rept. 5455, 520.

JOLLY, W. T. (1978), *Metamorphic history of the Archean Abitibi belt.* Geol. Surv. Canada, Paper 78–10, 367.

KENNEDY, L. P. (1985), *Geology and geochemistry of the Archean Flavarian Pluton, Noranda, Quebec, Canada.* Ph.D. thesis, University of Western Ontario, London, Ontario, Canada, 469.

KERRICH, R. (1983), *Geochemistry of gold deposits in the Abitibi greenstone belt.* CIM Special Paper 27.

KERRICH, R. (1985), *Archean lode gold deposits of Canada, Pt. I.* Proc. Symp. Archean Gold, Barberton, South Africa.

KERRICH, R. and ALLISON, I. (1978), *Vein geometry and hydrostatics during Yellowknife mineralization.* Canad. J. Earth Sci. *15*, 1653–1660.

KERRICH, R., ALLISON, I., BARNET, R. L., MOSS, S. and STARKEY, J. (1980), *Microstructural and chemical transformations accompanying deformation of granite in a shear zone at Mieville, Switzerland, with implications for stress corrosion cracking and superplastic flow.* Contrib. Mineral. Petrol. *73*, 221–242.

KERRICH, R. and FRYER, B. J. (1979), *Archean precious metal hydrothermal systems, Dome Mine, Abitibi greenstone belt: II. REE and oxygen isotope relations.* Canad. J. Earth Sci. *16*, 440–458.

KERRICH, R., FYFE, W. S. and ALLISON, I. (1977), *Iron reduction around gold-quartz veins, Yellowknife district, Northwest Territories, Canada.* Econ. Geol. *72*, 657–663.

KERRICH, R. and FYFE, W. S. (1981), *The gold-carbonate association: Source of CO_2, and CO_2 fixation reaction in Archaean lode deposits.* Chem. Geol. *33*, 265–294.

KERRICH, R. and FYFE, W. S. (1983), *The ^{18}O, temperature and ^{34}S of Archean ocean bottom water (2.8 Ga).* Geol. Surv. Canada Abstr., A37.

KERRICH, R. and HODDER, R. W. (1982), 'Archean lode gold and base metal deposits: Chemical evidence for metal separation into independent hydrothermal systems', in R. W. Hodder and W. Petruk (eds.), CIM Spec. Vol. 24, Geology of Canadian Gold Deposits, 144–160.

KERRICH, R. and HYNDMAN, D. (1986), *Thermal and fluid regimes in the Bitterroot Lobe–Sapphire Block detachment zone: Evidence from* $^{18}O/^{16}O$ *and geologic relations*. Geol. Soc. Amer. Bull., *97*, 147–155.

KERRICH, R., KISHIDA, A. and WILLMORE, L. M. (1984), Geol. Assoc. Canada Prog. with Abstr. *9*, 78.

KERRICH, R., LOBATO, L., FYFE, W. S. and WILLMORE, L. M. (1986), *Shear zone hosted uranium deposits in overthrust Archean gneisses, Bahia, Brazil: Evidence on U provenance from Rb–Sr isotopic data*. Econ. Geol., in press.

KERRICH, R. and REHRIG, W. (1986) *Fluid motion associated with Tertiary mylonitization and detachment faulting:* $^{18}O/^{16}O$ *evidence from the Picacho metamorphic core complex*. G.S.A., in press.

KERRICH, R. and WATSON, G. P. (1984), *The Macassa Mine Archean lode gold deposit, Kirkland Lake, Ontario: Geology, patterns of alteration and hydrothermal regimes*. Econ. Geol. *79*, 1104–1130.

KING, R. W. (1983), *Auriferous 'Porphyry Zone', Taylor Township, Ontario: Petrology and geochemical relations*. B.Sc. thesis, University of Western Ontario, London, Ontario, Canada, 128.

KISHIDA, A. (1984), *Hydrothermal alteration zoning and gold concentration at the Kerr-Addison Mine, Ontario, Canada*. Ph.D. Thesis, Univ. Western Ontario, London, Canada, 231 p.

KISHIDA, A. and KERRICH, R. (1986), *Hydrothermal alteration zoning and gold concentration at the Kerr-Addison, Archean lode gold deposit, Kirkland Lake, Ontario*. Econ. Geol., in press.

KROGH, T. E., DAVIS, D. W., NUNES, P. D., and KORFU, F. (1982), *Archean evolution from precise U-Pb isotopic dating*. GAC/MAC Joint Annual Meeting, Winnipeg, Manitoba, Prog. with Abst. 7, 61.

KWONG, Y. T. J. and CROCKET, J. H. (1978), *Background and anomalous gold in rocks of an Archaean greenstone assemblage, Kakigi Lake area, Northwestern Ontario*. Econ. Geol. *73*, 50–63.

LA TOUR, T. E. (1981), *Significance of folds and mylonites at the Grenville Front in Ontario*. Geol. Soc. Amer. Bull. *92*, Pt. II, 997–1038.

LEE, D. E., FRIEDMAN, I. and GLEASON, J. D. (1984), *Modification of D values in eastern Nevada granitoid rocks spatially relatd to thrust faults*. Contrib. Mineral. Petrol. *88*, 288–298.

LOBATO, L. M., FORMAN, J. M. A., FUZIKAWA, K., FYFE, W. S. and KERRICH, R. (1983a), *Uranium in overthrust Archean basement, Bahia, Brazil*. Canad. Mineral. *21*, 647–654.

LOBATO, L. M., FORMAN, J. M. A., FYFE, W. S., KERRICH, R. and BARNETT, R. L. (1983b), *Uranium enrichment in Archean crustal basement associated with overthrusting*. Nature *303*, 235–237.

LOBATO, L. M., FUZIKAWA, K., FYFE, W. S. and KERRICH, R. (1983c), *Uranium enrichment in Archean basement: Logoa Real, Brazil*. Rev. Brasileira Geociencias *12*, 484–486.

LUDDEN, J. N., DAIGNEAULT, R., ROBERT, F. and TAYLOR, H. P. (1984), *Trace element mobility in alteration zones associated with Archean Au lode deposits*. Econ. Geol. *79*, 1131–1141.

LYNCH, S. P. (1979), *Mechanisms of hydrogen assisted cracking*. Metals Forum *2*, 189–200.

MCNEIL, A. M. and KERRICH, R. (1986), *Archean lamprophyric dykes gold mineralization, Matheson, Ontario: The conjunction of LIL-enriched mafic magmas, deep crustal structures and Au concentration*. Canad. J. Earth Sci., *23–3*, 324–343.

MICHALSEK, T. A. and FREIMAN, S. W. (1982), *A molecular interpretation of stress corrosion in silica*. Nature *295*, 511–512.

NESBITT, R. W. and SUN, S. S. (1976), *Geochemistry of Archean spinifex textured peridotites and magnesian tholeiites*. Earth Planet. Sci. Lett. *31*, 433–453.

NUNES, P. D. and JENSEN, L. S. (1980), *Geochronology of the Abitibi metavolcanic belt, Kirkland Lake area – progress report*. In E. G. Pye (ed.) Summary of Geochronology Studies 1977–1979, Ont. Geol. Surv. Misc. Paper 92, 40–45.

NUNES, P. D. and PYKE, D. R. (1980), *Geochronology of the Abitibi metavolcanic belt, Timmins – Matachewan area* – progress report. In E. G. Pye (ed.) Ont. Geol. Surv. Misc. Paper 92, 34–39.

OHMOTO, H. and RYE, R. O., *Isotopes of sulphur and carbon*, in H. L. Barnes (ed.), *Geochemistry of Hydrothermal Ore Deposits*. 2nd ed., Wiley, New York, 1979, 509–562.

O'NEIL, J. R. and TAYLOR, H. P. (1967), *The oxygen isotope and cation exchange chemistry of feldspars*. Am. Mineral. *52*, 1414–1437.

O'NEIL, J. R., CLAYTON, R. N. and MAYEDA, T. (1969), *Oxygen isotope fractionation in divalent metal carbonates*. Journ. Chem. Phys. *51*, 5547–5558.

Percival, J. A. and Krogh, T. E. (1983), *U–Pb zircon geochronology of the Kapuskasing structural zone and vicinity of the Chapleau-Foleyet area, Ontario.* Canad. J. Earth Sci. *20*, 830–843.

Rehrig, W. A. and Keith, S. B. (1984), unpublished mapping.

Robert, F. and Brown, A. C. (1984), *Chemical exchanges between gold mineralizing fluids and wall rocks at the Sigma Mine, Abitibi region, Quebec.* GAC-MAC Joint Ann. Meeting Progr. with Abstr. *9*, 100.

Sibson, R. H., Moore, J. McM. and Rankin, A. H. (1975), *Seismic pumping—A hydrothermal fluid transport mechanism.* J. Geol. Soc. London *131*, 653–659.

Sibson, R. H. (1981), 'Fluid flow accompanying faulting: Field evidence and models', in D. W. Simpson and P. G. Richards (eds.), *Earthquake Prediction.* A.G.U. Maurice Ewing Series 4, 593–603.

Sibson, R. H. (1982), Bull Seism. Soc. Am. *72*, 151–163.

Stein, J. H., Netto, A. M., Drummond, D. and Angeiras, A. G. (1980), *Nota preliminar sobre os processos de albitizacao uranifera de Lagoa Real (Bahia) e sua comparacao com os da URSS e Suecia.* Anais do 22nd Congr. Bras. Geol. Santa Catarina *3*, 1758–1775.

Taylor, H. P. (1974), *The application of oxygen and hydrogen isotope studies to problems of hydrothermal alteration and ore deposition.* Econ. Geol. *69*, 843–883.

Taylor, H. P., 'Oxygen and hydrogen isotope relations in hydrothermal mineral deposits', in H. L. Barnes (ed.), *Geochemistry of Hydrothermal Ore Deposits*, Wiley, New York, 1979, 236–277.

Thomson, J. E. (1941a), *Geology of Gauthier Township, east Kirkland Lake area.* Ont. Dept. Mines Ann. Rept. *50*, Pt. 8.

Veizer, J. and Compston, W. (1974), *$^{87}Sr/^{86}Sr$ composition of seawater during the Phanerozoic.* Geochim. Cosmochim. Acta *38*, 1461–1484.

Wang, C.-Y. (1980), *Sediment subduction and frictional sliding in a subduction zone.* Geology *8*, 530–533.

Wenner, D. B. and Taylor, H. P. (1971), *Temperature of serpentinization of ultramafic rocks based on 0/0 fractionations between coexisting serpentine and magnetite.* Contrib. Mineral. Petrol. *32*, 165–185.

(Received 6th October 1985, revised 28th April 1986, accepted 30th April 1986)

PAGEOPH, Vol. 124, Nos 1/2 (1986) 0033-4553/86/020269-20$1.50+0.20/0

Fabric Development in Brittle-to-Ductile Shear Zones

CAROL SIMPSON[1]

Abstract—Brittle-to-ductile shear zones from two separate geological settings are shown to have nucleated on zones of predominantly brittle deformation. The shear zones are not simply foliated cataclasites, since they contain abundant evidence of dynamic recrystallization of constituent minerals. A small quartz diorite lens in the Borrego Springs shear zone, southern California, contains centimeter-scale cataclasite zones that exhibit a gradual transition into foliated rock. Alteration of magnesiohornblende to actinolite, feldspar to white mica plus quartz, and biotite to chlorite, produced elongate minerals that define the foliation. During the later stages of deformation, intracrystalline slip and dynamic recrystallization of quartz and feldspar were important deformation mechanisms.

The widespread occurrence of mineralized dilatant cracks predated the development of meter-to-decimeter-scale ductile shear zones in the Striped Rock granite, southern Virginia. Again, important deformation mechanisms in the final stages of deformation were intracrystalline slip and dynamic recrystallization of quartz.

In both field areas the role of fluids has been important from the onset of brittle deformation. Fluids may have enhanced early fracturing in addition to causing the alteration and hydrolytic weakening of host rock minerals and the introduction of new mineral species. Each of these processes is thought to have contributed to the later localization of crystal plastic deformation in the rocks.

Key words: Cataclasites, ductile shear zones, microstructures.

Introduction

The brittle-to-ductile transition in naturally deformed rocks commonly involves a gradual change from predominantly cataclastic to predominantly crystal plastic-deformation mechanisms. Experimental rock deformation studies have shown that this transition may be accomplished by an increase in temperature or in effective confining pressure (e.g., TULLIS and YUND, 1977, 1980; PATERSON, 1978), a decrease in strain rate (RUTTER, 1976; WHITE, 1976), or some combination of these variables. Rock ductility may also be enhanced by an increase in fluid pressure or by a decrease in grain size, which may be brought about by cataclasis (e.g., MITRA, 1984), crystal plasticity (e.g., SCHMID, 1975, 1976; WHITE, 1976; TULLIS and YUND, 1985), or metamorphic reaction (e.g., WHITE and KNIPE, 1978; KNIPE and WHITE, 1979).

---

[1] Department of Geological Sciences, Virginia Polytechnic Institute and State University, Blacksburg, Virginia 24061.

Ductile shear zones that are centimeters to many tens of meters wide are common in otherwise quite homogeneous rocks, such as granites and granodiorites (e.g., RAMSAY and GRAHAM, 1970; MITRA, 1978; SIMPSON, 1983). Their localization is often assumed to require some initial heterogeneity in the rock. The effects of this localized strain progressively weaken the rock and cause the material to support a progressively lower stress as deformation proceeds. An initial heterogeneity could include preexisting microscopic or macroscopic fracture systems (e.g., SEGALL and POLLARD, 1983; SEGALL and SIMPSON, 1986). These fracture systems are advantageous to strain localization for several reasons, including the following: (a) a reduction in grain size by cataclasis may occur along the fracture zone, (b) new hydrous minerals may precipitate within the fractures, and (c) a hydrolytic weakening effect on minerals of the wall rock may be caused by circulating water-rich fluids (GRIGGS and BLACIC, 1965; BLACK and CHRISTIE, 1984; KIRBY and KRONENBERG, 1984).

This paper describes naturally occurring shear zones that show a gradual transition from a completely brittle behavior of the constituent minerals to predominantly crystal plastic-deformation mechanisms. The term *brittle* is used when a material's ability to resist a load decreases with increase in strain (JAEGER and COOK, 1979, p. 80) and when unstable fractures spontaneously propagate at high velocities in response to the applied stress (LAWN and WILSHAW, 1975; PATERSON, 1978). Zones of closely spaced, unstable crack arrays have been termed *brittle deformation zones* by MITRA (1984). A rock exhibits ductile behavior when it can sustain permanent deformation without any loss in ability to resist a load (JAEGER and COOK, 1979). Commonly the material will undergo stable fracturing of individual grains and build up an appreciable strain prior to macroscopic failure (LAWN and WILSHAW, 1975; PATERSON, 1978; MITRA, 1984). The deformation mechanisms that can accomplish bulk ductile flow of a material include cataclastic flow, dislocation glide, dislocation creep, diffusional creep, and pressure solution (RUTTER, 1976). For the purposes of the present paper, deformation zones that exhibit a progression from predominantly brittle behavior to predominantly ductile, regardless of the precise deformation mechanisms operative, are referred to as brittle-to-ductile shear zones. The shear zones described here are from two geological settings that are temporally and spatially unrelated: the Eastern Peninsular Ranges mylonite zone of southern California (SHARP, 1966, 1979) and the Striped Rock granite of southern Virginia (STOSE and STOSE, 1957). Despite the differences between these two settings, the shear zones have many features in common, suggesting that shear zones which have formed in a brittle-to-ductile mode may be more widespread in granitoid rocks than has hitherto been documented.

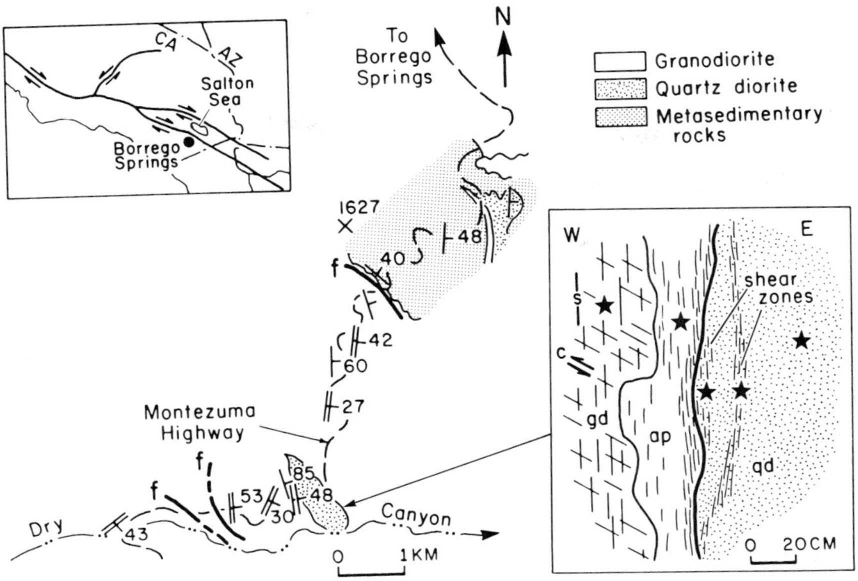

Figure 1
(a) Locality of quartz diorite lens in Borrego Springs section of Eastern Peninsular Ranges mylonite zone, southern California. Single bar strike symbols are S planes; double bar strike symbols are C planes. (b) Field sketch of brittle-to-ductile shear zones in quartz diorite (qd) at margin of aplite dike (ap); foliations in granodiorite (gd) strike 158° and dip 85° E (S planes) and strike 167° and dip 48° E (C planes); foliations in aplite strike 180° and dip 86° W; stars are sample localities.

Brittle-to-ductile deformation in quartz diorite, Borrego Springs

Geologic setting

The first examples of brittle-to-ductile shear zones occur in a 10-m-wide quartz diorite lens in the Borrego Springs section of the Eastern Peninsula Ranges mylonite zone (SIMPSON, 1984, 1985). In this locality the predominant rock type is mid-Cretaceous granodiorite of the Peninsular Ranges Batholith (TODD and SHAW, 1979) that has been deformed into a Type I S (schistosity) and C (cisaillement, shear) mylonite (LISTER and SNOKE, 1984) during west-directed thrusting in the late Cretaceous (SIMPSON, 1984, 1985). The mylonitic rocks have been cut in several places by later brittle normal faults of possible Miocene age (ENGEL and SCHULTEJANN, 1984).

An aplite dike separates the quartz diorite lens from the granodiorite (Fig. 1). Neither aplite nor quartz diorite exhibit the pervasive S–C mylonitic foliations, trending north to north-northeast, of the surrounding granodiorite. In thin sections taken from the aplite adjacent to the granodiorite contact, however, a very weakly developed quartz ribbon fabric is observed that is concordant with the S planes of the granodiorite. Sense-of-shear indicators in the granodiorite include S–C bands, asymmetrical recrystallized feldspar porphyroclasts, and oblique grain-shape foliations (SIMPSON and SCHMID, 1983; LISTER and SNOKE, 1984; PASSCHIER and SIMPSON, in press) and give an east-over-west movement sense (SIMPSON, 1984, 1985). Development of maximum intensity of foliation in the aplite occurs along and subparallel to the north-northwest-trending contact with the quartz diorite. Small-scale (up to 5 cm wide) brittle-to-ductile shear zones occur adjacent and subparallel to this contact in the otherwise undeformed quartz diorite. Oblique quartz grain-shape fabrics and rotated porphyroclasts in the highly strained aplite–quartz-diorite contact zone consistently point to an east-side down movement. This change in movement sense and in orientation of foliation with respect to the rest of the Borrego Springs mylonite suggests that the brittle-to-ductile shears in the quartz diorite did not develop synchronously with the foliations in the surrounding granodiorite. No unambiguous evidence of the timing of deformation in the shear zones has been found, but it is possible that they are related to the Miocene normal faults (ENGEL and SCHULTEJANN, 1984).

Microstructures

The undeformed quartz diorite is approximately 50% magnesiohornblende, 30% plagioclase (An 42), 10% quartz, 5% biotite, and 5% sphene, and has very minor amounts of epidote and white mica (<1%). Original igneous textures are well preserved. Some reaction of hornblende to biotite and sphene is evident, as is a minor alteration of plagioclase to epidote and white mica. Quartz grains show extremely slight undulatory extinction and, rarely, healed microcracks.

Undeformed quartz diorite and cataclastically deformed rock are not separated by a single planar fracture. Rather, the transition occurs across a 2-cm-wide crush microbreccia zone (SIBSON, 1977), with randomly oriented, dilatant cracks. There is a very slight tendency of the cracks to follow grain boundaries or cleavage planes in the amphiboles (intergranular fracture), but the majority are transgranular fractures (LAWN and WILSHAW, 1975) that show no lateral offset of grain boundaries (Fig. 2a). Within these dilatant cracks are angular fragments of 5 to 100 μm size, of hornblende, plagioclase, quartz, and sphene. Figure 2a illustrates the transition from crush microbreccia to cataclasite. All fragments are angular, and no preferred grain-shape orientation exists.

There is a slight but significant change in composition from undeformed quartz diorite to the cataclasite. Biotite is absent in the cataclasite, and as much as 5% magnesian calcite is present. The latter grains are generally anhedral and range in size from less than 5 to 200 μm. The proportions of mineral species present were difficult to estimate in the finer-grained matrix, but there is no apparent increase in epidote, sphene, or white mica content. Plagioclase grains in the clasts are unchanged in composition (An 42), although they commonly show undulatory extinction, as do quartz grains. Cores of brecciated amphibole grains show no change in composition

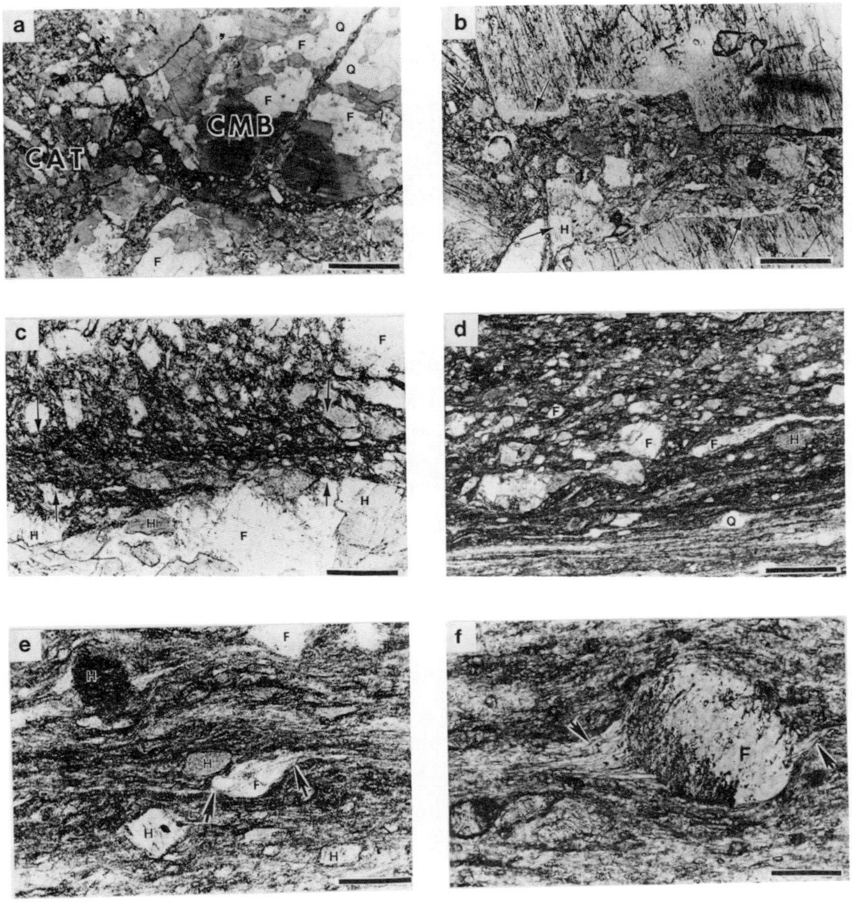

Figure 2

Transition from brittle to ductile deformation in Borrego Springs quartz diorite: H, actinolitic hornblende; F, feldspar; Q, quartz; E, epidote. All with plane-polarized light. (a) Transition from crush microbreccia (CMB) to cataclasite (CAT); transgranular dilatant cracks produce angular fragments that show no shape alignment; scale bar, 1500 μm. (b) Overgrowths of new actinolitic hornblende (arrows) onto cataclasite fragments; scale bar, 300 μm. (c) Narrow zone (between arrows) of aligned elongate fragments at boundary with large, undeformed clast; scale bar, 300 μm. (d) Relict subangular clasts in fine-grained matrix of aligned recrystallized amphibole, white mica, and quartz; scale bar, 300 μm. (e) Feldspar porphyroclast with elongated tails (arrows) of dynamically recrystallized grains in well-foliated matrix; all amphibole clasts have beards of white mica and new actinolitic hornblende; scale bar, 300 μm. (f) Feldspar porphyroclast with asymmetrical beards of white mica (arrows) in well-foliated matrix; scale bar, 150 μm.

from the undeformed rock (see Table 1 and Fig. 3). Almost all of the amphibole grains, however, have narrow irregular rims of inclusion-free magnesian to actinolitic hornblende (Figs. 2b and 3 and Table 1). These overgrowths occur preferentially on grain margins that face the (001) direction; they also occur along the margins of transgranular microcracks in the amphiboles (Fig. 4a). No evidence for nucleation and growth of

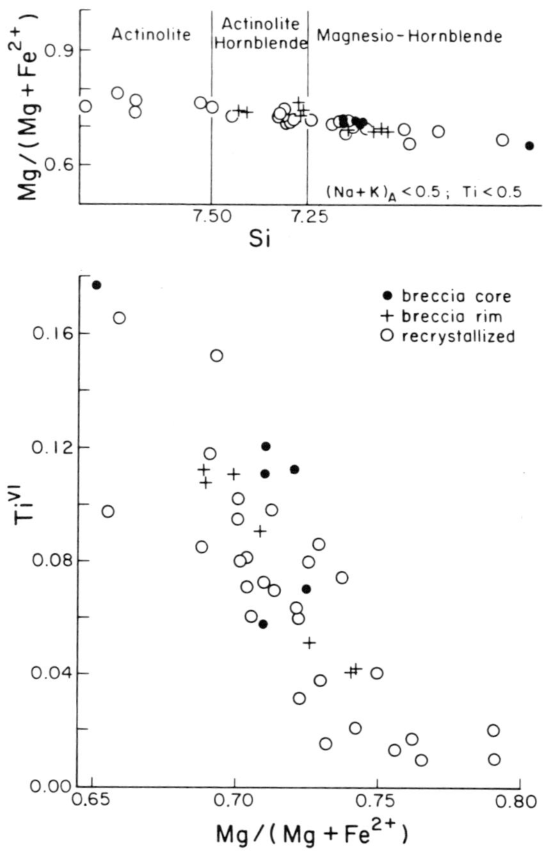

Figure 3

(a) Variation of Mg/Mg + Fe^2 with Si content of the amphiboles in quartz diorite, Borrego Springs, California; note tendency of overgrowths on breccia grain rims and recrystallized grains to be more actinolitic; nomenclature after LEAKE (1978). (b) Variation of Ti^{VI} with Mg/Mg + Fe^2 content of amphiboles in (a): symbols as in (a).

entirely new amphiboles in the cataclasite was found, but this may be due in part to the difficulty of an optical resolution of grains in the very fine-grained matrix material.

Along the margins of some of the larger clasts in the cataclasite there are thin, discontinuous zones of foliated rock (Fig. 2c). These zones are commonly 100 to 200 μm wide and <1000 μm long and are defined by a grain-shape alignment of phyllosilicates, fibrous actinolitic hornblende, and elongate angular fragments. The average grain size within these foliated zones ranges from 5 to 50 μm. With increasing distance into the cataclasite and away from the contact with the crush microbreccia, the zones of foliated rock become longer, wider, and more common.

Approximately 1 cm inside the crush microbreccia–cataclasite boundary a well-developed foliation is apparent, and most of the clasts are subangular to subrounded

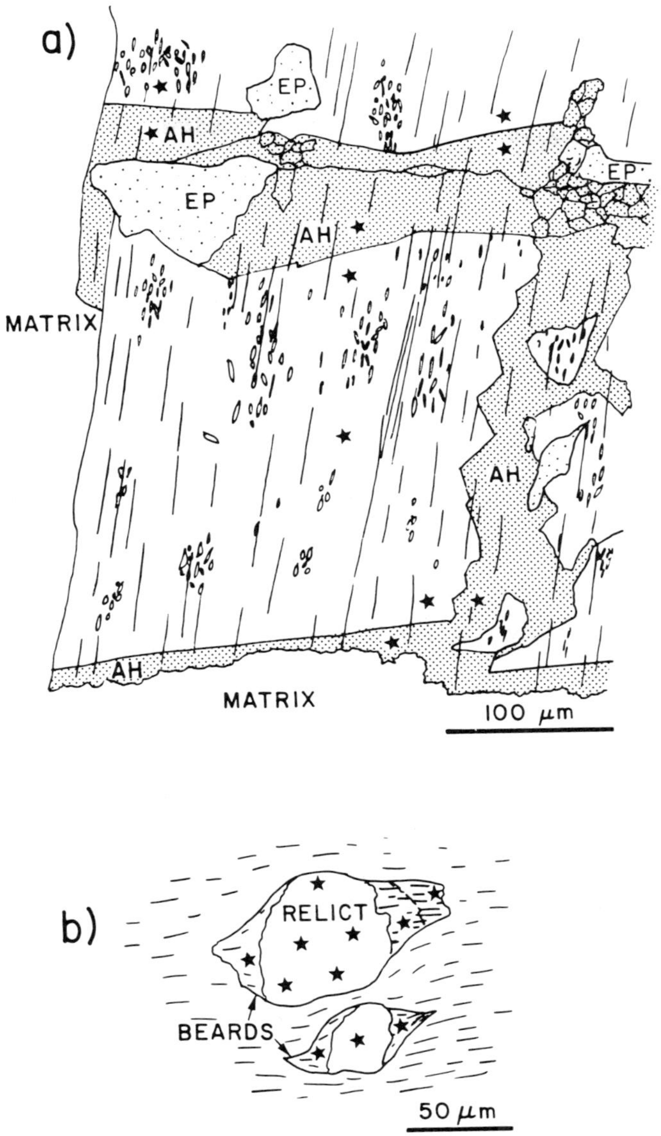

Figure 4
Sketches from thin sections of Borrego Springs sheared quartz diorite to show microprobe analysis localities (stars). (a) Grain of cataclastically deformed magnesiohornblende with paler, inclusion-free actinolitic hornblende rims along microcracks: EP, epidote; AH, actinolitic hornblende. Grains in matrix (not illustrated) are all angular and unsorted. (b) Two actinolitic hornblende porphyroclasts with distinctive beards of newly recrystallized actinolitic hornblende in a fine-grained, well-foliated matrix.

in outline (Fig. 2d). The foliation is defined by aligned elongate clasts, actinolitic hornblende, minor chlorite, white mica, and epidote. Subround to subangular grains,

< 10 μm in diameter, of calcite, sphene, and epidote, are also present. The extremely fine grain size (< 2 μm) of much of the matrix for the clasts inhibited an exact estimation of the mineral proportions. Dynamically recrystallized grains of quartz and plagioclase (An 38), 2 to 4 μm in diameter, occur in narrow ribbons that help to define a weak compositional banding (Fig. 2d). The quartz C-axes in these ribbons show a random orientation distribution. Shear bands that cut across the foliation at a low angle occur in regions with many larger porphyroclasts; they have not been observed in zones with porphyroclast sizes of less than 100 μm.

Most of the porphyroclasts are either amphibole or feldspar, and occasionally larger sphene and epidote grains are present. The amphibole porphyroclasts usually have 'beard' overgrowths of fine-grained white mica, chlorite, and actinolitic hornblende (Fig. 2e, 4b). These beards are almost always asymmetrically developed with respect to the mean orientation of foliation; the sense of the asymmetry is always consistent with the sense of the shear, as indicated by the shear bands and oblique recrystallized grain-shape fabrics (SIMPSON and SCHMID, 1983; LISTER and SNOKE, 1984). Plagioclase (An 42–45) porphyroclasts have either dynamically recrystallized tails (An 38) that are asymmetric to the mean foliation (Fig. 2e) or asymmetric 'beard' overgrowths of white mica (Fig. 2f). No detectable compositional difference was found between those grains with white mica beards and those with dynamically recrystallized tails.

Deformation history, Borrego Springs zones

The sequence of events in the deformation of the quartz diorite appears to have commenced with the formation of planar zones of randomly oriented, dilatant fractures and crush microbreccia that are now preserved on the margins of the brittle-to-ductile shear zones. Further comminution reduced the overall grain size to that of a cataclasite. The formation of undulatory extinction in quartz and plagioclase clasts probably is due to lattice bending below the macroscopic fracture stress (PATERSON, 1978, p. 181). Fluids circulating in the fault zone introduced calcite and altered the biotite to chlorite and some of the feldspar to white mica and epidote. Optically continuous overgrowth of actinolitic hornblende on the original magnesiohornblende grains occurred during this stage.

The next stage in the deformation sequence was the formation of shear bands. These occurred at sites of high stress concentration, such as along breccia clast boundaries and within the cataclasite. Some of the small, fibrous actinolitic hornblende and white-mica grains may have been passively rotated and aligned within the shear zones ('March model'; see reviews by TULLIS, 1976, MEANS, 1977, OERTEL, 1983). None of the grains are bent or kinked, and most appear to have grown with a preferred grain-shape alignment in response to the prevailing stress orientations. This is especially evident among those grains that form the asymmetric beard-like structures adjacent to porphyroclasts.

In the final stages of the deformation sequence a weak compositional layering was developed, aided by dynamic recrystallization of feldspar and quartz into new grains 2 to 4 μm in size. The absence of a crystallographic preferred orientation pattern in recrystallized quartz-rich areas suggests that grain-boundary sliding ('geometrical softening,' WHITE et al., 1980) predominated in these fine-grained regions.

Brittle-to-ductile deformation in Striped Rock Granite

Geologic setting

A transition from predominantly cataclastic to predomiantly crystal plastic deformation occurs in the Striped Rock granite of southern Virginia on a scale significantly greater than in the example above. First mapped by STOSE and STOSE (1957), the Striped Rock granite is one of the late Proterozoic plutons in the Crossnore plutonic-volcanic group of the southern Appalachians; it is situated in the Blue Ridge thrust sheet, near the town of Independence, Virginia (Fig. 5). The Striped Rock granite is really a composite pluton containing several phases in the range of syenite to granite (RIECKEN, 1966). Only those features affecting the granitic rocks are described in detail here. The granite has a Rb-Sr whole-rock age of 681 ±5 Ma (ODOM and FULLAGER, 1984) and is surrounded by Cranberry gneiss with an age of 1227 ±44 Ma (Rb-Sr whole-rock data of FULLAGAR and ODOM, 1973). The existing K-Ar whole-rock ages for the Cranberry gneiss (CARR and KULP, 1957) and K-Ar biotite ages for the Striped Rock granite (RIECKEN, 1966) are 406 ±20 Ma and 438 ±12 Ma, respectively. These data suggest that a significant metamorphic event could have occurred in the region about 400 or 420 Ma ago.

About 2 km to the southeast the Fries thrust fault system juxtaposes metasedimentary rocks of the Late Proterozoic Ashe formation and the Cranberry gneiss. Both the Striped Rock granite and the Cranberry gneiss are cut by numerous brittle-to-ductile shear zones that are subparallel to the Fries fault system and that vary in width from a few meters to several tens of meters (STOSE and STOSE, 1957). The age and geological significance of these shear zones is at present under investigation.

Between the shear zones completely undeformed Striped Rock granite is found only in very few localities. Elsewhere the Striped Rock plutonic rocks are pervasively, although heterogeneously, fractured. Millimeter-wide mineralized transgranular cracks are millimeters to centimeters apart in the more intensely fractured regions, and up to several centimeters in length (Fig. 6a). Locally the Striped Rock granite contains potassium feldspar phenocrysts up to 2 cm in length. Quartz-filled fractures along (001) cleavage planes (intergranular fractures) in these large grains commonly extend up to three grain diameters into the adjacent rock, to form

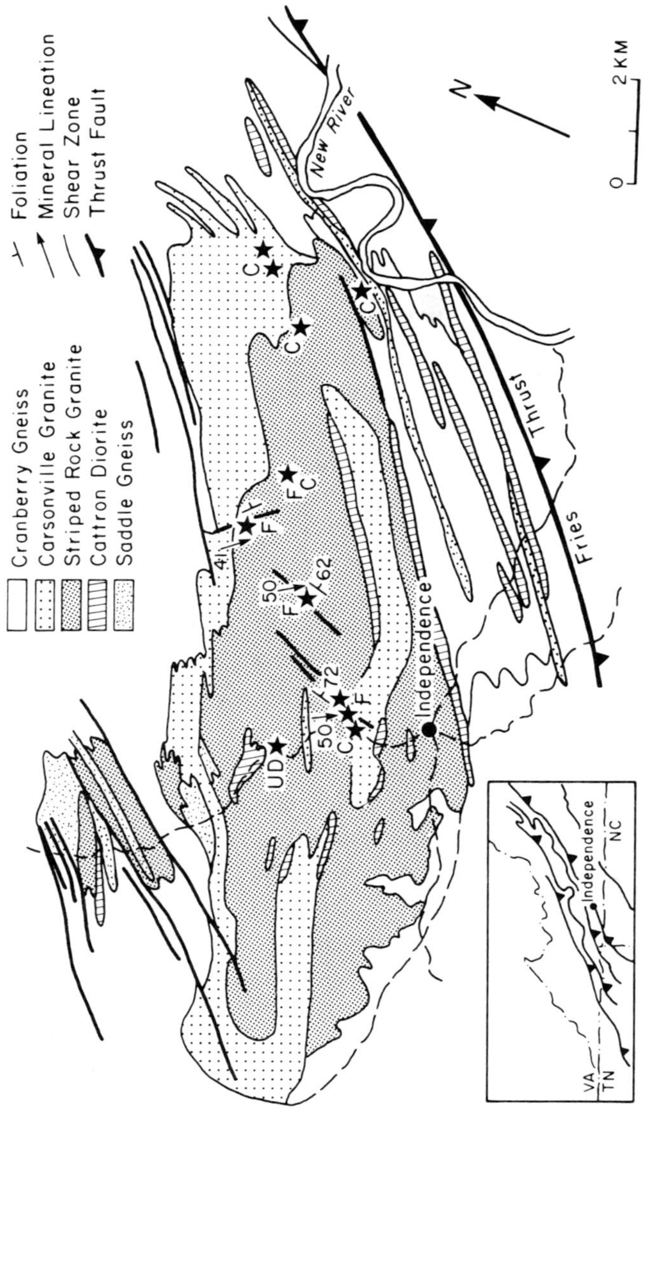

Figure 5

Map of sample sites (stars) Striped Rock granite: UD, undeformed; C, cataclastically deformed; F, foliated. Modified from STOSE and STOSE (1957)

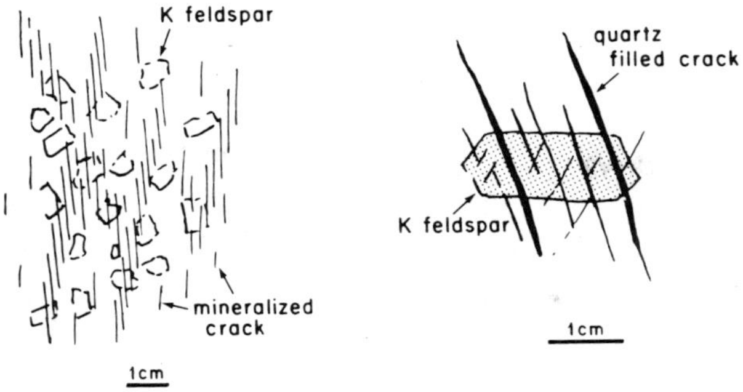

Figure 6
(a) Sketch of transgranular cracks in Striped Rock granite approximately 150 m south of major brittle-to-ductile shear zone located 1.5 km NNW of Independence (see Fig. 5). (b) Sketch of quartz-filled intergranular fractures in feldspar phenocrysts that extend into the adjacent rock to form transgranular cracks.

transgranular cracks (Fig. 6b). Preliminary results show highly complex geometrical relationships among intergranular and transgranular fractures over the pluton as a whole. In general, the transgranular fractures tend to lie at low angles to, and do not cross-cut, the brittle-to-ductile shear zones. In addition, the most intensely fractured rock occurs in the vicinity of the major brittle-to-ductile shear zones. The precise geometrical relationships among all the fracture sets and the shear zones is the subject of current research.

Microstructures

In its least deformed state the Striped Rock granite is approximately 50% microcline–microperthite, 25% quartz, 20% plagioclase (An 13), 5% green biotite, and minor amounts (<2%) of allanite, zircon, apatite, epidote, and white mica, the latter as an alteration product of plagioclase (see also Stose and Stose, 1957; Riecken, 1966). The original igneous textures are well preserved only adjacent to feldspar grains; nowhere is the rock completely undeformed. Almost all feldspar grains contain abundant intergranular fractures (Fig. 7a), usually dilatant and usually along the (010) or (001) cleavage planes. The fractures are sealed with the same mineral as the host grain, in optical continuity, or they are filled with quartz or green biotite, or both, or they contain both host grain mineral and green biotite (Fig. 7a). In addition, it is not uncommon to find intergranular fractures in microcline grains that are filled with plagioclase (≈An 10). Cross-cutting relationships among filled fractures that contain different mineral species are commonly observed (Fig. 7a), but no systematic sequence of mineralization could be ascertained.

In regions of intense transgranular fracturing the bulk-rock mineralogy is changed to approximately 35% microcline, 30% quartz, 10% relict plagioclase, 10% green biotite, 8% epidote, 5% white mica, <2% magnetite and ilmenite, and <1%

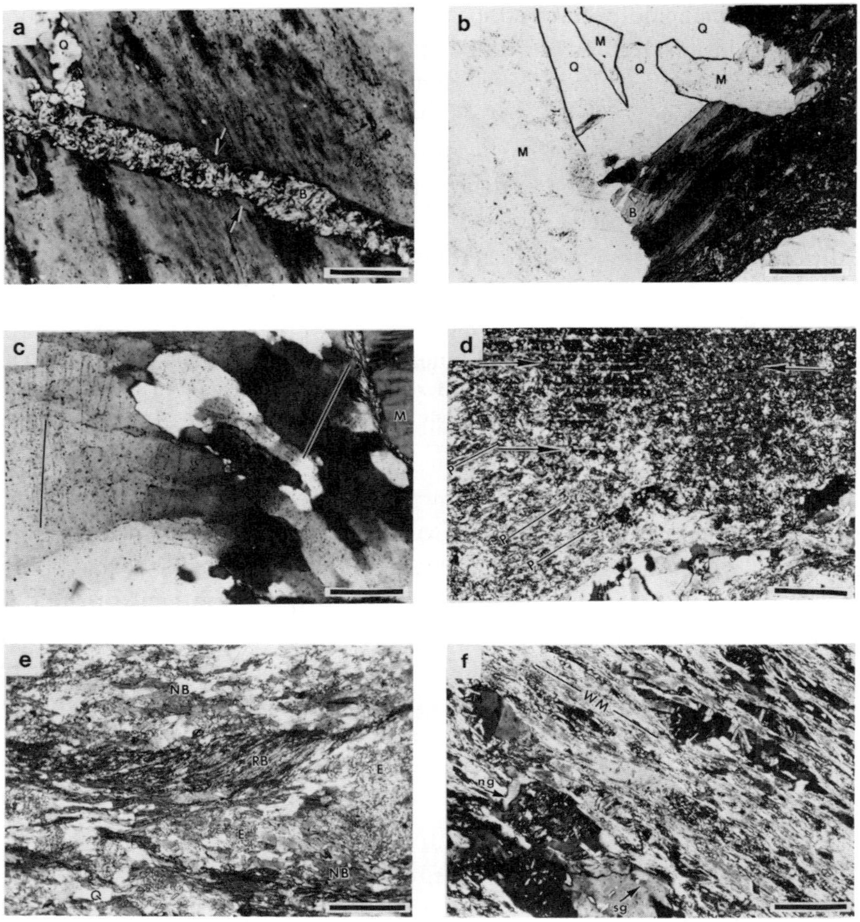

Figure 7

Brittle-to-ductile transition in Striped Rock granite, southern Virginia. All with crossed nicols except (b) and (e). (a) Dilatant cracks in microline filled with quartz (Q) and green biotite (B); margins of main crack show overgrowths (arrows) of optically continuous microcline; scale bar, 150 μm. (b) Fractured microline (M) with quartz (Q) and green biotite (B) filling; fractures highlighted for clarity; note tendency of biotite (001) to track opening direction; scale bar, 300 μm. (c) Progressive rotation of elongate subgrains and recrystallized quartz grains at margin of microline grain (M) resulting in rotation of earlier healed microcracks; after ca. 25° of rotation microfractures disappear; scale bar, 150 μm. (d) Alteration of plagioclase to unoriented sericite and epidote obscures albite twins (arrows); where alteration is complete, phyllosilicates (P) show strong alignment; scale bar, 300 μm. (e) Quartz (Q), new green biotite (NB), and relict sheared green biotite (RB) define fluxion structure around epidote clots (E); scale bar, 300 μm. (f) Strong foliation defined by white mica (WM) and dispersed, fine-grained epidote; quartz grain aggregates show subgrains (sg) and dynamically recrystallized new grains (ng); scale bar, 150 μm.

chlorite, calcite, apatite, and zircon. Original igneous green biotite grains are clearly distinguished by their relatively large size (>1000 μm) and by numerous inclusions of zircon and apatite <2 μm in diameter. These older biotites are commonly rimmed by 10 to 20 μm epidote grains. Biotite grains within transgranular fractures are commonly inclusion-free and have a grain size that varies with fracture width. In the narrower transgranular microcracks new biotite has a grain size of 30 μm or less; in wider fractures the new grains may be up to 300 μm long (Fig. 7b) but are commonly 80 to 100 μm long. The new biotite grains usually have their (001) planes parallel to the apparent opening direction of the cracks (Fig. 7b), which suggests that they grew concurrently with crack opening.

All interstitial quartz grains (i.e., those that are not apparently associated with fracture filling) show undulatory extinction, deformation bands, and minor development of 80 to 100 μm equidimensional subgrains and recrystallized grains around grain margins. Almost all of these grains contain dense arrays of planar, healed intergranular microcracks (Fig. 7c) that are subparallel to intergranular fractures in adjacent feldspar grains. Bubble trails and submicron-sized inclusions define the healed microcracks. Quartz within transgranular fractures, on the other hand, is generally inclusion-free, shows no evidence of microcracking, and is completely recrystallized to a polygonal mosaic structure.

The onset of plastic deformation along the margins of ductile shear zones in the Striped Rock granite is manifested differently by each mineral type present. Elongate aggregates of recrystallized quartz grains define a weak foliation that anastomoses around the more rigid feldspar grains. Figure 7c shows a quartz grain at the margin of a microcline grain. In the core of the quartz grain arrays of planar, healed microcracks are clearly visible, and few deformation band boundaries or subgrains occur. Adjacent to the microcline grain, elongate (70 × 30 μm) subgrains and recrystallized grains locally define a quartz grain-shape foliation. With progressive misorientation of the subgrains away from the core of the quartz grain, a concomitant rotation of the healed microcrack arrays occurs. After about 25° of rotation (measured on a Universal Stage by the change in orientation of C-axes of recrystallized grains), the planar nature of the healed microcracks is no longer discernible, and the new grains contain randomly scattered bubbles and inclusions. No transgranular fractures have been observed to cross-cut the recrystallized quartz grains. Thus the microcracks in the quartz and, by inference, the subparallel intergranular fractures in the feldspars, were precursory to crystal plastic deformation of the quartz.

Plagioclase grains always show some alteration to sericite and epidote in the Striped Rock granite, but along shear zone margins the alteration is almost complete (Fig. 7d). No apparent orientation of the white micas within the host plagioclase crystals was found. However, where alteration reactions appear to have gone to completion, the fine-grained white-mica quartz and epidote grains show a strong preferred grain-shape orientation (Fig. 7d) that is subparallel to the margins of adjacent microcline grains or to elongate recrystallized quartz grain aggregates.

Few relict grains of the original green biotite persist into the shear zones. Figure 7e shows one such grain that has been sheared in a right-lateral sense at the margin of a larger aggregate of epidote. Very minor alteration of the biotite to chlorite and opaque minerals occurs along cleavage planes. A poorly developed fluxion structure wraps around relict microcline grains and epidote grain aggregates. This fluxion structure is defined by moderately well-aligned white mica and recrystallized green biotite and by elongate recrystallized grains and grain aggregates of quartz.

Apart from the intergranular fractures, microcline–microperthite grains show very little internal strain at the shear zone margins. In the cores of the grains, bent and misaligned twins and patchy extinction are most commonly seen. Very minor recrystallization to grains <2 μm in size occurs along some microfractures and grain margins; these tiny grains proved too small to analyze with the electron microprobe. Epidote always occurs as aggregates of grains 10 to 20 μm in size; the high relief and granular nature of this mineral precluded positive identification of its deformation mechanisms.

Within ductile shear zones of width greater than 1 m a well-developed foliation wraps around relict fractured microline porphyroclasts >1 cm in diameter, producing an augen gneiss. No transgranular fractures have been observed in these augen gneisses. Although plagioclase is almost completely altered to white-mica quartz and epidote, the overall chemical composition of the sheared rock is not significantly different from that of the densely fractured rock. Aligned recrystallized green biotite, white mica, and dispersed epidote grains define the foliation (Fig. 7f). A true ribbon structure in quartz grain aggregates is rare, but lenticular aggregates that are 100 to 200 μm long and contain subgrains and recrystallized grains of quartz or microcline are common (Fig. 7f). These aggregates tend to have their long dimensions effectively pinned by adjacent phyllosilicates.

Deformation history, Striped Rock zones

Deformation in the Striped Rock granite appears to have commenced with widespread intergranular fracturing, especially of feldspar and quartz grains. Cross-cutting relationships (Fig. 7a) show that some cracks were mineralized prior to the formation of others. This suggests that fluids were present in significant quantities during the fracturing event. Such fluids may have enhanced crack growth ('stress corrosion cracking,' ATKINSON, 1979), particularly as the ionic species present would cause a reactivity greater than that of pure water (WIEDERHORN and JOHNSON, 1973; LAWN and WILSHAW, 1975; ATKINSON and MEREDITH, 1981). Compositional changes from relatively undeformed to highly fractured rock include the introduction of quartz and green biotite, with minor amounts of calcite. The presence of fluid may also have promoted the severe alteration of plagioclase grains, although the fluid's temperatures must have been sufficiently elevated to preclude the alteration of biotite to chlorite.

At a later stage ductile deformation localized along the earlier zones of more intense fracturing, possibly in response to movement on the major thrust systems to the southeast. The absence of any major change in bulk-rock chemical composition from fractured into sheared rock suggests that most of the fluid was introduced at the fracturing stage. The mechanisms of deformation within the shear zones included (a) slip on basal planes of original biotite grains, (b) recrystallization of new biotite and white-mica grains (the latter from altered plagioclase) with a preferred grain-shape orientation, probably accompanied by significant mechanical rotation of existing phyllosilicates, (c) intracrystalline plasticity and dynamic recrystallization of quartz grains, resulting in elongate aggregates of subgrains and new grains, and (d) continued intergranular fracturing of microcline grains together with minor crystal plastic deformation, producing bent twins and patchy extinction. That cracking was a precursor to the crystal plastic deformation in quartz is shown by the rotation of previously formed microcracks with subgrain rotation.

Discussion

In each of the two field areas discussed above there is good evidence of intergranular and transgranular fracturing of the rocks prior to the onset of crystal plastic deformation. In the Borrego Springs quartz diorite the earlier brittle event produced discrete zones of cataclasite, whereas in the Striped Rock granite, mineralized dilatant fractures pervaded the rock. Several factors could explain this difference in the mechanisms of brittle deformation. The most obvious factor is the potential variation in each of the extrinsic parameters that governed rock deformation in the two regions. Such parameters include strain rate, confining pressure, fluid pressure, fluid composition, applied stress, and temperature. Even if the extrinsic parameters were identical in each field area (an admittedly unlikely possibility), variations in intrinsic properties, such as modal composition and grain size, would play an important role in controlling the deformation mode. For example, under brittle conditions quartz-poor rocks are generally weaker than quartz-rich rocks and, although a granite may be able to support the applied stress and respond to it by distributed stable cracking, a quartz diorite would be more likely to fail along discrete zones of cataclasis.

Fluids that passed through the fractured quartz diorite introduced calcite and probably enhanced the reaction from magnesian to actinolitic hornblende. In the Striped Rock granite the fluids introduced new quartz, green biotite, and minor calcite. In both areas plagioclase grains were severely altered, but only in the Borrego Springs zones were fluid temperatures low enough to permit the biotite-to-chlorite reaction to proceed significantly. Pressure–temperature conditions of deformation cannot be estimated with any degree of certainty in either field area. In the Borrego Springs brittle-to-ductile zones the growth of new actinolitic amphiboles, both in the cataclasite and in the foliated, sheared rock, suggests that conditions of uppermost greenschist to lowest amphibolite facies prevailed throughout the deformation sequence. In the Striped Rock granite the deformation conditions are even less well constrained, but the presence of dynamically recrystallized quartz and the absence of

Table 1

Averaged Amphibole Analyses [OH calculated] from Borrego Springs Quartz Diorite

No. analyses:	Undeformed, 5	Breccia clasts, 4	Breccia overgrowths, 9	Recrystallization, 28
SiO_2	48.78	49.20	49.44	50.14
TiO_2	1.04	0.63	0.72	0.58
Al_2O_3	7.03	6.80	6.62	6.21
FeO	10.45	10.51	10.68	10.38
MnO	0.26	0.27	0.26	0.28
MgO	14.99	15.25	15.22	15.50
CaO	12.27	12.41	12.30	12.36
Na_2O	0.81	0.76	0.77	0.67
K_2O	0.56	0.49	0.52	0.46
H_2O	2.05	2.05	2.05	2.06
TOTAL	98.24	98.35	98.58	98.64
Formulae (24 oxygen atoms)				
Si	7.138	7.187	7.207	7.287
Al^{IV}	0.861	0.813	0.793	0.717
Al^{VI}	0.351	0.358	0.309	0.346
Ti	0.115	0.069	0.079	0.063
Fe^{2+}	1.280	1.284	1.303	1.261
Mn	0.032	0.033	0.032	0.034
Mg	3.270	3.320	3.307	3.356
Ca	1.924	1.943	1.922	1.924
Na	0.231	0.216	0.217	0.189
K	0.104	0.092	0.096	0.085

dynamically recrystallized feldspar suggest that deformation conditions in the greenschist facies range prevailed during the formation of the foliated rock.

Foliations in rocks that have been deformed by cataclastic flow have long been recognized in low-temperature experiments (e.g., JACKSON and DUNN, 1974; LOGAN et al., 1979, 1981) and in naturally occurring low-temperature and low-pressure fault gouge (CHESTER et al., 1985). Such foliations have been attributed to the mechanical rotation of rigid, inequant grains by cataclastic flow on a fine scale, along with grain-scale frictional sliding and the formation of discrete shear offsets (CHESTER et al., 1985). Foliated cataclasites may also be produced in very fine-grained fault gouge by late-stage pressure solution (MITRA, 1984). In each case these foliations are the result of a switch in the dominant deformation mechanism as the grain size is reduced. Microstructurally, one of the most significant differences between the ductile shear zones described here and foliated cataclasites is the absence of dynamically recrystallized grains in the latter.

The reason for the transition from cracking as the predominant mode of deformation to crystal plasticity is worthy of discussion. PATERSON (1978, p. 181) considers how an increase in confining pressure can increase the fracture stress so that the operation of intragranular slip systems is favored over microcracking. That such an increase in confining pressure could have occurred in either of the two field areas is doubtful, except as a result of tectonic loading by emplacement of an overlying thrust sheet. While this may have been possible in the Striped Rock granite area, it is unlikely in the Borrego Springs region, especially if the shear zones are related to Miocene normal faulting. Another consideration is the role of fluid pressure rather than confining pressure *per se*. It is well known that an increase in fluid pressure may cause a decrease in the effective normal stress and that rupture will tend to occur at lower applied stress levels. If a significant decrease in fluid pressure were to occur—if, for example, the fluids were allowed access to the surface or to a more permeable rock body, then the effective stress would be raised and both the strength and ductility of the rock would be increased. Microcrack growth could then become stable and crystal mechanisms predominate.

During brittle deformation of an originally unstrained grain, a minor amount of dislocation glide may occur prior to fracturing (PATERSON, 1978). The crack serves as a sink for the dislocations, so that for the most part the resultant smaller grains are relatively strain-free. When conditions are such that dislocation glide may begin to play an important role in the deformation, these small grains will be relatively weak and able to undergo dynamic recrystallization fairly easily (cf. TULLIS and YUND, 1985). Thus a switch to crystal plastic-deformation mechanisms is favored as the grain-size is reduced, and the shear strain will tend to localize and be contained within the finer-grained areas.

The onset of dynamic recrystallization in both field areas may also be in part the result of hydrolytic weakening due to increased water uptake along microcracks in the grains (e.g., KIRBY and KRONENBERG, 1984). Although definitive evidence of hydrolytic weakening in these naturally deformed rocks is lacking, water was certainly sufficient in the deformational environment to permit its participation on the lattice scale as well as the grain scale. As shown by TULLIS and YUND (1985), fine-grained minerals undergo recrystallization relatively readily, because the migrating grain boundaries do not have far to travel. In the presence of trace amounts of water trapped along grain boundaries and in the crystal lattice, diffusional mass transfer or pressure solution (RUTTER, 1976) also could contribute significantly to the plastic deformation of the small grains.

Conclusions

The vast majority of ductile shear zones described in the literature have been attributed to deformation at mid-crustal levels and any mineral or chemical changes within them attributed to the influence of syn or post-deformational fluid flow. It is suggested here that fluids may in fact play an even more significant role in brittle-to-ductile shear-zone formation by promoting early fracturing and alteration of the rock and thereby locally weakening the rock for subsequent crystal plastic deformation. In many shear zones evidence of earlier brittle deformation may not be preserved through the later plastic deformation. A close examination, however, of both along-strike and across-strike boundaries of ductile shear zones may reveal the signature of a precursor brittle event.

References

ATKINSON, B. K. (1979), *A fracture mechanics study of subcritical tensile cracking of quartz in wet environments.* Pure Appl. Geophys. *117*, 1011–1024.

ATKINSON, B. K. and MEREDITH, P. G. (1981), *Stress corrosion cracking of quartz: A note on the influence of chemical environment.* Tectonophys. *77*, T1–T11.

BLACIC, J. D. and CHRISTIE, J. M. (1984), *Plasticity and hydrolytic weakening of quartz single crystals.* J. Geophys. Res. *89*, B6, 4223–4239.

CARR, D. R. and KULP, J. L. (1957), *The K-Ar method of geochronology.* Bull. Geol. Soc. Amer. *68*, 763–784.

CHESTER, F. M., FRIEDMAN, M. and LOGAN, J. M. (1985), *Foliated cataclasites.* Tectonophys. *111*, 139–146.

ENGEL, A. E. J. and SCHULTEJANN, P. A. (1984), *Late Mesozoic and Cenozoic tectonic history of south central California.* Tectonics, *3*, 659–675.

FULLAGAR, P. D. and ODOM, A. L. (1973), *Geochronology of Precambrian gneisses in the Blue Ridge province of northwestern North Carolina and adjacent parts of Virginia and Tennessee.* Bull. Geol. Soc. Amer. *84*, 3065–3080.

GRIGGS, D. T. and BLACIC, J. D. (1965), *Quartz: Anomalous weakness of synthetic crystals.* Science, *147*, 292–295.

JACKSON, R. F. and DUNN, D. E. (1974), *Experimental sliding friction and cataclasis of foliated rocks.* Int. J. Rock Mech. Min. Sci. and Geomech. Abstr. *11*, 235–249.

JAEGER, J. C. and COOK, N. G. W., *Fundamentals of Rock Mechanics.* 3rd Ed., Chapman and Hall, London, 1979.

KIRBY, S. H. and KRONENBERG, A. K. (1984), *Hydrolytic weakening of quartz: Uptake of molecular water and the role of microfractures.* (abstr.), EOS (Amer. Geophys. Union Trans.) *65*, 277.

KNIPE, R. J. and WHITE, S. H. (1979), *Deformation in low grade shear zones in the Old Red Sandstone, S.W. Wales.* J. Struct. Geol. *1*, 53–66.

LAWN, B. R. and WILSHAW, T. R., *Fracture of Brittle Solids.* Cambridge Univ. Press, Cambridge, 1975, p. 112–114.

LEAKE, B. E. (1978), *Nomenclature of amphiboles.* Min. Mag. *42*, 533–563.

LISTER, G. S. and SNOKE, A. W. (1984), *S-C mylonites.* J. Struct. Geol. *6*, 617–638.

LOGAN, J. M., FRIEDMAN, M., HIGGS, N., DENGO, C. and SHIMAMOTO, T. (1979), *Experimental studies of simulated gouge and their application to studies of natural fault zones.* Proc. 8th Conf. Anal. Actual Fault Zones in Bedrock, U.S. Geol. Surv., Open File Rpt. *79–1239*, 305–343.

LOGAN, J. M., HIGGS, N. G. and FRIEDMAN, M. (1981), 'Laboratory studies on natural gouge from the U.S. Geological Survey Dry Lake Valley No. 1 well, San Andreas fault zone,' in *Mechanical Behavior of Crustal Rocks: The Handin Volume.* Geophys. Monogr. Am. Geophys. Union, *24*, 121–134.

MEANS, W. D. (1977), *Experimental contributions to the study of foliations in rocks: A review of research since 1960.* Tectonophys. *39*, 329–354.

MITRA, G. (1978), *Ductile deformation zones and mylonites: The mechanical processes involved in the deformation of crystalline basement rocks.* Amer. J. Sci. *278*, 1057–1084.

MITRA, G. (1984), *Brittle to ductile transition due to large strains along the White Rock Thrust, Wind River mountains, Wyoming.* J. Struct. Geol. *6*, 51–62.

ODOM, A. L. and FULLAGAR, P. D. (1984), 'Rb-Sr whole-rock and inherited zircon ages of the plutonic suite of the Crossnore complex, southern Appalachians, and their implications regarding the time of opening of the Iapetus Ocean,' in *The Grenville Event in the Appalachians and Related topics.* M. J. Bartholomew (ed.), Geol. Soc. Amer. Spec. Paper, *194*, 255–261.

OERTEL, G. (1983), *The relationship of strain and preferred orientation of phyllosilicate grains in rocks—A review.* Tectonophys. *100*, 413–447.

PASSCHIER, C. W. and SIMPSON, C., *Porphyroclast systems as kinematic indicators.* J. Struct. Geol., *in press.*

PATERSON, M. S., *Experimental Rock Deformation—The Brittle Field.* Springer-Verlag, Berlin, 1978, p. 161–185.

RAMSAY, J. G. and GRAHAM, R. H. (1970), *Strain variations in shear belts,* Canad. J. Earth Sci. *7*, 786–813.

RIECKEN, C. C., *Petrology of the Striped Rock Granite and surrounding rocks, Grayson County, Virginia.* Ph.D. Thesis (umpubl.), Virginia Polytech. Inst. and State Univ., 1966.

RUTTER, E. H. (1976), *The kinetics of rock deformation by pressure solution.* Phil. Trans. R. Soc. Lond. A, *283*, 203–219.

SCHMID, S. M. (1975), *The Glarus overthrust: Field evidence and mechanical model.* Eclog. Geol. Helv. *68*, 247–280.

SCHMID, S. M. (1976), *Rheological evidence for changes in the deformation mechanisms of Solnhofen limestones towards low stresses.* Tectonophys. *31*, T21–T28.

SEGALL, P. and POLLARD, D. D. (1983), *Nucleation and growth of strike-slip faults in granite.* J. Geophys. Res. *88*, B1, 555'–568.

SEGALL, P. and SIMPSON, C. (1986), *Nucleation of ductile shear zones on dilatant fractures.* Geology, *14*, 56–59.

SHARP, R. V. (1966), *Ancient mylonite zone and fault displacements in the Peninsular Ranges of southern California.* Geol. Soc. Amer. Spec. Paper, *101*, 333.

SHARP, R. V. (1979), *Some characteristics of the eastern Peninsular Ranges mylonite zone.* Proc. 8th Conf. Anal. Actual Faults in Bedrock, U.S. Geol. Surv., Open File Rpt. *79–1239*, 258–267.

SIBSON, R. H. (1977), *Fault rocks and fault mechanisms.* J. Geol. Soc. Lond. *133*, 191–213.

SIMPSON, C. (1983), *Strain and shape fabric variations associated with ductile shear zones,* J. Struct. Geol. *5*, 61–72.

SIMPSON, C. (1984), *Borrego Springs–Santa Rosa mylonite zone: A Late Cretaceous west-directed thrust in southern California.* Geology, *12*, 8–11.

SIMPSON, C. (1985), *Deformation of granitic rocks across the brittle–ductile transition.* J. Struct. Geol. *7*, 503–511.

SIMPSON, C. and SCHMID, S. M. (1983), *An evaluation of criteria to deduce the sense of movement in sheared rocks.* Bull. Geol. Soc. Amer. *94*, 1281–1288.

STOSE, A. J. and STOSE, G. W. (1957), *Geology and Mineral Resources of the Gossan Lead District and Adjacent Areas in Virginia.* Virginia Div. Mineral Resources Bull. *72*.

TODD, V. R. and SHAW, S. E. (1979), 'Structural, metamorphic and intrusive framework of the Peninsular Ranges batholith in southern San Diego County, California,' in *Mesozoic Crystalline Rocks.* P. L. Abbott and V. R. Todd (eds.) Geol. Soc. Amer. Annual Mtg. Field Trip Guidebook, San Diego State Univ., 177–231. Calif., San Diego.

TULLIS, J. A. and YUND, R. A. (1977), *Experimental deformation of dry Westerly granite.* J. Geophys. Res. *82*, 5705–5718.

TULLIS, J. A. and YUND, R. A. (1980), *Hydrolytic weakening of experimentally deformed Westerly granite and Hale albite rock.* J. Struct. Geol. *2*, 439–451.

TULLIS, J. A. and YUND, R. A. (1985), *Dynamic recrystallization of feldspars: A mechanism for ductile shear zone formation.* Geology *13*, 238–241.

TULLIS, T. E. (1976), *Experiments on the origin of slaty cleavage and schistosity.* Bull. Geol. Soc. Amer. *87*, 745–753.

WHITE, S. H. (1976), *The effects of strain on the microstructure, fabrics and deformation mechanisms in quartzite.* Phil. Trans. R. Soc. Lond. A, *283*, 69–85.

WHITE, S. H., BURROWS, S. E., CARRERAS, J., SHAW, N. D. and HUMPHREYS, F. J. (1980), *On mylonites in ductile shear zones.* J. Struct. Geol. *2*, 175–188.

WHITE, S. H. and KNIPE, R. J. (1978), *Transformation-and reaction-enhanced ductility in rocks.* J. Geol. Soc. Lond. *135*, 513–516.

WIEDERHORN, S. M. and JOHNSON, H. (1973), *Effect of electrolyte pH on crack propagation in glass.* J. Am. Ceram. Soc. *56*, 192–197.

(Received 1st September 1985, revised 24th April 1986, accepted 27th April 1986)

PAGEOPH, Vol. 124, Nos. 1/2 (1986) 0033-4533/86/020289-19$1.50+0.20/0

The Effect of Cyclic Operation of Brittle and Ductile Deformation on the Metamorphic Assemblage in Cataclasites and Mylonites

H. Stel[1]

Abstract—Cataclasites and mylonites, and the brittle–ductile processes that produce them, were studied at exposures along the northern rim of the Grong culmination, a transverse basement antiform in the central Scandinavian Caledonides. The rock suite studied is composed of gneisses, mylonites, and cataclasites which have a granodioritic composition. The microstructure of the rocks appears to be the result of repeated alternations of brittle faulting (associated with hydrothermal mineral growth) and ductile deformation of the crystallization products. During brittle faulting K-feldspar–chlorite veins are formed, probably by incongruent pressure solution of micas. During plastic deformation of the rocks the mineral association is transformed to white mica and green biotite, according to the reaction

$$5 \text{ biotite} + 3 \text{ white mica} + 9 \text{ quartz} + 4\ H_2O = 8 \text{ K-feldspar} + 9 \text{ chlorite}$$

During this reaction plagioclase is overgrown by mica and epidote, and K-feldspar crystals are replaced by albite. The reactions which involve K-feldspars are cyclic: K-feldspar that is generated in cracks tends to be removed by albitization during ductile deformation. It is concluded that the mylonites studied represent a movement zone in the Earth's crust in which seismic and aseismic slip alternated during a large part of the deformation history.

Key words: Brittle–ductile transitions, cyclic metamorphic reactions, feldspar deformation, Transmission Electron Microscopy, Scandinavian Caledonides.

Introduction

Fossil movement zones in the earth's crust show evidence of both brittle and ductile deformation. The circumstances that determine the operation of either process during the deformation history of a rock may vary. In mylonites, rigid minerals such as feldspars and garnet may have behaved in a brittle manner, while others (quartz, calcite) show microstructures indicative of plastic yielding (Mitra, 1978; Boullier, 1980). In these cases the divergence of deformation processes may be attributed to the differences in ductility of the various minerals at given metamorphic conditions. It can be shown that in some mylonite zones a period of 'steady state' ductile deformation is followed by a late stage of brittle faulting (e.g., Passchier, 1982). Such

[1] Institute of Earth Sciences of the Free University, 1085, de Boelelaan, 1081 HV Amsterdam, The Netherlands.

change in the deformation process may be attributed to declining metamorphic conditions during the deformation history, that is, by uplift of a rock suite, in which case the 'brittle–ductile transition field' is entered.

In a recent paper STEL (1981) suggested that the microstructure of some mylonites and cataclasites had been produced by a repetitive alternation of brittle and ductile deformation mechanisms. The former process was associated with the hydrothermal growth of quartz; in the latter the crystallization products were plastically deformed and transposed to produce a foliation in the rocks. This study did not assess, however, whether metamorphic conditions remained stable or not during the deformation history. Moreover, since both brittle and ductile deformation tend to destroy each other's effects, only the latest stages of the deformation history are reflected by the microstructure.

The present paper describes microstructures which indicate that a cyclic operation of brittle and ductile processes dominated the development of the (micro) structure of cataclasites and mylonites. It will be argued that both processes were active during a considerable part of the deformation history while the rocks remained at approximately the same level in the earth's crust. The paper also discusses the parameters that governed repetitive brittle–ductile transitions.

Material

The mylonite studied is found at the northern edge of the Grong culmination, a basement antiform that transects the central Scandinavian Caledonides. The rock unit is part of the Granite-mylonite nappe, a mobilized basement slice which was thrusted over its parent rock during the Caledonian orogeny (OFTEDAHL, 1956; GEE, 1975; AUKES et al., 1979). Whole rock chemical analyses of 16 samples of the mylonite zone were determined and the results are given in Table 1, CIPW normative values are given in Figure 1.

The texture of the rocks from the granite mylonite nappe is highly variable. Undeformed ingneous rocks are found in platy mylonites and cataclasites with chaotic texture. The diversity in texture is accompanied by differences in the mineralogical composition, due to the ongoing retrograde metamorphic reactions during the deformation of the rocks (AUKES et al., 1979).

In the following sections the interrelations of the different types of deformation to mineralogical reactions are described.

Table 1

*Whole-Rock Chemical Analyses of Samples of the Granite Mylonite Nappe**

Nr	00.78	79.26	81.1c	81.1b	81.1b′	81.1a	79.24′	80.39	81.21	78.04	81.56	81.56′	81.60	80.86	79.28	78.89
SiO_2	59.70	67.77	57.19	58.00	57.36	57.19	67.53	63.43	61.28	62.11	62.41	65.18	62.49	62.38	63.52	60.27
TiO_2	0.72	0.71	0.66	0.67	0.72	0.66	0.67	0.70	0.73	0.58	0.70	0.68	0.69	0.73	0.71	0.68
Al_2O_3	17.73	15.05	17.32	17.20	17.19	17.40	14.90	16.32	16.44	16.58	17.36	16.63	17.13	17.32	17.09	16.28
Fe_2O_3	0.50	0.17	2.68	2.80	2.89	2.20	1.31	1.05	0.78	0.63	1.04	0.43	0.51	1.06	0.45	0.81
FeO	4.57	2.91	3.21	3.25	3.25	3.40	1.78	3.56	4.02	4.43	3.51	3.27	3.28	3.62	3.22	3.88
MnO	0.09	0.08	0.08	0.05	0.07	0.07	0.02	0.06	0.03	0.06	0.07	0.08	0.09	0.06	0.10	0.04
MgO	2.33	0.94	2.75	2.89	2.90	2.91	0.91	2.63	3.41	2.27	2.86	2.43	1.36	2.24	1.24	3.52
CaO	5.00	2.96	6.00	6.44	6.44	6.00	1.92	4.32	4.36	4.79	4.35	3.57	3.75	3.68	2.88	4.58
Na_2O	3.77	3.54	3.01	4.27	4.16	3.94	3.70	3.41	4.02	3.98	3.37	3.62	4.18	3.75	4.21	3.88
K_2O	3.65	5.44	4.72	2.60	2.62	2.90	5.92	3.52	3.98	4.13	3.58	2.61	4.56	4.11	4.70	4.14
P_2O_5	0.37	0.25	0.38	0.40	0.39	0.37	0.24	0.28	0.27	0.36	0.26	0.24	0.22	0.32	0.24	0.38
H_2O	1.36	0.76	1.29	1.35	1.32	1.83	0.50	0.68	1.38	0.56	0.72	1.12	0.81	0.68	0.92	1.42
CO_2	—	—	—	0.08	0.20	0.10	—	—	0.10	—	—	—	—	—	—	—
Tot.:	99.79	100.58	99.29	100.10	99.51	98.97	99.40	99.96	99.98	100.48	100.23	99.97	99.07	99.94	99.48	99.88

* Specimens Nr 00.78, 79.26, 81.1, a–c, and 79.24′ are deformed granodiorites; 81.1 a, b, b′, and c are from the shear zone described in the paper; 81.1 a is the least deformed (igneous texture), 81.1 b and b′ are moderately deformed, and 81.1c is mylonitic; Nr 80.39, 81.21, 78.04, 81.56, 81.56′, 81.60, 80.86, 79.28, and 78.89 represent strongly deformed rocks.

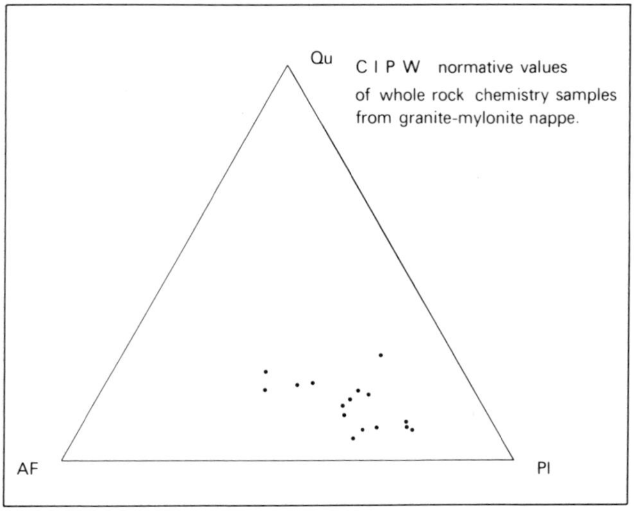

Figure 1
Plots of CIPW normative values of whole-rock chemistry of 16 samples from Granite-mylonite nappe: Qu, quartz; Pl, plagioclase; AF, alkali feldspar.

Textural and mineralogical transformations in sheared granodioritic rocks

Relatively undeformed granodioritic rocks show transitional relationships to mylonites in small-scale shear zones, in which a textural variation from igneous to mylonitic can be studied in relation to mineralogical changes. Figure 2 is a schematic drawing of the studied shear zone. From the edge towards the centre of the shear zone the rocks show a gradational increase in the development of the foliation, which progressively rotates in orientation. It is interpreted that this structure has been formed by a deformation history approximating progressive inhomogeneous simple shear (RAMSAY and GRAHAM, 1970). At the edge of the shear zone the rock is coarse-grained and unfoliated. Irregularly shaped feldspar-rich domains contain predominantly plagioclase crystals; quartz and alkali fedlspar are found locally.

Plagioclase crystals are subhedral and twinned; microprobe analyses reveal that these are compositionally An_{30}. Quartz occurs as blebs, and the mineral shows undulose extinction and local recrystallization. In one thin section is found alkali feldspar, which demonstrates a flame type of perthitic exsolution structure.

Biotite-rich domains are composed of aggregates of brown biotite in a decussate arrangement (VERNON, 1974). The mineral often includes rutile, sphene, and zircon.

Towards the centre of the shear zone the rocks are affected by metamorphism and deformation and show an incipient foliation, predominantly in biotite-rich

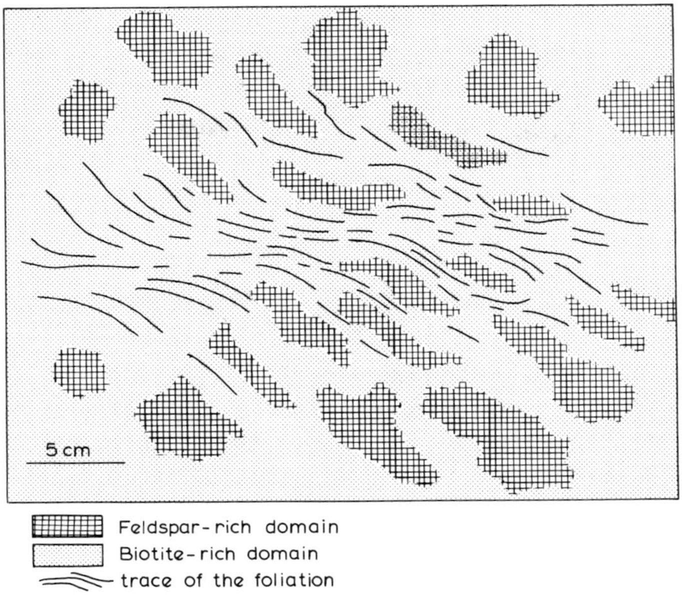

Figure 2
Diagram of small-scale shear zone in relatively weakly deformed granodiorite.

domains. Large biotite crystals are folded or kinked. Kink-band boundaries are decorated by aggregates of rutile and sphene. This mineral association is also found at grain boundaries of deformed biotite aggregates. Locally, epidote is found at such sites; it seems to have grown into the biotite. The relative amount of biotite versus epidote + sphene + rutile + opaques in deformed regions varies from 9:1 to 1:1.

An incipient foliation is defined by irregularly spaced bands 1 to 5 mm in width, composed of fine-grained epidote, sphene, rutile, and occasionally chlorite and green biotite. These bands run oblique to the trend of the shear zone. They may abut coarse, relatively undeformed biotite aggregates. Because the mineral association in the spaced bands is similar to that found at kink-bands and grain boundaries of deformed biotite (aggregates), a genetic relation is suggested. The locally high amount of Ti-bearing minerals relative to the amount of biotite suggests the operation of nonconservative processes: breakdown of biotite and selective transport of elements. The newly formed mineral association would indicate loss of K and introduction of Ca in these biotite-rich domains. BEACH (1979, 1980) and KERRICH et al. (1980) described similar microstructures in biotite and attributed them to the operation of pressure solution and stress corrosion cracking. The spaced bands are interpreted accordingly: as deformed biotite aggregates which were completely altered by solution and redeposition of elements.

Plagioclase crystals show further alteration to epidote-group minerals and white mica. The alteration products are also found in bands parallel to the foliation.

In the centre of the shear zone the rocks are composed completely of fine-grained epidote, phengite, green biotite, chlorite, albite, and quartz; locally there are elongate saussuritized plagioclase relicts (aspect ratio, 10:1) which are aligned with the foliation. The minerals are more or less evenly dispersed throughout the rock, and the texture is mesoscopically homogeneous. The foliation is defined by dimensionally oriented new micas and runs parallel to the trend of the shear zone.

Metamorphic reactions related to ductile shear

The textural variation of the rocks from the different structural domains of the shear zone is the result of a close interplay of deformation and metamorphism. Chemical analyses of specimens from different parts of the shear zone are similar, showing that the mineralogical transformation was isochemical (Table 1, samples 81 a, b, c). The microstructures suggest that the following mineral transformation has taken place:

brown biotite + plagioclase = epidote-group minerals + albite +
phengite + green biotite + chlorite + sphene + rutile

The newly formed mineral association points to middle greenschist facies conditions (MIYASHIRO, 1973; FREY et al., 1976; GRAHAM et al., 1983).

Metamorphism related to brittle faulting

The Grong culmination is transected by numerous large-scale faults, some of which have produced a displacement exceeding 1 km (AUKES et al., 1979). Within the fault zones are cataclasites that bear evidence of hydrothermal mineral growth (STEL, 1981). Brittle faults in the Granite-mylonite nappe are invariably associated with K-feldspar–chlorite veins (Fig. 3). The width of these veins varies from 1 mm to 0.5 m (observed lengths exceeding 200 m).

Microprobe analyses indicate that the K-feldspar mineral is nearly pure $KAlSi_3O_8$/<0.2 wt-% Na_2O), with a detectable amount of BaO (<0.1 wt-%). The structural state of the mineral is maximum microcline, as was determined by x-ray diffraction powder techniques (STEL, 1984).

K-feldspar crystals from undeformed veins show growth structures. Some crystals are built up by numerous microveins (Fig. 4). Overprinting relations indicate that numerous crack-seal events have taken place (RAMSAY, 1980). Other veins contain subgrains that are elongate in a direction normal to the vein wall and display a 'comb structure' (SPRY, 1969). Subgrain boundaries and microvein walls are stained by haematite particles and fluid inclusions. Microcline crystals do not show perthitic

Figure 3
Discordant K-feldspar vein in brittle fault in mylonitic rock. Note offset of mylonitic foliation.

Figure 4
Micrograph of composite K-feldspar vein. Vein composed of numerous microveins, as revealed by dust trails. Crossed polarizers. Scale bar, 1 mm.

Figure 5
Micrograph of folded K-feldspar vein showing segmentation into clasts. Crossed polarizers. Scale bar, 1.5 mm.

exsolution structures and are untwinned.

Most K-feldspar veins are deformed; some are folded and demonstrate breakup into clasts (Fig. 5). Several generations of veins are found, deformed ones being transected by younger, undeformed ones.

Throughout the entire rock suite K-feldspar megacrysts are found, which in some cases are concentrated in bands parallel to the foliation in mylonites (Fig. 6). The microstructure and chemical composition of the megacrystals are identical with those of clasts from deformed veins. It is thought that the megacrysts were generated as syntectonic veins in brittle structures and were subsequently broken up by plastic yielding of the matrix. The occurrence of different generations of veins, displaying a wide variety in the amount of strain, indicates that alternations of brittle and ductile deformation processes took place during a large part of the deformation history.

The mechanical behaviour of feldspar and its relation to the metamorphic processes

The repeated alternation of brittle and ductile deformation processes in the rocks is reflected in the microstructure of the alkali feldspars, which show a strong interrelation of mechanical behaviour and metamorphic processes.

Alkali feldspars occur in two microstructural arrangements: as megacrystals,

Figure 6
Outcrop. Concentration of K-feldspar megacrystals in bands parallel to a mylonitic foliation. Note also discordant veins.

which show little evidence of plastic deformation, and in fine-grained bands, the microstructure of which suggests that the mineral underwent ductile deformation associated with recrystallization. In the latter, the various microstructures are related to different metamorphic reactions that alternately took place in the rocks.

Most K-feldspar megacrysts show undulose extinction; lattice misorientations of about 5° have been measured. At grain boundaries and cracks the mineral is kinked. The minerals are embedded in a fine-grained matrix mainly composed of albite grains that form mantles and elongate tails around K-feldspar. Tails of various clasts may join to produce a compositional layering (Fig. 7).

By transmission electron microscopy the dislocation substructure of K-feldspar was revealed. The structure is characterized by slip bands showing a high dislocation density (Fig. 8a). The dislocation form loops, the different parts of which run parallel to crystallographic directions (Fig. 8b). The nature of these defects is discussed elsewhere (Stel, 1984). There is little evidence of recovery structures; observed dislocation walls, by correlation with optical microscopy, are thought to be growth structures. No recrystallization effects have been observed.

Nearly all the K-feldspar porphyroclasts show evidence of brittle deformation. Two types of crack, described in the following, have been distinguished on the basis of their 'post-crack' history.

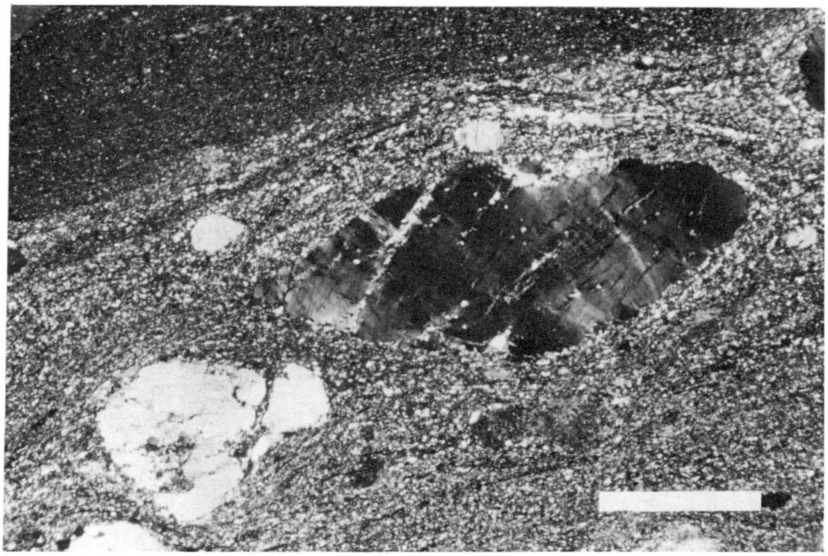

Figure 7
Micrograph of K-feldspar clasts in albite-rich matrix formed by joined tails and mantles which constitute a banding (light versus dark bands). Crossed polarizers. Scale bar, 2 mm.

1. Healed cracks

The cementing mineral is newly crystallized K-feldspar, which consists of microveins that are variable in width from 0.1 to 2 mm. The orientation of the veins is

A

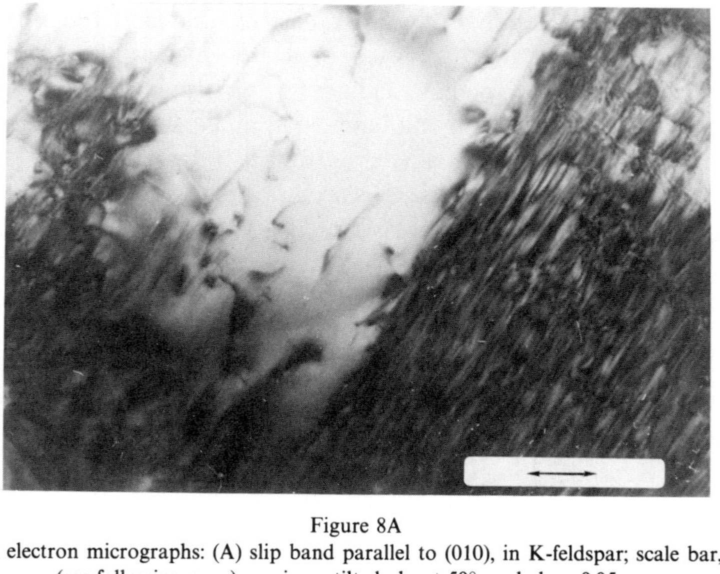

Figure 8A
Transmission electron micrographs: (A) slip band parallel to (010), in K-feldspar; scale bar, 0.1 μm; (B) (see following page) specimen tilted about 50°; scale bar, 0.05 μm

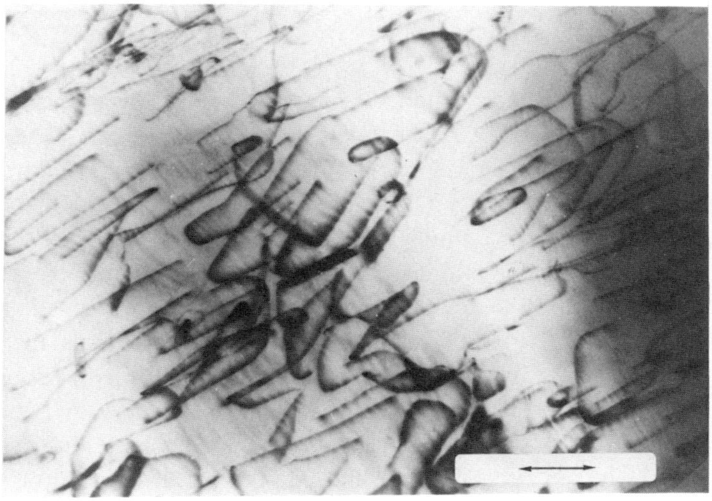

Figure 8B

such that in general an extension parallel to the local foliation may be deduced. In some cases the microveins occur in a conjugate pattern. This type of crack does not need to be confined to the K-feldspar clast; in general, the microveins are continuous in the matrix. In most cases the cracks contain composite chlorite–K-feldspar veins.

Transmission electron microscopy demonstrates the difference in defect structure between the healed crack and its host. The microstructure indicates that the cracks were opened before cementation took place (Fig. 9).

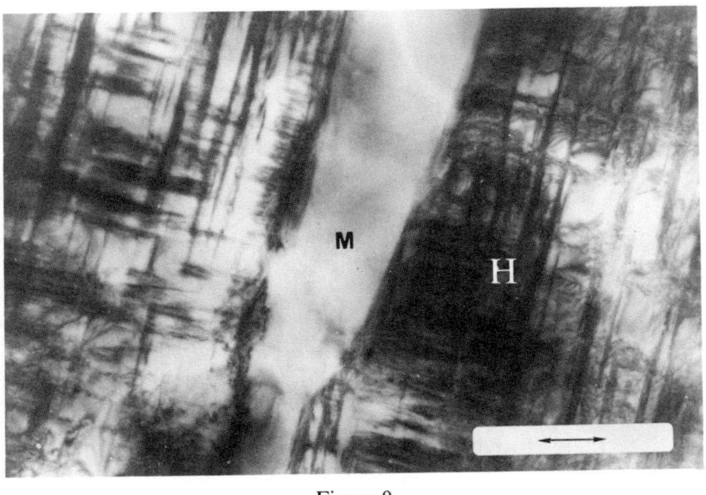

Figure 9
Transmission election micrograph of healed crack (H) = microvein (M) of neocrystallized K-feldspar in deformed K-feldspar host. Mineral analyses based on diffraction characteristics. Scale bar, 0.1 μm.

2. Cracks along which albite is found

This type of crack is always confined to one specific K-feldspar clast. The cracks are not persistent in the matrix. Two microstructural types can be distinguished, which show different relations of the albite phase to the K-feldspar host. In one type of crack the albite phase is in crystallographic continuity with the K-feldspar host. In this microstructure albite contains fluid inclusions very much exceeding those in the host mineral. In general, the number of cracks increases from the core to the rim of a clast; at the rim the albite domains join to produce a pure albite mantle around K-feldspar crystals (Fig. 10). The microstructure is interpreted as the result of a replacement of K-feldspar by albite, this process taking place by a cation exchange through the pore fluid, which finds access through the cracks (ORVILLE, 1963). The anomalously high density of fluid inclusions in the albite phase with respect to K-feldspar is thought to be the result of the shrinkage of the Si–Al–O lattice during the cation exchange (cf. HEIMANN, 1979).

The other microstructure is characterized by rows of small, equant albite grains along crack-like structures. The albite grains appear to corrode the K-feldspar host (Fig. 11). In this microstructure there is a crystallographic misorientation between albite and its host. The microstructures resemble recrystallized deformation bands. Because the host minerals show no associated relative displacements, however, an origin by microcracking is favoured.

Transmission electron microscopy demonstrates a low crystal-defect density of the albite phase with respect to the K-feldspar host (Fig. 12). The grain boundaries of albite are smooth or lobate, suggesting corrosion. In some cases the boundaries are decorated by bubbles. The microstructural relations of albite grains and the K-feldspar host in this type of crack are similar to those found at the rims of porphyroclasts. The microstructure is interpreted as being due to the replacement of K-feldspar by albite by a solution neocrystallization process, such as proposed by PETROVIC (1974). In the present case a correlation is found between the type of replacement and the defect structure of the K-feldspar host. Porphyroclasts showing albitization by a cation-exchange mechanism demonstrate no undulose extinction and in transmission electron microscopy demonstrate a relatively low defect density. Those porphyroclasts that show albitization related to solution neocrystallization are deformed, showing undulose extinction and a relatively high dislocation density.

The microstructures described above suggest that in conditions of greenschist facies plastic deformation of K-feldspar occurs. The mechanisms of deformation, such as the formation of slip bands, kinking, and brittle failure, are typical of a low-temperature environment in which diffusional processes are thought not to be rate-determining (cf. NICOLAS and POIRIER, 1976). Accordingly, the formation of a finely grained K-feldspar susceptible to plastic flow (as is typical of mylonites) is suppressed, because in most cases such a matrix formation is a result of diffusion-

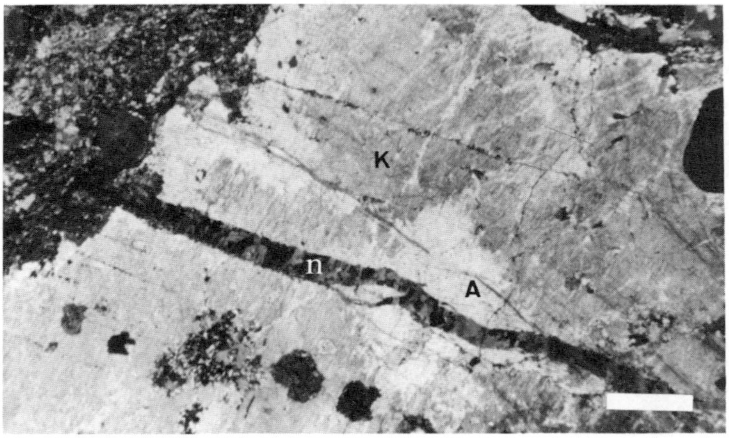

Figure 10
Micrograph. K-feldspar (K) showing albitization along cracks and grain boundaries. Note that albite phase (A) does not constitute new grains. Note also microvein of neocrystallized K-feldspar. Crossed polarizers. Scale bar, 1.5 mm.

controlled solid-state recrystallization (HOBBS et al., 1976; WHITE, 1976; POIRIER and GUILLOPÉ, 1979).

In the rocks studied the relation of megacrysts to the matrix seems to indicate that during deformation K-feldspar megacrysts were replaced by finely grained

Figure 11
Micrograph of K-feldspar host (clast from boudinaged vein) showing two types of crack: a—b, healed cracks; c—d, cracks along which albite grains appear to corrode K-feldspar. Different generations of each type of crack can be deduced. Crossed polarizers. Scale bar, 1.5 mm.

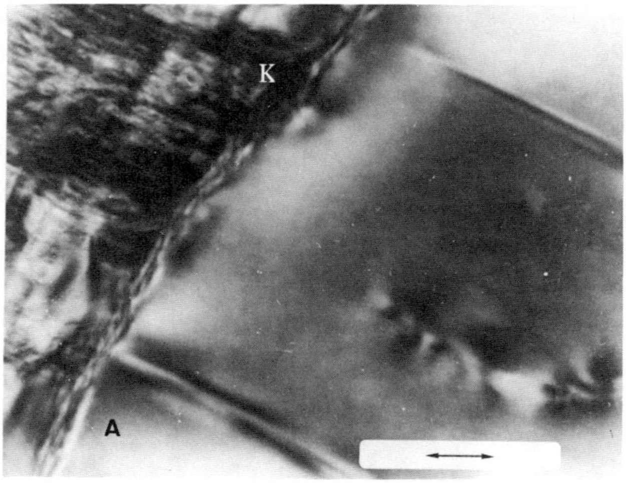

Figure 12
Transmission electron micrograph of strain-free albite grain in K-feldspar host (K). Note difference in defect density between albite (A) and host. Mineral determination based on diffraction pattern analyses. Scale bar, 0.1 μm.

feldspar aggregates of varying composition. Similar observations have been made with respect to granite mylonites from other areas; cf. ALLISON et al. (1979), KERRICH et al. (1980), and HANMER (1982). These authors attributed the apparent ductility to a flow-strength reduction by strain-induced recrystallization. The compositional variation of the recrystallized material with respect to the host was thought to be indicative of diffusive mass transfer during deformation, allowing grain-boundary sliding ("superplasticity"). Whereas in metals and most silicates the temperature required for the operation of diffusion-enhanced grain-boundary sliding is thought to be about $T = 0.5T_{\text{melt}}$, following KERRICH et al. (1980), in feldspars the process may operate at much lower temperatures ($T = 0.2T_{\text{melt}}$).

In the present study it is found that the interaction of the pore fluid and the K-feldspar crystals is of vital importance to the formation of new, strain-free grains. The microstructures suggest that strain-enhanced solution took place in the presence of an Na-rich fluid and neocrystallization of fine-grained albite took place reducing the strength of the rock. The nature of the pore fluid appears to be of major importance to the mechanical behaviour of the rocks and as such may be denoted an independent variable (cf. FERRY, 1983).

The microstructures indicate that the chemical properties of the pore fluid changed repeatedly during the deformation. In the next section these are described and discussed in relation to ongoing metamorphic reactions during brittle and ductile deformation of the rocks.

Relationships of microstructure, pore fluid, and large-scale deformation mechanism

Figure 11 shows various overprinting relations of two different types of crack. The younger type associated with neocrystalline K-feldspar, cross-cut the other, older type, along which albitization took place, which in turn overprinted the first type. It may be concluded that the composition of the pore fluid changed repeatedly. This phenomenon appears to be related to the manner of occurrence of albite and K-feldspar. K-feldspar is found in cracks that have been opened and are throughout the whole rock. Apparently the mineral grew in open space. Albite is never found in tension gashes or at other sites where crystallization in free space could have taken place; instead, it is found solely as a replacement for K-feldspar.

Because all the specimens analysed have chemical compositions close to that of granodiorite, which is assumed to be the parent rock of the mylonites, and no large variations in the contents of K and Na were found, it is concluded that the change in the compositions of the pore fluid did not result from an influx of material from external sources but is due to internal factors. It is thought that the source material of K-feldspar + chlorite veins is derived from micas undergoing pressure solution as follows (cf. WINTCH, 1975; MEYER and HEMLEY, 1967; BEACH, 1979),

$$\text{muscovite} + 6\ SiO_2 + 2K^+ = 3\text{K-feldspar} + 2H^+ \qquad (1)$$

or, more generally (cf. HOSCHEK, 1973, and GRAHAM et al., 1983),

$$5\ \text{biotite} + 3\ \text{white mica} + 9\ \text{quartz} + 4H_2O = 8\ \text{K-feldspar} + 9\ \text{chlorite} \qquad (2)$$

The replacement of K-feldspar by albite is thought to be related to the overgrowth of plagioclase by epidote-group minerals and white mica (see section shear zone) following Reaction 3 (MEYER and HEMLEY, 1967):

$$4\ \text{anorthite} + \text{K-feldspar} + 2\ H_2O = 2\ \text{clinozoisite} + \text{muscovite} + 2\ SiO_2 \qquad (3)$$

By reaction 3 the Na^+ is released to the pore fluid by breakdown of the albite component of plagioclase and is then available for the albitization of K-feldspar.

In an earlier section it is found that, during the plastic deformation of grandiorites under greenschist facies conditions, retrograde mineral reactions take place that are qualitatively expressed as follows:

$$\text{plagioclase} + \text{brown biotite} = \text{white mica} + \text{green biotite} + \text{epidote} + \text{albite}\ (+\ \text{rutile, sphene}) \qquad (4)$$

The transformation reaction 4 involves pressure solution of brown biotite and overgrowth of plagioclase by a new generation of mica. Accordingly, dissolved mica appears to be incorporated in the rocks in two distinct manners: first, by mica

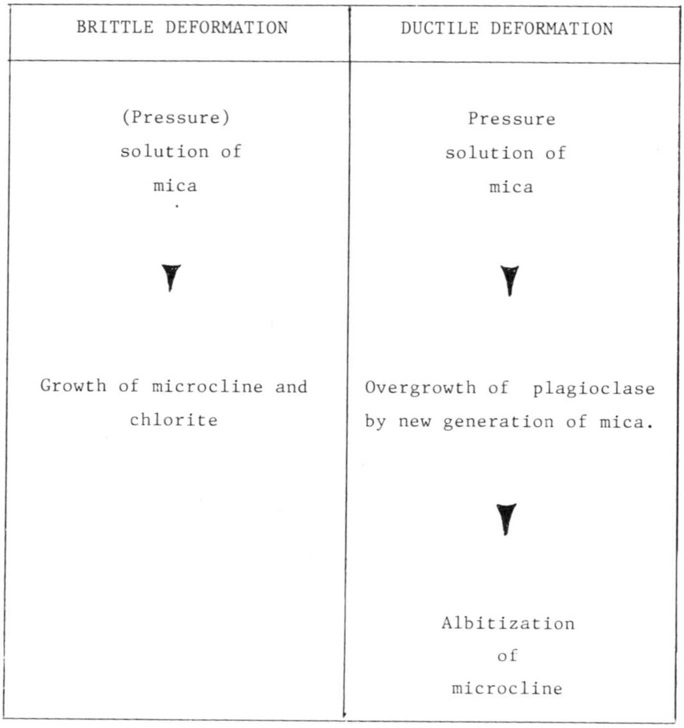

Figure 13
Schematic representation of relation of metamorphic processes and deformation mechanisms in mylonites studied.

overgrowth of plagioclase during ductile deformation, thereby releasing Na^+, which causes albitization of K-feldspar and secondly, as K-feldspar–chlorite veins following a brittle event.

The metamorphic reactions involving alkali feldspars appear to be cyclic. K-feldspar is generated in cracks, and by subsequent albitization during ductile deformation it is removed from the rocks, and in a renewed brittle phase it may be formed again. The mineral associations both before and after a brittle event being similar, it may be concluded that the metamorphic conditions remained essentially stable, that is, that deformation took place at approximately the same level in the Earth's crust. It appears that the different mineral reactions are related to distinct deformation mechanisms, as illustrated in Figure 13.

It is not known which factors governed the differences in mineral reaction associated with either brittle or ductile deformation. Three possible parameters are given: (a) pressure variations in the rocks, (b) variations in the temperature, and (c) a change in the chemical activity of the pore fluid. With respect to (a), if the microcracks in a rock form a fluid-filled network extending over a considerable distance, the fluid pressure within these structures is lower than the rock pressure (FYFE et al., 1978), thereby favouring K-feldspar growth over K-rich mica (MEYER and HEMLEY, 1967); then, after the cementation of the microcracks, the pressure will increase to a value equal to the rock pressure, favouring K-rich mica growth. With respect to (b), fluid migration in a rock through cracks may cause a fast heat transport, thereby governing a change in mineral reactions (FERRY, 1983); in the present case, a sudden lowering of the temperature would favour K-feldspar + chlorite growth at the expense of mica. Concerning (c), brittle deformation involves a migration of fluids, present as films at grain boundaries, to cavities in the rock; in this case the fluid might be buffered, thereby changing its chemical activity and, as a result, favouring different mineral reactions (GRAHAM et al., 1983).

Discussion

The granite mylonite nappe is mainly composed of mylonites that are well lineated and foliated and intensively folded, suggesting that the tectonic history involved ductile deformation (BELL and ETHERIDGE, 1973), that is, an aseismic movement. Cataclasites and breccia are found only in late-stage faults; this would indicate that seismic slip took place during the latest stages of the deformation history only. The microstructure of the mylonites, however, indicates that the deformation history involved repeated brittle–ductile transitions. In the movement zone cataclasites and breccia must have been formed repeatedly, their mesoscopic fabric having been destroyed by subsequent ductile deformation. It is concluded that the nappe represents a movement zone in which seismic and aseismic slip alternated. It is difficult to establish whether intrinsic or extrinsic parameters governed the operation of either brittle or ductile deformation. A possible external cause of a ductile–brittle transition could be that a different large-scale tectonic movement took place. For example, ductile deformation could be the result of permanent stress (as by loading), whereas a brittle event could be due to a suddenly induced movement. Alternatively, if intrinsic factors governed the changes in mechanism of deformation, it is likely that strain hardening played an important role. If it is assumed that this process took place, the microstructural development of the rocks may be imagined as follows. As a result of strain hardening during ductile deformation the differential stress rises until failure strength is reached, at which time brittle failure takes place. Faulting involves a sudden drop in the differential stress and the formation of a fluid-filled network of cracks and cavities in which hydrothermal mineral growth causes cementation of the clasts (cf. FYFE et al., 1978). After cementation the differential stress in the rock will rise again in order to allow renewed plastic deformation, and a new cycle of events may take place.

If the latter parameter determined the mechanical behaviour, it follows that the rocks underwent a deformation involving periodic stress releases. The seismic history of the movement zone would be similar to a stick-slip behaviour (cf. AUGEVIN et al., 1982).

Conclusions

There is a strong interaction between deformation mechanisms and metamorphic reactions. Incongruent pressure solution of micas yields material for the growth of K-feldspar–chlorite veins during and after brittle faulting. Intragranular cracking of K-feldspar clasts during ductile deformation of the matrix of mylonites is associated with albitization. The mylonites studied may represent an ancient seismic zone in which seismic and aseismic slip alternated. The various processes took place under approximately similar metamorphic conditions.

Acknowledgments

This work was carried out while the author benefited from a scholarship from the Netherlands Organization for the Advancement of Pure Research (ZWO). The research was performed at the Department of Structural Geology of the State University at Utrecht under the supervision of Prof. Dr. H. J. Zwart and Dr. J. N. Boland. Dr. I. van der Molen is thanked for his stimulating criticism.

REFERENCES

ALLISON, J. BARNETT, R. L. and KERRICH, R. (1979), *Superplastic flow and changes in crystal chemistry of feldspars.* Tectonophys. *53*, T41–46.

AUGEVIN, C. L., TURCOTTE, D. L. and FURNISH, M. D. (1982), *Pressure solution lithification as a mechanism for the stick-slip behaviour of faults.* Tectonics *1*, 151–160.

AUKES, P. G., REYMER, A. P. S., DE RUITER, G. W. M., STEL, H. and ZWART, H. J. (1979), *The geology of the Lierne district, north-east of the Grong culmination, central Norway.* Norges Geol. Unders. Publ. *354*, 115–129.

BEACH, A. (1979), *Pressure solution as a metamorphic process in deformed terrigenous sedimentary rocks.* Lithos *12*, 51–58.

BEACH, A. (1980), *Retrogressive metamorphic processes in shear zones with special reference to the Lewesian complex.* J. Struct. Geol. *2*, 257–263.

BELL, T. H. and ETHERIDGE, M. A. (1973), *Microstructure of mylonites and their descriptive terminology.* Lithos *6*, 337–348.

BOULLIER, A. M. (1980), *A preliminary study of the behaviour of brittle minerals in a ductile matrix.* J. Struct. Geol. *2*, 211–217.

FERRY, J. M. (1983), *Regional metamorphism of the Vassalboro formation, South Central Maine, USA: A case study of the role of fluid in metamorphic petrogenesis.* J. Geol. Soc. London *140*, 551–576.

FREY, M., HUNZIKER, J. C., O'NEILL, J. R. and SCHANDER, H. W. (1976), *Equilibrium disequilibrium relations in the Monte Rosa granite, western Alps: Petrological, Rb-Sr and stable isotope data.* Contrib. Mineral. Petrol. *55*, 147–179.

FYFE, W. S., PRICE, N. J. and THOMPSON, A. B. (1978), *Fluids in the Earth's Crust*. Elsevier, Amsterdam, 383 pp.

GEE, D. G. (1975), *A geotraverse through the Scandinavian Caledonides-Ostersund to Trondheim*. SGU *C717*, 66 pp.

GRAHAM, C. M., GREIG, K. M., SHEPPARD, S. M. and TURI, B. (1983), *Genesis and mobility of the H_2O–CO_2 fluid phase during regional greenschist and epidote amphibolite facies metamorphism: Petrological and stable isotope study in the Scottish Dalradian*. J. Geol. Soc. London *140*, 577–599.

HANMER, S. K. (1982), *Microstructure and geochemistry of plagioclase and microcline in naturally deformed granite*. J. Struc. Geol. *4*, 197–215.

HEIMANN, R. B. (1979), *Hydrothermale Verdrängungsreaktionen an Alkalifeldspäte der Zusammensetzungen Or93Ab7 und Ab9Or4*. Neues Jahrb. Min. Abh. *137*, 1–19.

HOBBS, B. E., MEANS, W. D. and WILLIAMS, P. F. (1976), *An Outline of Structural Geology*. Wiley, New York, 571 pp.

HOSCKEK, G. (1973), *Zur Stabilität metamorpher Biotit Paragenesen*. Tschermaks Mineral. Petrol. Mitt. *20*, 48–58.

KERRICH, R., ALLISON, I., BARNETT, R. L., MOSS, S. and STARKEY, J. (1980), *Microstructural and chemical transformations accompanying deformation of granite in a shear zone at Miéville, Switzerland; with implications for stress corrosion cracking and superplastic flow*. Contrib. Mineral. Petrol. *73*, 221–242.

MEYER, C. and HEMLEY, J. J. (1967), 'Wall rock alteration,' in L. Barnes (ed.), *Geochemistry of Hydrothermal Ore Deposits*. Holt, Rinehart, New York, 166–235.

MITRA, G. (1978), *Ductile deformation zones and mylonites, the mechanical processes involved in the deformation of crystalline basement rocks*. Am J. Sci. *278*, No. 8, 1057–1084.

MIYASHIRO, A. (1973), *Metamorphism and Metamorphic Belts*, Allen and Unwin, London, 492 pp.

NICOLAS, A. and POIRIER, J. P. (1976), *Crystalline Plasticity and Solid State Flow in Metamorphic Rocks*. Wiley, London, 444 pp.

OFTEDAHL, C. (1956), *Om Grongkulminasjonen og Grongsfeltet skyvedekken*. Norges Geol. Unders. *195*, 57–64.

ORVILLE, P. M. (1963), *Alkali metasomatism and feldspars*. Norsk Geol. Tidskr. *42*, 283–316.

PASSCHIER, C. W. (1982), *Mylonitic deformation in the Saint Barthélemy Massif, French Pyrenees, with emphasis on the genetic relationship between ultramylonite and pseudotachylyte*. GUA Papers Geol., Ser. 1, No. 16, 171 pp.

PETROVIC, R. (1974), 'Diffusion of alkali ions in alkalifeldspars,' in W. S. McKenzie and J. Zussman, (eds.), *The Feldspars*. Manchester Univ. Press, Manchester, p. 174–182.

POIRIER, J. P. and GUILLOPPÉ, M. (1979), *Deformation induced recrystallization of minerals*, Bull. Minéral. *102*, 67–74.

RAMSAY, J. G. (1980), *The crack-seal mechanism of rock deformation*. Nature *284*, 135–139.

RAMSAY, J. G. and GRAHAM, R. H. (1970), *Strain variations in shear belts*. Canad. J. Earth Sci. *7*, 786–813.

SPRY, A. (1969), *Metamorphic Textures*, Pergamon, Oxford, 350 pp.

STEL, H. (1981), *Crystal growth in cataclasites: Diagnostic microstructures and implications*. Tectonophys. *78*, 585–600.

STEL, H. (1984), *Hydrothermal processes in mylonitic rocks at the NW edge of the Grong culmination, Central Norway*. Geologica Ultraiectina, No. 27.

VERNON, R. H. (1974), *Metamorphic Processes*. Allen and Unwin, London, 247 pp.

WHITE, S. (1976), *The effect of strain on the microstructures, fabrics and deformation mechanisms in quartzites*. Phil. Trans. R. Soc. London Ser. A, *3283*, 69–86.

WINTCH, R. P. (1975), Solid-fluid equilibria in the system $KAlSi_3O_8$–$NaAlSI_3O_8$–$Al_2Si_2O_5$–HCl, J. Petrol. *16*, 57–69.

(Received 1st November 1985, revised 12th April 1986, accepted 14th April 1986)

PAGEOPH, Vol. 124, Nos 1/2 (1986) 0033-4533/86/020309-28$1.50+0.20/0

Earthquakes in the Ductile Regime?

B. E. HOBBS,[1] A. ORD[1] and C. TEYSSIER[2]

Abstract—Pseudotachylytes from a crustal scale shear zone in Central Australia have developed in a cyclical manner: once developed, an individual pseudotachylyte is deformed in a ductile manner, only to be overprinted at a later stage by a new generation of pseudotachylytes. Such cyclic generation and deformation of pseudotachylyte has been interpreted in the past as representing conditions at the brittle–ductile transition; a different interpretation, however, is presented here. It is proposed that psuedotachylytes and associated ultramylonites can develop entirely within the ductile regime as ductile instabilities. Such instabilities are different in nature to those previously discussed at length in the geophysical literature but are identical in principle with the instabilities that develop for velocity-weakening frictional behavior in spring-slider systems. At a given strain rate a critical temperature, T_c, is defined, at which the transient work hardening equals the product of stress relaxation due to a thermal fluctuation and the heat generated by shearing. A necessary condition for ductile instability at a given strain rate is that the temperature is below T_c; then the rate of change of stress with respect to strain is negative. An additional requirement is that this rate of change exceeds, in magnitude, the effective elastic stiffness of the loading system. Ductile instabilities are marginally possible at geological strain rates in quartzites but are possible at mid-crustal temperatures in other rock types. On the basis of these observations a new interpretation is presented for the base of the seismogenic zone in crustal regions.

Key words: Earthquakes, pseudotachylytes, ductile instabilities.

1. *Introduction*

Earthquakes in the crust are generally considered to be the result of stick-slip events on rock surfaces undergoing frictional sliding (BRACE and BYERLEE, 1966). Such a concept implies that the base of the seismogenic regime in crustal regions corresponds to the brittle–ductile transition, where pressure-sensitive brittle sliding is replaced by temperature-sensitive ductile flow (SIBSON, 1982, 1984). A different interpretation for the base of the continental seismogenic regime derives from experimentally determined velocity-dependent friction laws, where the steady-state frictional stress, τ^{ss}, is observed to vary with the velocity, V, of sliding at constant normal stress (DIETERICH, 1978, 1979a, 1979b, 1981; RUINA, 1980, 1983; TULLIS and WEEKS, 1985). Analyses of the stability of frictional sliding based on these laws (RUINA, 1980, 1983, RICE and GU, 1983; RICE and RUINA, 1983; GU et al., 1984) show that if $\partial\tau^{ss}/\partial V < 0$, then the system may become unstable if the elastic stiffness of the system is below some critical value or if the velocity perturbation is sufficiently large. If $\partial\tau^{ss}/\partial V < 0$ then the constitutive behavior is said to be velocity-weakening.

For velocity-strengthening behavior, $\partial \tau^{ss}/\partial V > 0$, frictional sliding is stable (although an exception to this has been noted by WEEKS and TULLIS, 1984, and by TULLIS and WEEKS, 1985). This line of argument leads to the base of the continental seismogenic zone being identified as the transition from velocity-weakening to velocity-strengthening behavior (SHIMAMOTO and LOGAN, 1985; TSE and RICE, 1985).

We present another interpretation of the base of the seismogenic zone. The transition from seismically active to seismically inactive behavior is identified with a critical temperature, T_c, below which the stress relaxation due to a thermal fluctuation exceeds the work hardening due to a fluctuation in strain rate. Provided that the elastic stiffness of the crustal loading system is low enough, T_c then marks the transition from unstable to stable ductile flow at a given strain rate.

Some models of earthquake events due to frictional sliding on pre-existing faults indicate that stress is transmitted downwards into the ductile lower crust and mantle at and after the stick-slip event, so that sliding is propagated downwards into the ductile regime (GOETZE, 1978; LEHNER et al., 1981; CHEN and MOLNAR, 1983; TSE and RICE, 1985). Workers such as SIBSON (1977) have interpreted pseudotachylytes as the products of such events. A pseudotachylyte zone is a zone of intensely deformed rock, sometimes composed of melted local rocks (PHILPOTTS, 1964; MADDOCK, 1983) and sometimes composed of brecciated material (WENK, 1978) that commonly displays intrusive relations with the host rocks. SIBSON (1980) also suggests that pseudotachylytes result from brittle frictional sliding in an otherwise ductile regime, due to concentrations of shear strain around inclusions of stronger rocks, or that they result from shear melting during the development of a ductile instability in much the same manner as OROWAN (1960) and GRIGGS and BAKER (1969) have suggested for the origin of deep mantle earthquakes.

The aim of this paper is to investigate the suggestion that some pseudotachylytes develop from ductile instabilities, by examining the structures within a large crustal shear zone (the Redbank Shear Zone of Central Australia) and considering the criteria for the development of ductile instabilities on the basis of what is known of the constitutive behavior of rocks under mid-crustal conditions. We do not mean to imply that all pseudotachylytes develop from ductile-instability events; other mechanisms have been suggested (e.g., WEISS and WENK, 1983).

The development of localized zones of high strain during the ductile deformation of metals is well known and has been studied for many years (ZENER and HOLLOMON, 1944; DRUCKER, 1960; RECHT, 1964; RUBIN and DRUCKER, 1969; STAKER, 1981, and review therein). The effects, which resemble the development of ductile shear zones in rocks (see ROGERS, 1979, Figs. 1 and 2), are particularly well developed in metals at fast strain rates such as those involved in impact loading or machining. This is because the thermal conductivity of metals is so high compared to rocks that fast strain rates are required to approach adiabatic conditions. In such situations the heat generated during deformation can lead to stress relaxation that exceeds work hardening so that there is an overall decrease in stress with continued deformation. Intuitively, one would expect such effects at slower strain rates in rocks because the thermal conductivity of rocks is three orders of magnitude lower than that of metals.

The main emphasis in the geophysical literature has been on the possibility that shear instabilities develop in fluids simply because of the heat developed during deformation (see, for instance, SCHUBERT and YUEN, 1978, BRUN and COBBOLD, 1980, and FLEITOUT and FROIDEVAUX, 1980). By the term *fluid* it is implied that no work hardening due to a fluctuation in strain rate occurs, so that any instability that develops is due to the thermal feedback between the heat generated by deformation and the relaxation of stress due to an increase in temperature. Stability analyses for such situations (see YUEN and SCHUBERT, 1977) are particularly sensitive to the boundary conditions. In some instances, notably those in which the velocity at the boundary of a shear zone is fixed, perturbations in the system do not grow, even for velocity-weakening behavior. If the shear zone is driven by constant stress conditions, then perturbations may amplify under velocity-weakening conditions, but the conditions for such amplification are strongly model-dependent.

In this paper we emphasize that rocks are not fluids, so that, first, transient work hardening is to be associated with any fluctuation in strain rate, and stress relaxation due to a thermal fluctuation must outweigh work hardening for instability to be possible. This allows the delineation of temperature–strain rate fields in which ductile instabilities may develop. Second, stored elastic-strain energy in the surrounding rock mass is envisaged as the driving force for the instability. It is the interaction of strain-softening behavior with the elastic stiffness of the loading system that determines whether an instability will develop. We propose that, for fast enough deformation rates during the instability event, shear melting can develop in rocks. For slower deformation rates, however, zones of cataclastic deformation could form during the instability event and, at slower rates again, ultramylonites or high-strain, ductile shear zones.

2. *Geological evidence for ductile instabilities*

2.1 *Structures within the Redbank Shear Zone, Central Australia*

The Central Australian continental crust is composed of medium-to-high-grade metamorphic rocks exposed in the Lower Proterozoic Arunta basement (Fig. 1), unconformably overlain by the Upper Proterozoic-Palaeozoic Amadeus sedimentary basin (RUTLAND, 1976). A characteristic of the Arunta basement is the presence of numerous long, narrow shear zones which display a complex structural and metamorphic history (WARREN, 1983) reflecting successive uplift of deep, high-grade crust. In particular, the Early Carboniferous Alice Springs Orogeny (FORMAN,

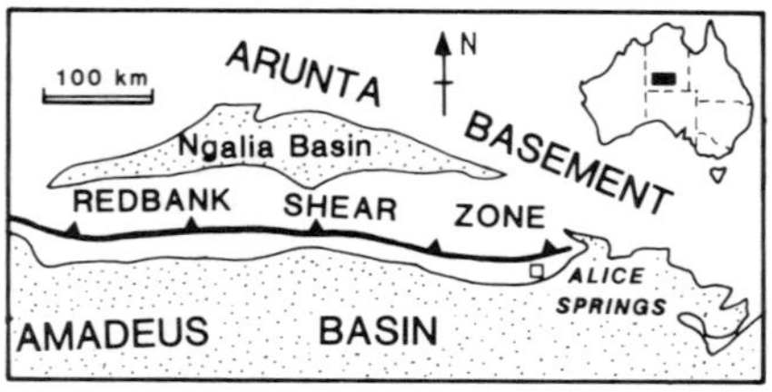

Figure 1
Locality map of the Redbank Shear Zone. The Upper Proterozoic-Paleozoic Amadeus and Ngalia Basins are stippled. The Precambrian Arunta basement is blank.

1966) is responsible for crustal-scale thrust tectonics which resulted in shear-zone development in the Arunta basement and nappe formation in the North Amadeus Basin (Teyssier, 1985). The Redbank Shear Zone is the most impressive of these basement shear zones. This north-dipping structure extends for 500 km along strike and can be several kilometers thick. It is characterized by extensive development of a mylonitic series (Sibson, 1977), described in Teyssier and Hobbs (1985).

This ductile fault zone separates terrains of different lithologies and structural and metamorphic histories. South of the zone the region is dominated by quartzofeldspathic gneisses with rare interlayered pelitic and quartzose horizons, intruded by elongate bodies of granitoids. The mylonitic equivalents of all these rock types are present within the Redbank Shear Zone. To the north of the zone, mafic and felsic granulites are the dominant rock type. These metagabbros to meta-anorthosites do not contain the strong mylonitic foliation, except for local shear zones which are characterized by recrystallized plagioclase and clinopyroxene; orthopyroxene is kinked and fractured. Zones of brittle fracturing and ultramylonitic deformation contain pseudotachylyte.

The metamorphic conditions of mylonitic deformation in the Redbank Shear Zone evolved from amphibolite to upper greenschist facies. An upper limit of 500 to 550 °C is determined by garnet–biotite thermometry on protomylonites. In mylonites and ultramylonites the syntectonic development of green biotite, muscovite, blue-green hornblende, epidote, rare garnet, and plagioclase is compatible with upper greenschist conditions and temperatures of 350 to 400 °C (Miyashiro, 1973; Vernon, 1976). Consequently, the exposed structures were probably produced at a depth of 12 to 15 km.

Mylonite zones are numerous in the Redbank Shear Zone and anastomose around relatively less deformed rock volumes (Fig. 2a). The mylonitic foliation is north-dipping (30° to 80°), and the mylonitic lineation is north-plunging. The mylonites are typical S–C tectonites (Berthe et al., 1979; Lister and Snoke, 1984) in which meso- to microscopic kinematic indicators (S–C relations, asymmetric or rotated porphyroclasts, oblique foliation in recrystallized quartz ribbons, and

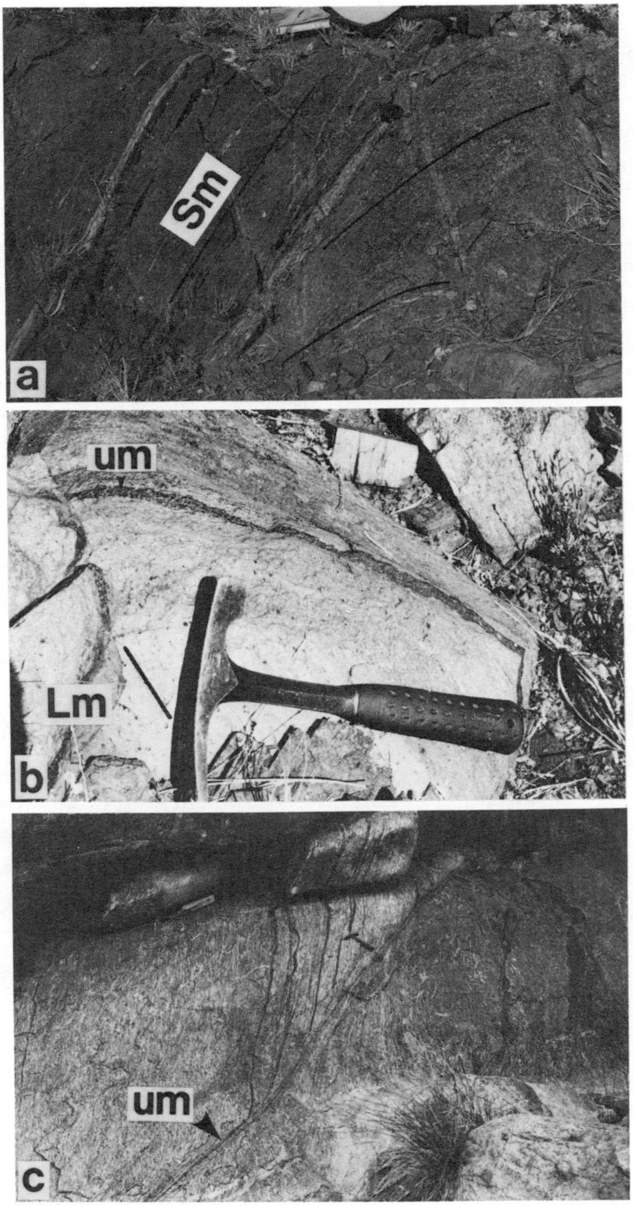

Figure 2
Mylonitic deformation in the Redbank Shear Zone: (a) shear zone development in a metagranodiorite, (b) ultramylonite band (um) concordant with mylonitic foliation in mylonitized granitoid. Lm is the mylonitic lineation; (c) ultramylonite band (um) cross-cutting foliation in mylonitized migmatitic gneiss.

crystallographic fabrics) statistically indicate a southward-directed shear parallel to the mylonitic lineation (see TEYSSIER and HOBBS, 1985, for details).

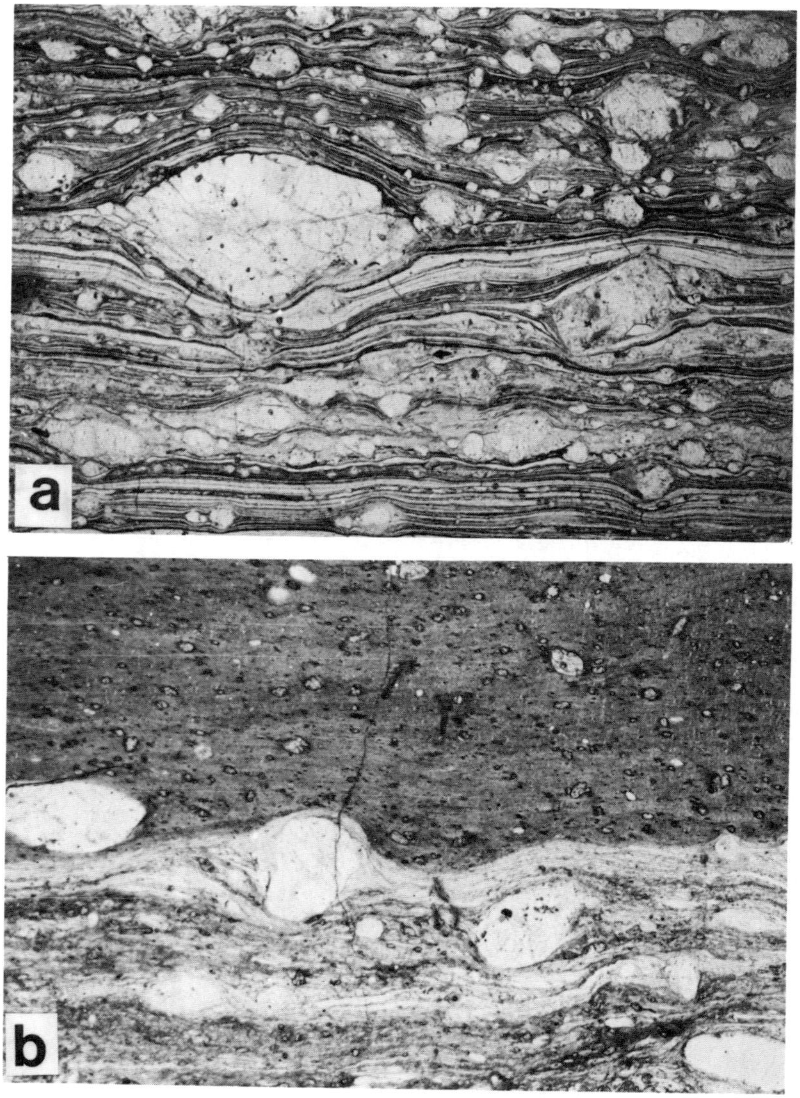

Figure 3
Photomicrographs of (a) mylonitic foliation in quartzofeldspathic gneiss and (b) contact ultramylonite–mylonite. The ribbons around feldspar porphyroclasts are composed of recrystallized quartz or feldspar. Width of each photomicrograph is 4 mm.

Bands of ultramylonites (a few millimeters to a few meters thick) are concordant (Fig. 2b) or discordant (Fig. 2c) with the regional E–W mylonitic foliation. These rocks are characterized by a very fine grain size (a few microns in diameter) and a dark, locally glassy and structureless aspect. However, thin sections (Fig. 3a and b) reveal an extremely strong penetrative foliation delineated by the alignment of fine-grained sheet silicates (usually green biotite) and by a strong crystallographic preferred orientation of isolated recrystallized quartz ribbons. The internal fabric of the ultramylonites, as well as their relation with the surrounding mylonites, also indicates a southward sense of shear.

Pseudotachylyte is commonly found in the mafic to felsic granulites of the hanging wall, and it has been studied in detail in a kilometer-long mafic granulite inclusion surrounded by the quartz feldspathic mylonites of the Redbank Shear Zone (TEYSSIER and HOBBS, 1985). It is only rarely observed in the gneissic terrains of the footwall. Pseudotachylyte appears in the field as a dark rock with a glassy aspect, which develops in centimeter-wide veins, usually showing intrusive relations controlled by several sets of microfractures. Clasts of the host rock, which range in diameter from a few millimeters to a few centimeters, may form a large component of the pseudotachylyte veins (Fig. 4a). These clasts have generally been transported within the fine-grained matrix, which shows a microscopic flow banding. The opposite walls of the veins can often be geometrically matched, supporting a dilational character for the veining.

In the Redbank Shear Zone some pseudotachylyte veins within the granulite inclusion are plastically deformed by ultramylonite bands which contain a strong biotite foliation (Fig. 4b). These deformed veins are in turn cross-cut by a later generation of pseudotachylyte veins (Fig. 5, a and b), indicating a cyclic occurrence of pseudotachylyte formation followed by ductile deformation and recrystallization.

In summary, the Redbank Shear Zone may be regarded as a mid-crustal example of a ductile reverse-fault zone. Strain variation within the mylonitic series is considerable and, at least locally, there is evidence of alternation of pseudotachylyte production and mylonitic and ultramylonitic deformation. As SIBSON (1980) pointed out, such a phenomenon may be common in mylonitic terrains, but shear discontinuities are likely to be obliterated by subsequent ductile deformation.

2.2 *Other geological evidence of pseudotachylyte–mylonite alternation*

Alternation in the development of psuedotachylyte and mylonite in the same shear zone has been described by a number of authors, including SIBSON (1980), STEL (1981; 1986), WENK and WEISS (1982), and PASSCHIER (1984). We concentrate on two of these below.

SIBSON (1980) has described the occurrence of pseudotachylyte within a ductile deformation shear zone of the Outer Hebrides Thrust Zone, northwest Scotland. SIBSON invokes either downward-migrating seismic fracturing from the brittle-frictional regime in the crust or local amplification of stress and strain rate, with attendant production of pseudotachylyte, around lensoid bodies of metabasites situated deep in the crust (15 km and deeper).

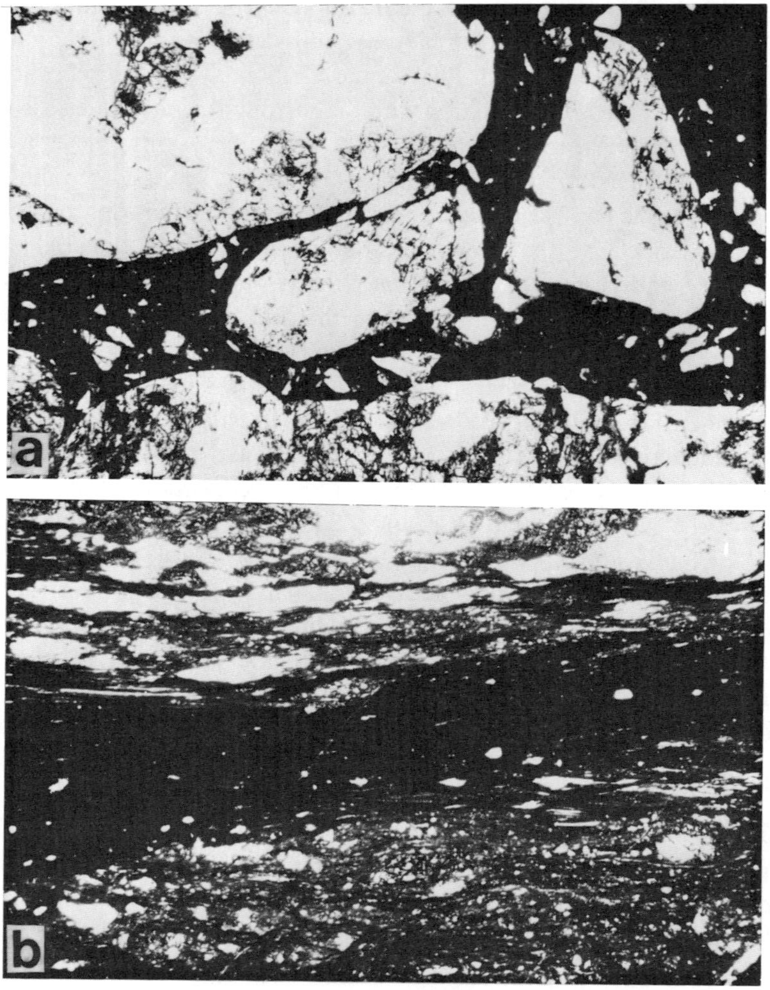

Figure 4
Photomicrographs of undeformed (a) and plastically deformed (b) pseudotachylyte veins. Width of each photomicrograph is 4 mm.

PASSCHIER (1984) described the cyclic development of pseudotachylyte and ultramylonite in the Saint-Barthelemy Massif, French Pyrenees, and proposed that this phenomenon is restricted to the brittle–ductile transition.

An alternative to these interpretations is developed below.

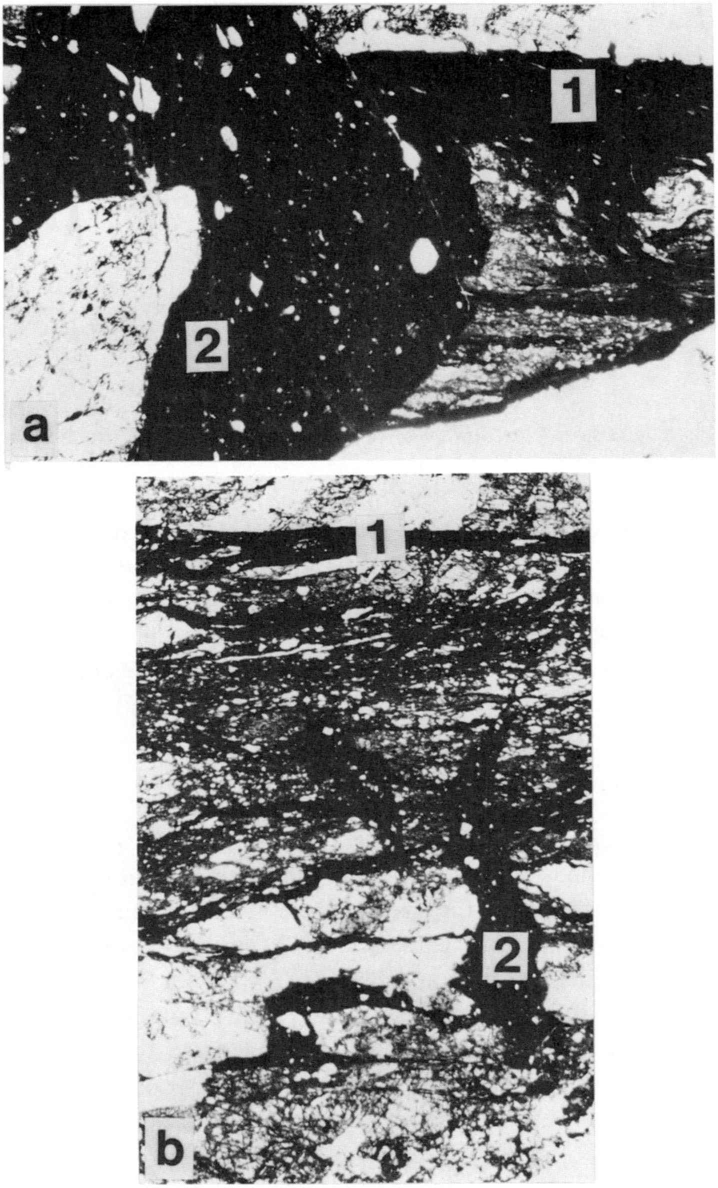

Figure 5
Cross-cutting relations between ductilely deformed pseudotachylyte (1) and second-generation undeformed pseudotachylyte (2). Width of each photomicrograph is 2.5 mm.

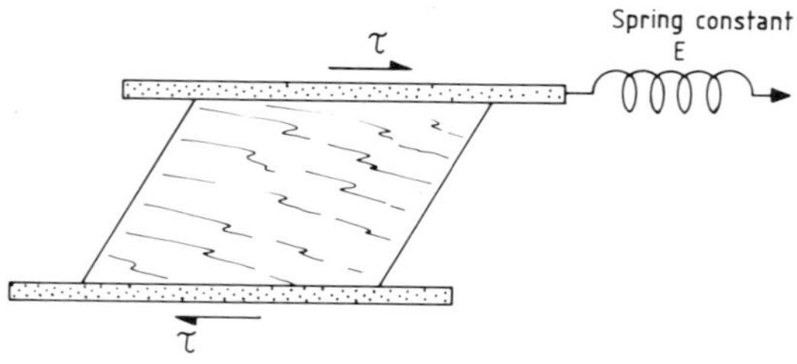

Figure 6
Model for deformation. Material deforms in ductile manner under shear stress, τ. The spring, with stiffness E, represents the elastic response of the surrounding crustal material.

3. Processes leading to instability of deformation in material undergoing creep

The processes that lead to the development of instabilities in the deformation of a ductile solid have been considered by a large number of workers (see, for instance, DRUCKER, 1960, GRUNTFEST, 1963, RUBIN and DRUCKER, 1969, BACKOFEN, 1972, ARGON, 1973, POIRIER et al., 1979, POIRIER, 1980, and NICHOLSON and KIDDY, 1984). At present the experimental data necessary for discussing the problem in detail are not available for rocks (as will become apparent below), so we adopt a fairly simple-minded approach in order to emphasize the principles involved.

3.1 Conditions for instability

Consider a solid creeping in simple shear at a shear stress τ as shown in Figure 6. The spring in this system is meant to represent the elastic behavior of the surrounding mass. We ask, what happens to the stress if there is a small fluctuation in variables such as the strain rate, $\dot{\gamma}$. or the temperature, T? Can the system become unstable with attendant rapid increases in τ and $\dot{\gamma}$?

We consider that the motion is essentially steady but that, for a fluctuation in strain rate, transient strain hardening occurs owing to the generation and re-arrangement of dislocations to accommodate the new strain-rate conditions. We suppose that the constitutive law may be expressed as

$$\tau = \tau(\gamma, \dot{\gamma}, T, f\mathrm{H_2O}) \tag{1}$$

where γ is the shear strain and $f\mathrm{H_2O}$is the fugacity of water. During transient fluctuations in strain rate or stress the constitutive behavior follows a law of the form considered by CARTER and KIRBY (1978):

$$\tau^{\text{transient}} = \beta^{-1/n}\gamma^{1/n}t^{-m/n} \exp\{Q/nRT\} \tag{2}$$

where β is a constant strongly dependent on rock type, t is time, n and m are constants typically with values of 1.5 and 0.5 respectively, Q is an activation energy

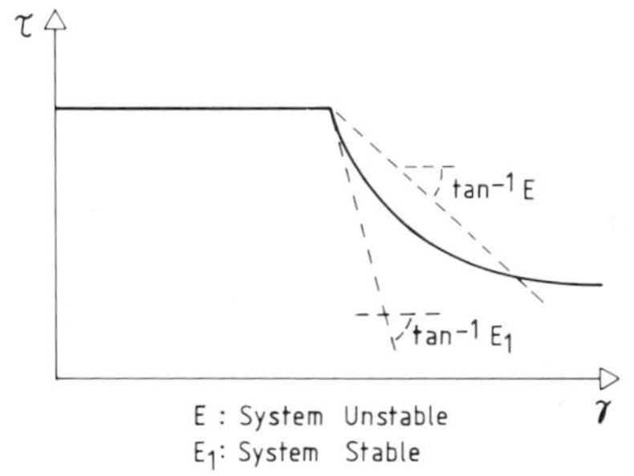

Figure 7
Strain-weakening response. If $|d\tau/d\gamma| > |E|$, the response is unstable. For $|d\tau/\gamma| < |E_1|$, the response is stable. E and E_1 are the effective elastic stiffness of the surrounding crustal material.

similar in magnitude to that for steady-state creep, and R is the gas constant. The dependence of transient creep upon fH_2O is unknown. Equation (2) is true only for strains up to about 10^{-2}.

During steady-state creep the material obeys a standard power-law constitutive law, and we consider a specific case of such laws suggested for quartzites by the work of Ord and Hobbs (1986a):

$$\tau = A^{-1/N}\dot{\gamma}^{1/N} \exp\{Q/NRT\} \cdot fH_2O^{-1/3N} \tag{3}$$

where A and N are material constants and Q is the activation energy for steady-state creep.

If, for a small fluctuation in any or all of γ, $\dot{\gamma}$, T, and fH_2O, the shear stress decreases with continued deformation—that is, $d\tau/d\gamma < 0$—then the potential exists for the system to become unstable. Specifically, for the system shown in Figure 6, if $|d\tau/d\gamma| > |E|$, where E is stiffness of the spring (Fig. 7), then for a thin enough shear zone the situation is formally similar to that discussed by Rice and Ruina (1983), Gu et al. (1984), and Rice and Tse (1985) for the spring-slider system with velocity-weakening frictional constitutive behavior. The stress may oscillate with continued deformation, or τ may undergo rapid decreases with time in a manner similar in principle to stick-slip behavior on sliding surfaces.

From equation (1),

$$\frac{d\tau}{d\gamma} = \left(\frac{\partial\tau}{\partial\gamma}\right) + \left(\frac{\partial\tau}{\partial\dot{\gamma}}\right)\left(\frac{d\dot{\gamma}}{d\gamma}\right) + \left(\frac{\partial\tau}{\partial T}\right)\left(\frac{dT}{d\gamma}\right) + \left(\frac{\partial\tau}{\partial fH_2O}\right)\left(\frac{d fH_2O}{d\gamma}\right) \tag{4}$$

We consider each of these terms in order; positive terms have a stabilizing influence on the deformation, negative terms are destabilizing.

$\left(\frac{\partial \tau}{\partial \gamma}\right)$ is the strain hardening and in general, except for some materials such as single crystals of synthetic quartz (HOBBS et al., 1972), some anisotropic rocks, and perhaps some cases in which dynamic recrystallization accompanies continued deformation (TULLIS and YUND, 1985), this parameter is positive. It is important to point out that the strain-weakening effects mentioned above are associated with relatively large increases in strain. For a (positive) fluctuation in strain rate, even for these materials, the strain hardening would probably be positive, but no experimental data are available.

$\left(\frac{\partial \tau}{\partial \dot{\gamma}}\right)$ is the change in stress with strain rate, and for power-law materials this is positive.

$\left(\frac{d\dot{\gamma}}{d\gamma}\right)$ is the change in strain rate with continued straining. For homogeneous steady-state deformations this is zero, but for deformations in which the strain becomes localized this parameter may become nonzero (see BACKOFEN, 1972, p. 215–217).

$\left(\frac{\partial \tau}{\partial T}\right)$ is negative for most materials of interest, since the stress tends to relax with increase in temperature. It clearly has a marked destabilizing influence.

$\left(\frac{dT}{d\gamma}\right)$ is the temperature increase due to straining; in general, the parameter is positive. Considerable attention has been paid to calculating the influence of this effect in fluids (see for instance, the review by BRUN and COBBOLD, 1980).

$\left(\frac{\partial \tau}{\partial f H_2O}\right)$ is the stress response to changes in the fugacity of H_2O. Increases in fH_2O lead to a decrease in stress, so this parameter is destabilizing. The effect is similar to an increase in temperature. Fluctuations in fH_2O would arise from local fluctuations in pressure or temperature, or both.

$\left(\frac{d f H_2O}{d\gamma}\right)$ is the water-fugacity response to changes in strain. This could arise from local regions of dilation developing during deformation where a transient pressure decrease could arise in the vapor phase. This leads to a decrease in fH_2O and associated increase in τ. The development of local regions of dilation, from this viewpoint, would have a stabilizing influence on the deformation. Fluctuations in fH_2O are responsible for the local deposition or dissolution of SiO_2 in dilating sites (see ETHERIDGE et al., 1983, 1984), so the effect of this parameter in deforming rocks is well known.

The balance of all of these processes must be such that the right-hand side of equation (4) is negative for an instability to develop.

One can see, therefore, that the development of an instability in the general case is a complicated process, and we have little understanding of the magnitudes of the various terms in equation (4). A discussion of some of these terms and of other mechanisms by which ductile instabilities may develop in rocks is given by EVANS and WONG (1985). In what follows we examine a simplified, but realistic situation in which the deformation remains homogeneous at least up to the time that the instability forms, that is, $\partial\dot{\gamma}/\partial\gamma = 0$. Throughout we assume that pressure, P, is constant ($dP/d\gamma = 0$) and that fH_2O remains constant with changes in strain ($dfH_2O/d\gamma = 0$).

Expression (4) then reduces to

$$\frac{d\tau}{d\gamma} = \left(\frac{\partial\tau}{\partial\gamma}\right) + \left(\frac{\partial\tau}{\partial T}\right)\left(\frac{dT}{d\gamma}\right) \tag{5}$$

Since $\partial\tau/\partial T$ is always negative whilst $dT/d\gamma$ is positive, the necessary condition for instability becomes

$$\left(\frac{\partial\tau}{\partial\gamma}\right) < \left|\left(\frac{\partial\tau}{\partial T}\right)\left(\frac{dT}{d\gamma}\right)\right| \tag{6}$$

Since we are considering only small fluctuations from steady state, we take $\tau^{\text{transient}} \approx \tau^{\text{steady state}}$, then from (3),

$$\left(\frac{\partial\tau}{\partial T}\right) = -\frac{\tau Q}{NRT^2} - \frac{\tau}{3N} fH_2O^{-1}\frac{dfH_2O}{dT} \tag{7}$$

and $dT/d\gamma = k\tau/c\rho$ (see RUBIN and DRUCKER, 1969), where c is the specific heat, ρ is the density, and k is the fraction of heat retained in the shear zone and not removed to the surrounding rocks by conduction or by advection in a moving fluid phase.

In order to gain some feeling for the influence of fH_2O in expression (7) we tabulate the two terms on the right-hand side of (7) in Table 1 at various temperatures for a confining pressure of 500 MPa. The table is compiled for a strain rate of 10^{-12} sec^{-1} in a 'wet' quartzite of JAOUL et al. (1984) with data for fH_2O from BURNHAM et al. (1969) and DELANY and HELGESON (1978). For temperatures of

Table 1

Comparison of First and Second Terms on Right-Hand Side of Expression (7)

T, °C	$\tau Q/NRT^2$, Pa·K^{-1}	$(\tau/3N)fH_2O^{-1}$. dfH_2O/dT, Pa·K^{-1}
200	1.94×10^8	1.35×10^7
300	3.07×10^6	1.84×10^5
400	1.70×10^5	9.18×10^3
500	1.85×10^4	8.68×10^2
600	3.33×10^3	1.31×10^2
700	8.00×10^2	2.81×10^1
800	2.45×10	7.64

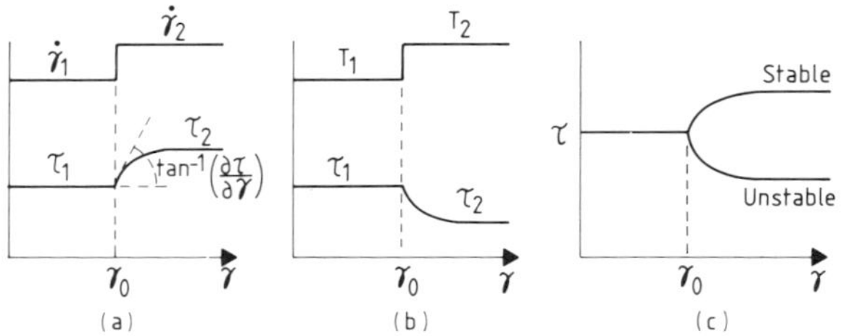

Figure 8

(a) Increase in shear stress, τ, for a stepwise increase in strain rate, $\dot{\gamma}$, at a strain, γ_0, under isothermal conditions. The transient work hardening is $\partial\tau/\partial\gamma$. (b) Decrease in shear stress, τ, for a stepwise increase in temperature, T, at a strain, γ_0. (c) Competition between work hardening in (a) and thermal relaxation in (b) results in strain-hardening behavior for stable situations and strain-weakening behavior for unstable situations (adiabatic conditions).

200 °C and above the second term is at least an order of magnitude smaller than the first. Thus, the effect of the $\mathrm{d}fH_2O/\mathrm{d}T$ term is small. In view of the uncertainty involved in incorporating the fH_2O dependence into the constitutive equation of JAOUL et al. (1984), we neglect the effect completely here. Overall, the effect of the $\mathrm{d}fH_2O/\mathrm{d}T$ term is to increase the likelihood that a ductile instability will develop.

Thus, the necessary condition for instability is taken to be

$$\left(\frac{\partial\tau}{\partial\gamma}\right) < \frac{\tau Q}{NRT^2} \cdot \frac{k\tau}{c\rho} \tag{8}$$

The interpretation of this expression is given in Figure 8. For a small increase in strain rate under isothermal conditions ($k = 0$), the stress changes transiently from one steady state to another (Fig. 8a). The transient stress states show hardening, and the slope of the hardening curve gives the value of $\partial\tau/\partial\gamma$. A small increase in temperature gives a decrease in τ, as shown in Figure 8b. In the adiabatic case ($k = 1$), if the product $(\partial\tau/\partial T)(\mathrm{d}T/\mathrm{d}\gamma)$ exceeds the strain hardening $\partial\tau/\partial\gamma$, then the stress decreases with further straining (Figure 8c). This is a necessary condition for instability. If $\mathrm{d}\tau/\mathrm{d}\gamma$ is negative and exceeds the effective elastic stiffness of the loading system, then we propose here that the system is unstable.

3.2 Calculation of strain hardening, $\partial\tau/\partial\gamma$

The data for transient creep of rocks are particularly unsatisfactory, and in what follows we assume that n and m in expression (2) are independent of rock type, with values of 1.5 and 0.5, respectively, and that Q in (2) is identical with the activation energy for steady-state creep (see CARTER and KIRBY, 1978). From (2):

$$\left(\frac{\partial\tau}{\partial\gamma}\right) = \frac{1}{n}\gamma^{-1}\tau \tag{9}$$

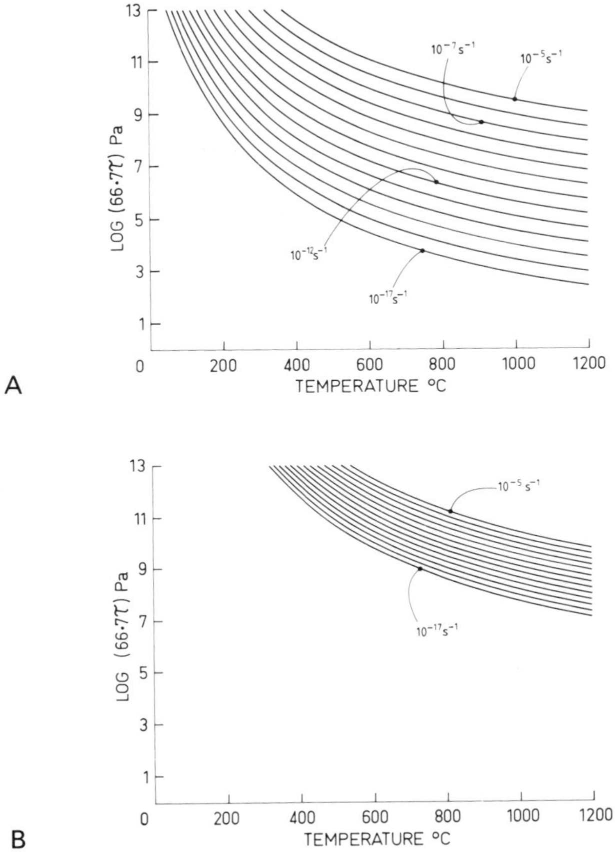

Figure 9

Plot of log of the work hardening = 66.7τ against temperature, T, for various strain rates: (A) 'wet' quartzite of JAOUL et al. (1984); (B) Åheim dunite of CHOPRA and PATERSON (1981).

We are concerned with that part of the transient stress–strain curve close to steady-state conditions, so that a strain of approximately 10^{-2} is relevant in expression (9). Thus, we have

$$\left(\frac{\partial\tau}{\partial\gamma}\right) = 66.7\tau \tag{10}$$

where τ is to be read as equal to the steady-state shear stress. Plots of the strain hardening, $\partial\tau/\partial\gamma$, against T for various strain rates for quartzite and dunite are given in Figure 9. Figure 9 shows that 66.7 τ as given in Equation (10) must be an upper limit to an estimate for the strain hardening, since this cannot exeed the elastic modulus of approximately 10^{11} Pa in rocks. However, since smaller values of strain hardening from (8) only tend to increase the likelihood of ductile instability, this estimate is retained here. The influence of assuming lower values for strain hardening is considered in Figure 12.

3.3 *Application to quartzite*

We take the results of JAOUL et al. (1984) for creep of 'wet' Heavitree quartzite as representative of wet quartzite rheology and assume (for want of better data) that the fH_2O dependence is already included in the quoted temperature dependence for their experiments. Then $N = 1.8$ and $Q = 151$ kJ·mol^{-1}. Also, we take c to be a constant, independent of strain and, for conditions in the crust, independent of P and T. BOTT (1971, p. 185) gives a value for c under crustal conditions of 1130 J·kg^{-1}·K^{-1}. The density is 2.65×10^3 kg·m^{-3}. With α for the right-hand side of expression (8), then log α may be plotted against temperature as in Figure 10a for strain rates ranging from 10^{-5} sec^{-1} to 10^{-17} sec^{-1}. We take the strain hardening as given by expression (10); then stable and unstable ductile fields are delineated as in Figure 10b. Figure 10b indicates that for a strain rate of 10^{-12} sec^{-1} quartzite undergoing ductile shear would be unstable to temperatures less than 200 °C for adiabatic situations. Departures from adiabatic conditions reduce the critical temperature, T_c, below which ductile instabilities can develop, as shown in Figure 10b. The general rule is that for a given value of $\partial\tau/\partial\gamma$ instability is favored by lower temperatures or faster strain rates (higher stresses). Despite the uncertainties which exist for values of $\partial\tau/\partial\gamma$ it would seem that the likelihood is very high that ductile shearing in quartzite should become unstable under geological conditions at moderate temperatures. Even at experimentally accessible strain rates of 10^{-6} sec^{-1} instability should occur below 430 °C under adiabatic conditions (Fig. 10b).

Figure 11 is a plot of $d\tau/d\gamma$, given by equation (5), against temperature for quartzite. The transition from positive to negative values occurs at T_c, as is also defined by Figure 10b. As indicated earlier, the necessary, but not sufficient, condition for ductile instability is that $d\tau/d\gamma$ be negative. It is also required (Fig. 7) that $d\tau/d\gamma$ be greater in magnitude than the effective elastic stiffness of the loading system. An upper limit for the stiffness of the mid-crust is a value for Young's modulus of approximately 10^{11} Pa, although, because of inhomogeneities on a large scale, one might expect the effective elastic stiffness to be substantially below this value. Figure 11 shows that this estimate of 10^{11} Pa for a given strain rate coincides almost exactly with T_c, so in this case T_c may be taken as defining the necessary and sufficient conditions for ductile instability.

It has been mentioned that 66.7τ must be an upper limit for an estimate for $\partial\tau/\partial\gamma$. Accepting lower values for strain hardening given by $\partial\tau/\partial\gamma = a\tau$, where $a = 30, 20, 10, 1, 0.1$, leads to progressively higher values of T_c, as shown in Figure 12. Thus, a decrease in a from 66.7 to 1 leads to an increase in T_c from 200 to 300 °C for the adiabatic case.

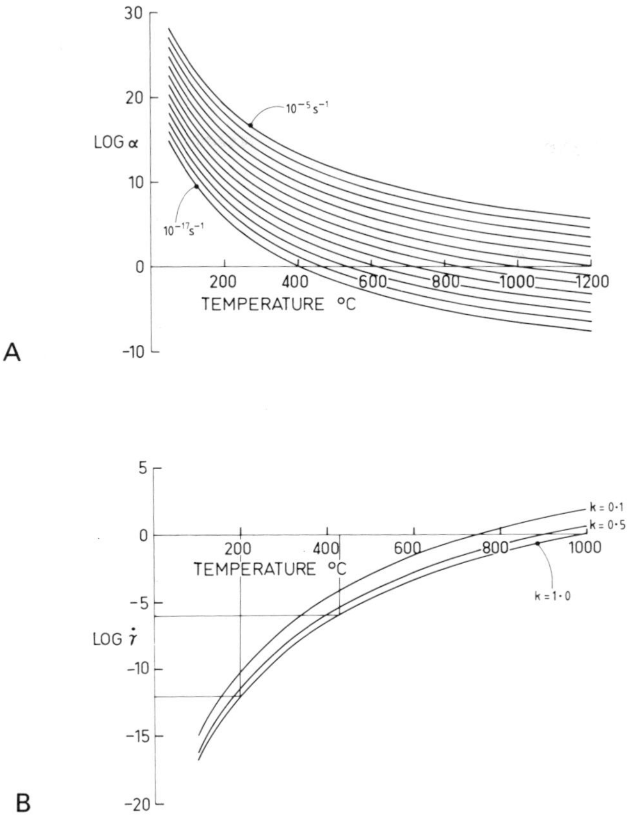

Figure 10

(A) Plot of $\alpha = k\tau^2 Q/NR\rho cT^2$ against temperature, T, for various strain rates. Flow law is for 'wet' quartzite of JAOUL et al. (1984). (B) Stability field for ductile shearing of 'wet' quartzite of JAOUL et al. (1984) for various values of k; $k = 1$ represents adiabatic conditions. The curves are plots of $\dot{\gamma} = (66.7NR\rho cT^2/kQ)^N A \exp\{-Q/RT\}$. Conditions above the curves are unstable; those below the curves are stable. The temperature below which ductile instabilities are possible, T_c, may be read off for a particular strain rate as indicated.

3.4 Deformation of quartzites within a shear zone

In order that heat losses may be taken into account, the geometry of the system must be considered, and in general this will require some form of numerical modelling. Here we consider a geometry proposed by Recht (1964), where a shear zone of thickness L develops a thin zone of high shear strain of thickness h (Figure 13). This high strain zone is nucleated in a layer which is slightly weaker than the remainder of the shear zone, and for the period of time that the high strain zone is forming it is supposed that the strain rate within it is the average for the shear zone as a whole; that is, all of the deformation is concentrated in the high strain zone (Fig. 13b). The

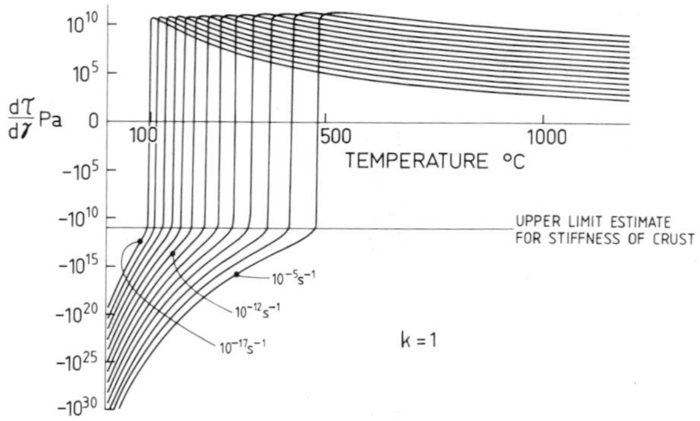

Figure 11
Plot of $d\tau/d\gamma$ against temperature, T, for various strain rates. Adiabatic conditions: $k = 1$. The intersection of the curve for a particular strain rate with the temperature axis is T_c for that strain rate.

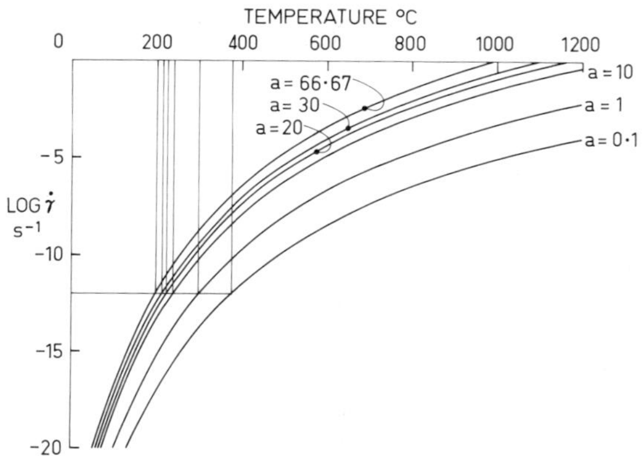

Figure 12
Influence of changes in work-hardening rate, $\partial\tau/\partial\gamma$, on the stability of ductile shearing for 'wet' quartzite of Jaoul et al. (1984). The curves represent various values of a in the expression $\partial\tau/\partial\gamma = a\tau$.

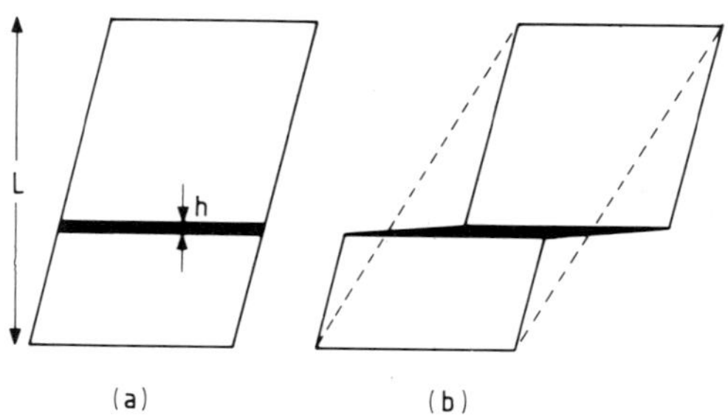

Figure 13
Geometry of shear zone of thickness L which develops a thin shear strain zone of thickness h.

temperature at this thin zone, which is embedded in an infinite medium, due to the heat generated by constant uniform shearing, is

$$T = \tau L\dot{\gamma}(t/\pi K\rho c)^{1/2} \tag{11}$$

where K is the thermal conductivity (see RECHT, 1964, p. 190). Differentiating equation (11) with respect to time and using $\gamma = \dot{\gamma} t$ gives

$$\frac{dT}{d\gamma} = \frac{1}{2}\tau L\left(\frac{\dot{\gamma}}{\pi K\rho c\gamma}\right)^{\frac{1}{2}} \tag{12}$$

which, from expression (6), gives

$$\frac{\partial\tau}{\partial\gamma} < \left|\frac{1}{2}\tau L\left(\frac{\dot{\gamma}}{\pi K\rho c\gamma}\right)^{\frac{1}{2}}\frac{\partial\tau}{\partial T}\right| \tag{13}$$

as the instability condition.

Expression (7) now gives, neglecting the fH_2O effect,

$$\frac{\partial\tau}{\partial\gamma} < \left|\frac{L}{2}\left(\frac{\dot{\gamma}\rho c}{\pi K\gamma}\right)^{\frac{1}{2}}\frac{\tau^2 Q}{R\rho c T^2 N}\right| \quad \text{for instability} \tag{14}$$

This expression is to be compared with expression (8), in which k is replaced by

$$k' = \frac{L}{2}\left(\frac{\dot{\gamma}\rho c}{\pi K\gamma}\right)^{\frac{1}{2}} \tag{15}$$

Taking $K = 2.51\ \mathrm{W{\cdot}m^{-1}K^{-1}}$ and ρ and c as before, this reduces to

$$k' = 308.12\,L\left(\frac{\dot{\gamma}}{\gamma}\right)^{\frac{1}{2}} \tag{16}$$

As defined in (15), k' is dimensionless and is a measure of the proportion of the heat retained in the shearing solid. Figure 14 is a plot of log L, the width of the shear zone, against log time, t. This plot is independent of both strain rate and temperature. Lines for $k' = 0.1$, 0.5, and 1.0 are shown. The line $k' = 1.0$ corresponds to adiabatic conditions, and as one proceeds to lower k' values, conditions approach isothermal. Situations to the left of $k' = 1$ are physically impossible, unless heat is added to the shear zone by circulating fluids or by some other mechanism.

The influence of amount of strain and the width of the shear zone are explored in Figure 15, by using the constitutive law for 'wet' quartzite, published by JAOUL et al. (1984). The amount of strain does not have a great influence on the critical temperature, T_c, below which ductile instabilities are possible (Figure 15b), and T_c increases as the shear zone width and the strain rate increase (Figure 15). Thus, at 10^{-12} sec^{-1} a shear strain of 10, and a shear zone width of 1 km, the critical temperature below which instability is possible is 150 °C. Notice that this temperature is 50 °C below that for the adiabatic situation (L infinite) under otherwise identical situations, shown in Figure 10b.

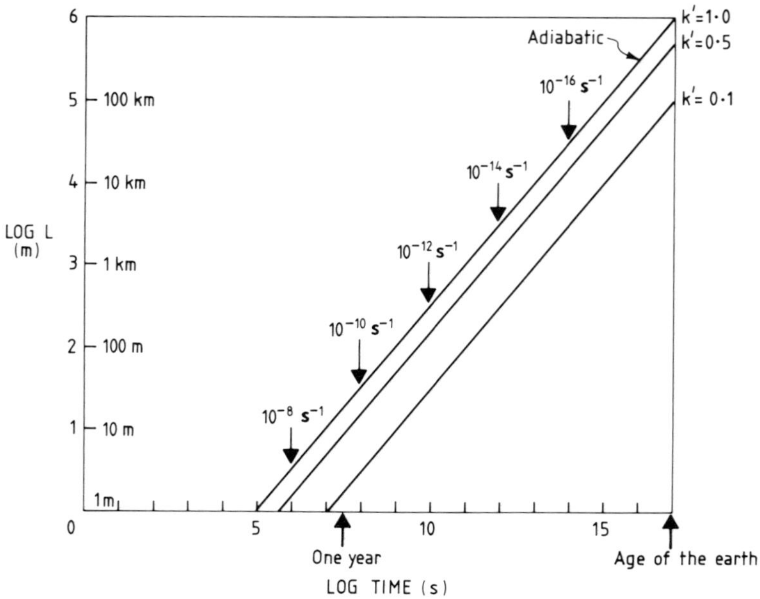

Figure 14
Plot of log L against log t for shear zone: $k = 1$ represents adiabatic conditions, and lower values of k approach isothermal conditions. The marks for each strain rate are the times required to reach a strain of 10^{-2} under steady-state conditions.

3.5 *Other rock types*

For other rock types some representative plots of stability fields are shown in Figure 16. The results are summarized in Table 2. Temperature T_c increases in the order quartzite, anorthosite, diabase, clinopyroxenite, diopsidite, and dunite. Under the conditions shown in Table 2. (strain rate $= 10^{-12}$ sec^{-1}, $L = 1$ km, $\gamma = 10$)

Table 2

Critical Temperature T_c *Below Which Ductile Instabilities are Possible for Various Rock Types**

Rock type	Reference	T_c °C
Quartzite	Jaoul et al. (1984)	150
Anorthosite	Shelton and Tullis (1981)	225
Diabase (average)	Shelton and Tullis (1981)	250
Clinopyroxenite	Kirby and Kronenberg (1984)	305
Diopsidite	Shelton and Tullis (1981)	385
Åheim dunite	Chopra and Paterson (1981)	410

* Shear zone thickness L is 1 km, strain is 10, and strain rate is 10^{-12} s^{-1}.

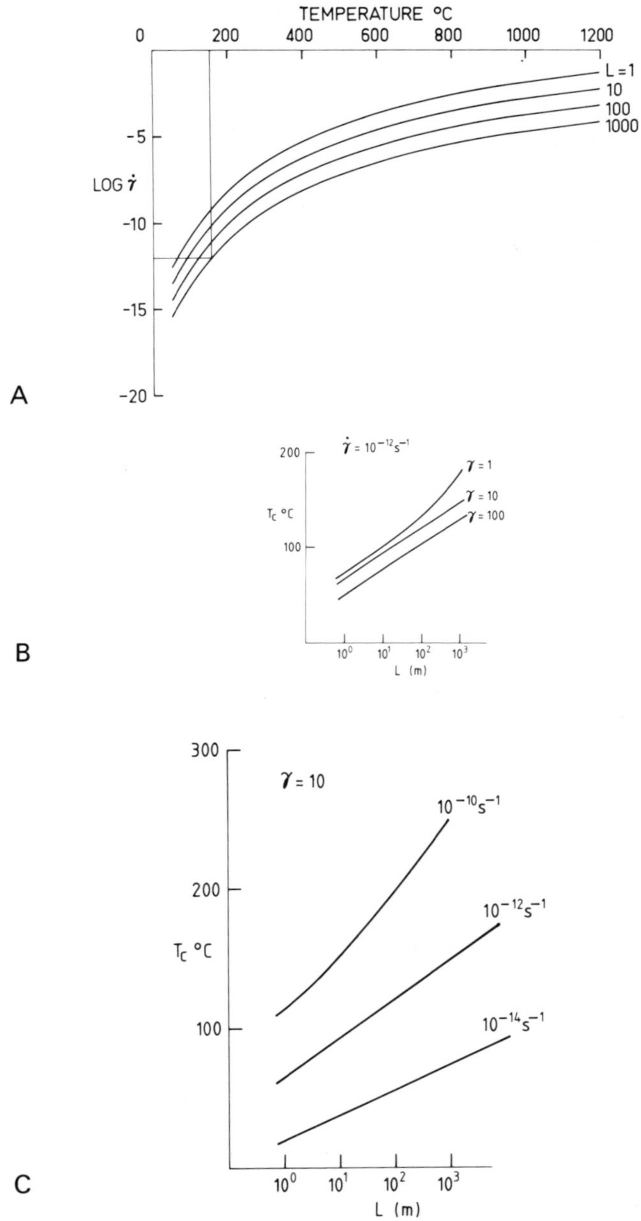

Figure 15
(A) Stability field for ductile shear of 'wet' quartzite of JAOUL et al. (1984): Shear Zone model; $\gamma = 10$. (B) Plot of T_c against log L for various strains at a fixed strain rate of 10^{-12} s^{-1}. 'Wet' quartzite of JAOUL et al. (1984). (C) Plot of T_c against log L for various strain rates at a fixed shear strain of 10.

clinopyroxenite would show ductile instabilities below 305 °C, whereas dunite would show instabilities below 410 °C.

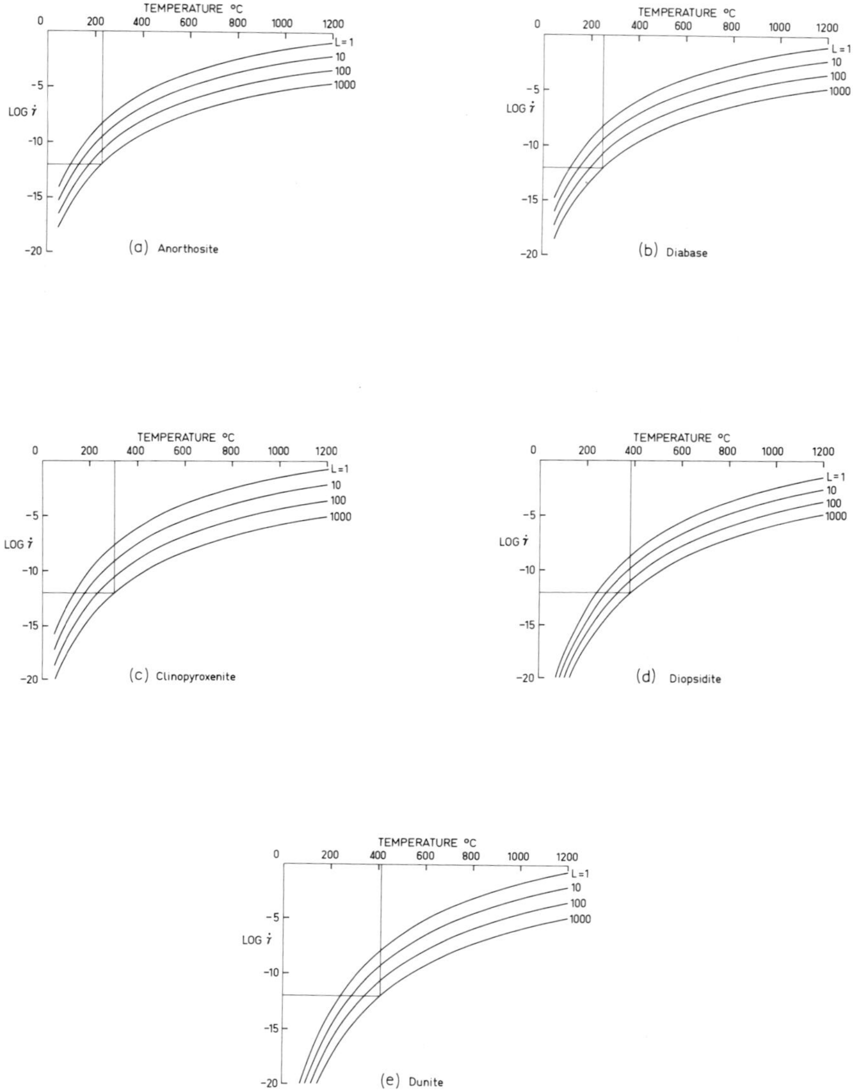

Figure 16

Stability fields for various rock types, with $\gamma = 10$: (a) anorthosite of Shelton and Tullis (1981); (b) diabase of Shelton and Tullis (1981), mean value; (c) clinopyroxenite of Kirby and Kronenberg (1984); (d) diopsidite of Shelton and Tullis (1981); (e) Åheim dunite of Chopra and Paterson (1981).

4. *Discussion*

The pseudotachylytes within the Redbank Shear Zone have a complex history of development in which periods of essentially ductile behavior (mylonite development) alternated with periods of pseudotachylyte formation. These periods of pseudotachylyte formation are interpreted as episodes of instability involving relatively rapid strain rates in an otherwise steady ductile-deformation regime. The conditions under which such unstable ductile shearing should develop have been considered, and if one assumes that the transient strain hardening accompanying a fluctuation in strain rate within an otherwise steady shearing environment does not exceed 66.7τ, where τ is the steady-state shearing stress, then unstable ductile shearing is to be expected in a range of rock types, as summarized in Table 2. Reference to Table 2 and Figures 10b and 15 indicates that unstable ductile shear in quartzites is marginally possible at the lower limits of metamorphism (ca. 200 °C) unless the strain rates are high (10^{-12} $\sec^{-1}$ or greater) and the shear zones are wide (1 km or more). In mafic and felsic rocks, however, the critical temperature above which only stable shearing is possible is higher than in quartzites (Table 2), so that ductile instabilities should be more common in mafic and felsic rocks than in quartzites. This is the situation in the Redbank Shear Zone, where evidence of such instabilities is better developed in the mafic rocks than in the quartzites.

The important point addressed in this paper is that instabilities potentially capable of leading to seismic events can develop in materials undergoing ductile deformation so long as the constitutive law embodies stress relaxation which is temperature-dependent. These shear instabilities are similar in principle to stick-slip events on sliding surfaces with velocity-weakening frictional constitutive behavior, and as such they require the interaction of the ductile constitutive law with the springiness of the surrounding loading system for their complete specification (cf. Gu et al., 1984). We interpret the pseudotachylytes in the Redbank Shear Zone to be examples of such ductile instabilities. Thus, they do not necessarily occur at the brittle–ductile transition in the crust, as interpreted by many authors (Sibson, 1977; Passchier, 1982a, 1982b, 1984; Macaudiere and Brown, 1982), although such an interpretation is not precluded by our discussion; pseudotachylytes may equally be the product of very rapid ductile instabilities. The pseudotachylyte-generating ductile instabilities envisaged here are those that potentially develop in any viscoplastic materials with thermomechanical coupling, as discussed more than twenty years ago by Drucker (1960) and by Rubin and Drucker (1969). Moreover, if pseudotachylytes are the product of ductile instabilities as outlined above, then their formation should be quasi-periodic with time, in much the same way that stick-slip events which result from velocity-weakening frictional behavior are quasi-periodic. Thus, it is expected that various members of the mylonite series (Sibson, 1980) should overprint each other. That is, at any one place within a shear zone one would expect to see mylonites overprinted by ultramylonites, cataclastic pseudotachylytes, or melt

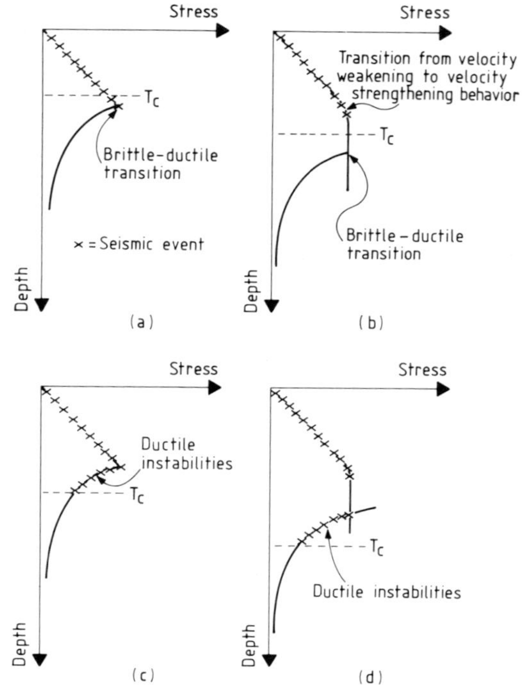

Figure 17
Models for the base of the seismogenic zone in crustal situations. (a) Base of seismogenic zone is the brittle–ductile transition. The temperature, T_c, for the transition from unstable to stable ductile shearing corresponds to depths above the brittle–ductile transition. (b) Base of seismogenic zone is defined by the transition from velocity-weakening to velocity-strengthening frictional-sliding constitutive laws (see ORD and HOBBS, 1985b, for discussion). The position of T_c lies below this transition but above the brittle–ductile transition. (c) Base of seismogenic zone defined by T_c in the ductile regime. Seismic events down to the brittle–ductile transition are due to frictional stick-slip behavior. Seismic events below the brittle–ductile transition are the result of unstable ductile shearing. (d) Base of the seismogenic zone defined by the transition from unstable to stable frictional sliding, but other deeper seismic events occur, owing to unstable ductile shear in the region where the temperature is less than T_c.

pseudotachylytes, and themselves overprinted by mylonites, as stable ductile shearing gives way to an unstable event to be followed again by stable ductile shearing. The complex internal geometries of most mylonite zones (BELL and HAMMOND, 1984) may result from such complicated cycles.

Thus, three different physical processes potentially define the base of the seismogenic zone in continental regions, as follows.

1. The base of the seismogenic zone is defined by the brittle–ductile transition, as discussed by BRACE and KOHLSTEDT (1980), KIRBY (1980), and SIBSON (1982); under such circumstances the temperature at the brittle–ductile transition would have to be higher than T_c, the temperature below which ductile shearing is unstable (see Fig. 17a).

2. The base of the seismogenic zone is defined by the transition from velocity-weakening to velocity-strengthening behavior, as discussed by SHIMAMOTO and LOGAN (1985), TSE and RICE (1985), and ORD and HOBBS (1986b); again, this view supposes that this transition from velocity-weakening to velocity-strengthening behavior occurs at temperatures where ductile behavior is unstable, but the brittle–ductile transition is deeper than the T_c isotherm (see Fig. 17b).

3. The base of the seismogenic zone is defined by T_c; seismic events due to unstable ductile shearing could occur at shallower depths (see Fig. 17c), and at even shallower depths seismic activity due to frictional stick-slip events would occur.

In addition to these three possibilities, deep crustal seismic events could occur anywhere within the ductile regime so long as the temperature is below T_c (Fig. 17d) for the particular rock type involved.

Acknowledgments

We wish to thank Jim Boland and Chris Beaumont for helpful discussion. The contents of this paper were developed under a grant from the U.S. Geological Survey, Department of the Interior. Those contents do not, however, necessarily represent the policy of that agency, and endorsement by the Federal Government should not be assumed.

REFERENCES

ARGON, A. S. 'Stability of plastic deformation', in *The Inhomogeneity of Plastic Deformation.* ASM Publication, 1973.

BACKOFEN, W. A., *Deformation Processing.* Addison Wesley, London, 1972.

BELL, T. H. and HAMMOND, R. L. (1984), *On the internal geometry of mylonite zones.* J. Geol. *92*, 667–686.

BERTHE, D., CHOUKROUNE, P. and JEGOUZO, P. (1979), *Orthogneiss, mylonite and non coaxial deformation of granites: The example of the South-American Shear Zone.* J. Struct. Geol. *1*, 31–42.

BOTT, M. H. P., *The Interior of the Earth.* Edward Arnold, 1971.

BRACE, W. F. and BYERLEE, J. D. (1966), *Stick-slip as a mechanism for earthquakes.* Science *153*, 990–992.

BRACE, W. F. and KOHLSTEDT, D. L. (1980), *Limits on lithospheric stress imposed by laboratory experiments.* J. Geophys. Res. *85*, 6248–6252.

BRUN, J. P. and COBBOLD, P. R. (1980), *Strain heating and thermal softening in continental shear zones: A review.* J. Struct Geol. *2*, 149–158.

BURNHAM, C. W., HOLLOWAY, J. R. and DAVIS, N. F. (1969), *Thermodynamic properties of water to 1000 °C and 10,000 bars.* Special Paper No. 132. Geol Soc. America.

CARTER, N. L. and KIRBY, S. H. (1978), *Transient creep and semibrittle behavior of crystalline rocks.* Pure Appl. Geophys. *116*, 807–839.

CHEN, W. P. and MOLNAR, P. (1983), *Focal depths of intracontinental and intraplate earthquakes and their implications for the thermal and mechanical properties of the lithosphere.* J. Geophys. Res. *88*, 4183–4214.

CHOPRA, P. N. and PATERSON, M. S. (1981), *The experimental deformation of dunite.* Tectonophys. *78*, 453–473.

DELANY, J. M. and HELGESON, H. C. (1978), *Calculation of the thermodynamic consequences of dehydration in subducting oceanic crust to 100 kb and >800 °C.* Amer. J. Sci. *278*, 638–686.

DIETERICH, J. H. (1978), *Time-dependent friction and the mechanics of stick slip.* Pure Appl. Geophys. *116*, 790–806.

DIETERICH, J. H. (1979a), *Modeling of rock friction: 1. Experimental results and constitutive equations.* J. Geophys. Res. *84*, 2161–2168.

DIETERICH, J. H. (1979b), *Modeling of rock friction: 2. Simulation of preseismic slip.* J. Geophys. Res. *84*, 2169–2175.

DIETERICH, J. H., 'Constitutive properties of faults with simulated gouge, in *Mechanical Behavior of Crystal Rocks*, Handin Volume, N. L. Carter, M. Friedman, J. M. Logan, and D. W. Stearns (eds), 1981. Geophys. Monograph Ser. *No. 24*, AGU, 103–120.

DRUCKER, D. C. (1960), *Extension of the stability postulate with emphasis on temperature changes*, Proc. 2nd Symp. Naval Struct. Mech. Brown Univ., E.-H. Lee and P. S. Symonds (eds), Pergamon, 1960.

ETHERIDGE, M. A., WALL, V. J., COX, S. F. and VERNON, R. H. (1984), *High fluid pressures during regional metamorphism and deformation: Implications for mass transport and deformation mechanisms.* J. Geophys. Res. (*Special Issue on Chemical Effects of Water on Rock Deformation*) *89*, 4344–4358.

ETHERIDGE, M. A., WALL, V. J. and VERNON, R. H. (1983), *The role of the fluid phase during regional metamorphism and deformation.* J. Metamorph. Geol. *1*, 205–226.

EVANS, B., and WONG, T.-F., 'Shear localization in rocks induced by tectonic deformation' chap. 10, in *Mechanics of Geomaterials*, Z. Bazant (ed), John Wiley, 1985, 189–210.

FLEITOUT, L. and FROIDEVAUX, C. (1980), *Thermal and mechanical evolution of shear zones.* J. Struct. Geol. *2*, 159–164.

FORMAN, D. J. (1966), *Regional geology of the south-west margin, Amadeus Basin, central Australia.* Bur. Miner. Resour. Aust. *Rept. 87*.

GOETZE, C. (1978), *The mechanisms of creep in olivine.* Phil. Trans. R. Soc. Lond. A *22*, 99–119.

GRIGGS, D. T. and BAKER, D. W. (1969), 'The origin of deep focus earthquakes,' in *Properties of Matter under Unusual Conditions,* H. Mark and S. Fernbach (eds), Interscience, 1969, p. 23–42.

GRUNTFEST, I. J. (1963), 'A note on thermal feedback and the fracture of solids,' in *Fracture of Solids.* Metallurg. Soc. Conf. *20*, Interscience, 189–193.

GU, J.-C., RICE, J. R., RUINA, A. L. and TSE, S. T. (1984), *Slip motion and stability of a single degree of freedom elastic system with rate and state dependent friction.* J. Mech. Phys. Solids *32*, 167–196.

HOBBS, B. E., MCLAREN, A. C. and PATERSON, M. S. 'Plasticity of single crystals of synthetic quartz, in *Flow and Fracture of Rocks.* Geophys. Monograph Ser. *16* H. C. Heard et al. (eds), AGU Washington 1972, p. 29–53.

JAOUL, O., TULLIS, J. A. and KRONENBERG, A. K. (1984), *The effect of varying water contents on the creep behavior of Heavitree quartzite.* J. Geophys. Res. (Special Issue on Chemical Effects of Water on Rock Deformation) *89*, 4298–4312.

KIRBY, S. H. (1980), *Tectonic stress in the lithosphere: Constraints provided by the experimental deformation of rocks.* J. Geophys. Res. *85*, 6353–6363.

KIRBY, S. H. and KRONENBERG, A. K. (1984), *Deformation of clinopyroxenite: Evidence for a transition in flow mechanisms and semibrittle behavior.* J. Geophys. Res. *89*, 3177–3192.

LEHNER, F. K., LI, V. C. and RICE, J. R. (1981), *Stress diffusion along rupturing plate boundaries.* J. Geophys. Res. *86*, 6155–6169.

LISTER, G. S. and SNOKE, A. W. (1984), *S-C mylonites.* J. Struct. Geol. *6*, 617–638.

MACAUDIERE, J. and BROWN, W. L. (1982), *Transcrystalline shear fracturing and pseudotachylyte generation in a meta-anorthosite* (*Harris, Scotland*). J. Struct Geol. *4*, 395–406.

MADDOCK, R. H. (1983), *Melt origin of fault generated pseudotachylytes demonstrated by textures.* Geology *11*, 105–108.

MIYASHIRO, A., *Metamorphism and Metamorphic belts*, Allen and Unwin, London, 1973.

NICHOLSON, D. W., and KIDDY, K. C. (1984), *A large deformation plasticity model with rate sensitivity and thermal softening*, Trans. ASME *106*, 388–392.

ORD, A. and HOBBS, B. E. (1986a), *Experimental control of the water-weakening effect in quartz, Geophysical* Monograph *36*, B. E. Hobbs and H. C. Heard (eds.) American Geophysical Union, 51–72.

ORD, A. and HOBBS, B. E. (1986b), *The strength of the continental crust, detachment zones and the development of plastic instabilities,* submitted to *Tectonics.*

OROWAN, E. (1960), *Mechanism of seismic faulting.* Geol. Soc. Amer. Mem. *79,* 323–345.

PASSCHIER, C. W. (1982a), *Mylonitic deformation in the Saint-Barthelemy Massif, French Pyrenees, with emphasis on the genetic relationship between ultramylonite and pseudotachylyte,* GUA Papers Geology, Ser. 1 *16,* 1–173.

PASSCHIER, C. W. (1982b), *Pseudotachylyte and the development of ultramylonite bands in the Saint-Barthelemy Massif, French Pyrenees* J. Struct. Geol. *4,* 69–79.

PASSCHIER, C. W. (1984), *The generation of ductile and brittle shear bands in a low-angle mylonite zone.* J. Struct. Geol. *6,* 273–281.

PHILPOTTS, A. R. (1964), *Origin of pseudotachylytes* Amer. J. Sci. *262,* 1008–1035.

POIRIER, J. P. (1980), *Shear localization and shear instability in materials in the ductile field* J. Struct. Geol. *2,* 135–142.

POIRIER, J. P., BOUCHEZ, J. L. and JONAS, J. J. (1979), *A dynamic model for seismic ductile shear zones* Earth and Planet Sci. Lett. *43,* 441–453.

RECHT, R. F. (1964), *Catastrophic thermoplastic shear,* Trans. ASME 31E (J. Appl. Mech.) *2,* 189–193.

RICE, J. R. and GU, J.-C. (1983), *Earthquake aftereffects and triggered seismic phenomena* Pure Appl. Geophys. *121,* 187–219.

RICE, J. R. and RUINA, A. L. (1983), *Stability of steady frictional slipping,* Trans. ASEM J. Appl. Mech. *50,* 343–349.

RICE, J. R. and TSE, S. T. (1986), *Dynamic motion of a single degree of freedom following a rate and state dependent friction law,* J. Geophys. Res. *91,* 521–530.

ROGERS, H. C. (1979), *Adiabatic plastic deformation,* Ann. Rev. Mat. Sci. *9,* 283–311.

RUBIN, D. and DRUCKER, D. C. 'On stability of viscoplastic systems with thermo-mechanical coupling, in *Contributions to Mechanics,* D. Abir (ed), Pergamon, 1969, p. 171–179.

RUINA, A. L., *Friction laws and instabilities: A quasistatic analysis of some dry frictional behavior,* Ph.D. Thesis, Brown Univ., 1980.

RUINA, A. L. (1983), *Slip instability and state variable friction laws* J. Geophys. Res. *88,* 10359–10370.

RUTLAND, R. W. R. (1976), *Orogenic evolution of Australia* Earth Sci. Rev. *12,* 161–196.

SCHUBERT, G. and YUEN, D. A. (1978), *Shear heating instability in the Earth's upper mantle* Tectonophys. *50,* 197–205.

SHELTON, G. and TULLIS, J. (1981), *Experimental flow laws for crustal rocks* (Abstr.), EOS, Trans. AGU *62,* 396.

SHIMAMOTO, T. and LOGAN, J. M. (1985), *Velocity-dependent behavior of simulated halite shear zones: An analog for silicates* unpub. ms. presented 1985 Ewing Symp.

SIBSON, R. H. (1977), *Fault rocks and fault mechanisms* J. Geol. Soc. (Lond.) *133,* 191–214.

SIBSON, R. H. (1980), *Transient discontinuities in ductile shear zones* J. Struct Geol. *2,* 165–171.

SIBSON, R. H. (1982), *Fault zone models, heat flow, and the depth distribution of earthquakes in the continental crust of the United States* Bull. Seismol. Soc. Am. *72,* 151–163.

SIBSON, R. H. (1984), *Roughness at the base of the seismogenic zone: Contributing factors* J. Geophys. Res. *89,* 5791–5799.

STAKER, M. R. (1981), *The relation between adiabatic shear instability strain and material properties* Acta Met. *29,* 683–689.

STEL, H. (1981), *Crystal growth in cataclasites: Diagnostic microstructures and implications* Tectonophys. *78,* 585–600.

STEL, H. (1986), *The effect of cyclic operation of brittle and ductile formation on the metamorphic assemblage in cataclasites and mylonites* Pure Appl. Geophys. *124,* 000–000.

TEYSSIER, C. (1985), *A crustal thrust system in an intracratonic tectonic environment* J. Struct Geol. *7,* 689–700.

TEYSSIER, C. and HOBBS, B. E. (1985), *Deformation and softening mechanisms in a major crustal fault zone, Redbank deformed zone (Central Australia)* submitted to J. Struct. Geol.

TSE, S. T. and RICE, J. R. (1985), *Crustal earthquake instability in relation to the depth variation of frictional slip properties* (unpub. ms presented 1985 Ewing Symp.).

TULLIS, T. and WEEKS, J. (1985), *Frictional sliding of dolomite: A variation in constitutive behavior* J. Geophys. Res. *90,* 7821–7826.

TULLIS, J. A. and YUND, R. A. (1985), *Dynamic recrystallization of feldspar: A mechanism for ductile shear zone formation* Geology *13*, 238–241.

VERNON, R. H., *Metamorphic Processes*, Allen and Unwin, London, 1976.

WARREN, R. G. (1983), *Granulite and the tectonic evolution of the Arunta Block* Geol. Soc. Aust. 6th Aust. Geol. Conv. Abstr. Ser. *9*, 59.

WEEKS, J. D. and TULLIS, T. (1984), *Frictional behavior of dolomite* (Abstr), EOS, Trans. AGU *65*, 1077.

WEISS, L. E. and WENK, H. R. (1983), *Experimentally produced pseudotachylyte-like veins in gabbro* Tectonophys. *96*, 299–310.

WENK, H. R. (1978), *Are pseudotachylytes products of fracture or fusion?* Geology *6*, 507–511.

WENK, H. R. and WEISS, L. E. (1982), *Al-rich calcic pyroxene in pseudotachylyte: An indicator of high pressure and high temperature?* Tectonophys. *84*, 329–341.

YUEN, D. A. and SCHUBERT, G. (1977), *Asthenospheric shear flow: Thermally stable or unstable?* Geophys. Res. Let. *4*, 503–506.

ZENER, C. and HOLLOMON, J. H. (1944), *Effect of strain-rate upon plastic deformation of steel* J. Appl. Phys. *15*, 22–32.

(Received 20th October 1985, revised 20th February 1986, accepted 23rd February 1986)

PAGEOPH, Vol. 124, Nos. 1/2 (1986) 0033–4533/86/020337–28$1.50 + 0.20/0

Ductile Deformation and Regional Strain Field in the Southern Segment of the Tancheng–Lujiang Fault Zone, Eastern China

JIAWEI XU,[1] PING WANG,[1] RENGAO CHING[1] and ZHOUJIE YE[1]

Abstract—Studies in a segment of the Tancheng–Lujiang fault zone show that it is a major transcurrent ductile shear zone with a considerable sinistral displacement. The formations ($Pt \rightarrow T_{1+2}$) at depths of 5 to 15 km may have cropped out in this area during subsequent erosion. Many ductilely deformed structures are in it. The deformed zone was formed during the Indo-Sinian orogeny. On the basis of measurements and analyses of many deformed pebbles within the shear zone, it is suggested that the deformation in the zone studied may be catagorized as a variation of simple shear and plane strain with a constant volume. The intensely deformed belt is generally 40 to 50 km wide, with the average strain ratio 27.68 and the maximum greater than 87.37. From analysis of elongation strain, we estimate that the deformed belt has been elongated by 164.23 km and narrowed by 222.5 km. The internal belt was formed by a high shear strain γ, calculated to be more than 11.34. The deformed belt is associated with syntectonic dynamothermal metamorphism, represented by greenschist facies and retrogressive metamorphism on previous amphibolite facies.

Key words: Tancheng-Lujiang Fault zone, ductile shearing, regional strain field, strain ratio, tectonic levels.

Introduction

The Tancheng–Lujiang fault zone, a major wrench fault to the northwest of the Pacific Ocean, cuts the eastern continental crust of Asia, crossing the Heilungjiang, Jilin, Liaoning, Shandong, Jiangsu, Anhui, and Hubei provinces from north to south in northern China and extending intermittently into southern China. The northern segment of this fault comes into the Kurtuhin fault zone in the easternmost areas of the Soviet Union. The end of the fault enters the Sakhalin Gulf in the Sea of Ochotsk. It may be traced for more than 3600 km, with a strike from northeast to north-northeast. This fault is 10 to 50 km wide and presents a fractural downwarping zone which seems to be a rift valley.

The Tancheng–Lujiang fault zone has been studied, by many Chinese geologists. Significant post-Mesozoic sinistral horizontal displacements along it have recently been brought to light (XU, 1980a, b, 1984; XU et al., 1985). XU presented convincing evidence that the maximum horizontal displacement found in the middle segment of

[1] Department of Geology, Hefei Polytechnical University, People's Republic of China.

the fault zone, from north Lieoning to west Shandong, is more than 740 km. The sinistral transcurrent faulting probably started in the Late Triassic (Indo-Sinian orogeny) and ceased in the middle Cretaceous (Early Yenshan orogeny). Intense dislocation took place in the Late Jurassic. The fault zone turned into a tension graben during the Late Cretaceous to Paleogene. The compression and small dextral shearing occurred after the Neogene.

In the southern segment of the fault zone, from south of the Huaiho River to north of the Yangtze River, especially along the Zhangbaling belt on the eastern side of the fault zone between Hungtse Lake and Chaohu Lake, there is a major ductile transcurrent fault (MATTAUER and MERCIER, 1980) characterized by strong ductile deformation at deep tectonic levels. The upper crustal structures of the Yangtze Platform appear in this segment and lie in the intensely sheared area between the Dabie Mountain Uplift and the Jiaonan Uplift; see Figure 1. Owing to intense uplifting and subsequent erosion since the Cretaceous, a series of formations, from old to young, cropped out on the eastern side of the fault zone from west to east: the Feidong group (Pt_1), the Zhangbaling group (Pt_2), and Sinian, Cambrian-Ordovician, and Silurian-Triassic. The deeper Dabie group ($Ar–Pt_1$) is exposed to the west of Tongcheng County. The western side of the fault zone had subsided 3 or 4 km. As a result, in the segment studied various deformed structures appeared at different tectonic levels, from deep to shallow, from ductile to brittle, rendering this fault zone a geological 'museum'; see Figure 2. These deformed formations have thicknesses of more than 15 km and are covered unconformably by Early and Middle Jurassic basins which have not suffered intense deformation. Studies of these formations indicate that an unconformity may exist between the Feidong group and the Zhangbaling group, and that a nonsequence or possibly weak unconformity may exist between the Zhangbaling group and the Sinian, without a clear deformation event between them. Here intense deformation within the group occurred chiefly in the Mesozoic, by shear along the Tancheng–Lujiang fault zone. In the Sinian rocks ductile structures developed. The depths of their formation, from 5 to 15 km, are determined from the calculated thicknesses of strata.

Main Structures of Ductile Deformation and their Locations in Vertical Section

The main ductile deformation structures are as follows: (1) tectonites of mylonite series and their structures, such as mylonite, ultramylonite, phyllonite, and schistomylonite; (2) pseudotachylytes; (3) mineral lineations, appearing in the lens-shaped massif consisting of the Feidong group near the fault zone (Fig. 2); slip schistosity and stretching lineations (Fig. 3), ubiquitous in the region (Fig. 4); (5) flattened intrusive rock bodies with the same foliation (schistosity) as in the adjacent formations, such as the Tonggan rock body and the Qiaotoujie rock body in the Qiaotoujie Massif; (6) syntectonic gash veins cutting the stretch lineations obliquely or normally,

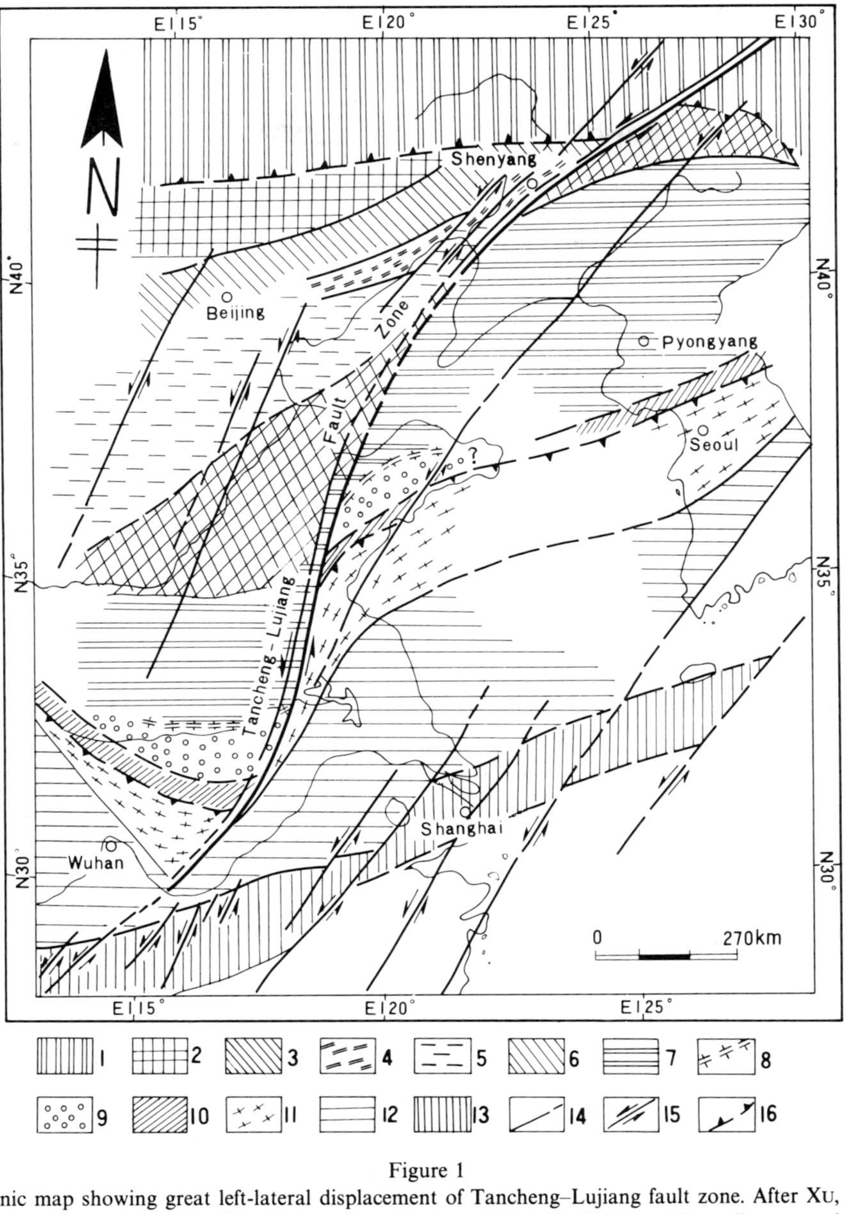

Figure 1

Tectonic map showing great left-lateral displacement of Tancheng–Lujiang fault zone. After Xu, 1980a. Legend: (1) Inner Mongolia–Jilin–Heilungjiang Variscan fold belt, (2) Inner Mongolia upwarping belt (Ar–Pt_2), (3) Yenshan fold belt (Z–Pz_2), (4) Shanhaiguan uplift belt (Ar–Pt_2), (5) Huanghua subsidence belt (Z-Pz_2), (6) West Shandong–North Liaoning uplift belt (Ar), (7) North Huaiho–South Liaoning–South Pyongan Massif (Z–Pz_2), (8) Banpu-Qixia anticlinorium, (9) Hefei–Jiaolai downwarping belt (J_3), (10) North Huaiyang–North Jiaonan–North Kyonggi fold belt (Pz), (11) Dabie–Jiaonan–Kyonggi uplift belt (Pt), (12) Lower Yangtze–Okcheon subsidence belt (Pz–T_2), (13) Jiangnan uplift belt (Pt_{1-2}). Note that areas indicated by legend numbers 2 to 9 are on the Sino-Korean Platform, and 11 to 13, on the Yangtze Platform. (14) Boundaries of tectonic unit, (15) wrench fault lines, showing sense of shear, (16) possible subduction zone within continent.

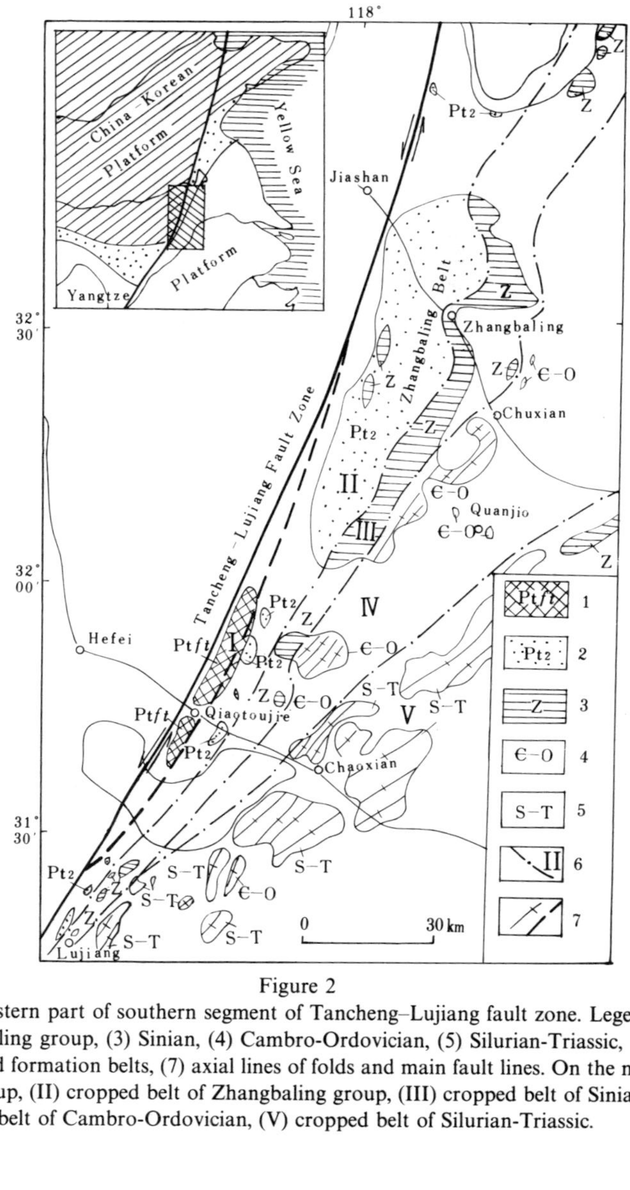

Figure 2

Tectonic map of eastern part of southern segment of Tancheng–Lujiang fault zone. Legend: (1) Feidong group, (2) Zhangbaling group, (3) Sinian, (4) Cambro-Ordovician, (5) Silurian-Triassic, (6) number and boundaries of folded formation belts, (7) axial lines of folds and main fault lines. On the map: (I) cropped belt of Feidong group, (II) cropped belt of Zhangbaling group, (III) cropped belt of Sinian, (IV) cropped belt of Cambro-Ordovician, (V) cropped belt of Silurian-Triassic.

and formed by polygenetic quartz fibers which are perpendicular to both walls of the vein (Durney, D. W. and Ramsay, 1973), Figure 5; (7) pebbles kinked, flattened and elongated, a great number resembling a fish in shape (*a* axis $\gg$ *b* axis $\gg$ *c* axis), some slip lines appearing on their surfaces (Fig. 6a), and a few competent pebbles not obviously deformed and with oblique gashes (Fig. 6b); (8) boudinage and similar structures, the latter associated with some brittle fractures; (9) flow cleavage and

Figure 3
Stretch lineations on slip schistosity plane; Zhangbaling group, Yaujie, Chuxian County.

fracture cleavage, their orientations stable in all areas (Fig. 4); (10) flattened linear folds; (11) strain-slip cleavage and ductile shear curings, etc.

A large number of flattened and elongated bodies are found within the formations in all the areas studied, including quartz and feldspar grains in the intrusives, volcanic breccia and conglomerate in the Zhangbaling group, gas cavities in andesitic layers of the Zhougan formation, boulders and pebbles of glacial drift and nodules in the Sujiawan formation (Z_1), and oölites in the Upper Sinian. There are also deformed fossil trilobites in the Cambrian formation. To the east of the fault zone center there gradually crop out the mylonite zone, mineral lineation zone, stretching lineation zone, flow cleavage zone, fracture cleavage zone, flattened linear fold zone, and weak linear fold zone. See Figure 7. The orientations of these deformation features lie between 0 and 20 degrees from that of the main fault plane of the Tancheng–Lujiang fault zone. The angle becomes larger with increasing distance from the fault zone center. On the whole, the various deformation characteristics represent different tectonic levels, from deep to shallow in the upper Earth's crust; see Table 1.

Principal Characteristics of Regional Strain Field

The authors have measured directly the values of the *a*, *b*, and *c* axes of deformed pebbles on the outcrops, which are related to the *X*, *Y*, and *Z* axes of a finite strain ellipsoid. At each location 50 to 100 deformed pebbles were measured, and a total of 1956 principal axes were obtained. The authors also measured the widths of syntectonic gash veins and a few deformed oölites and trilobite fossils. Using the methods of strain analysis suggested by BILLINGS (1972), HILLS (1972), MEANS (1976), HOBBS et al. (1976), RAMSEY (1967, 1980), RAMSAY and GRAHAM (1970), MATTAUER and

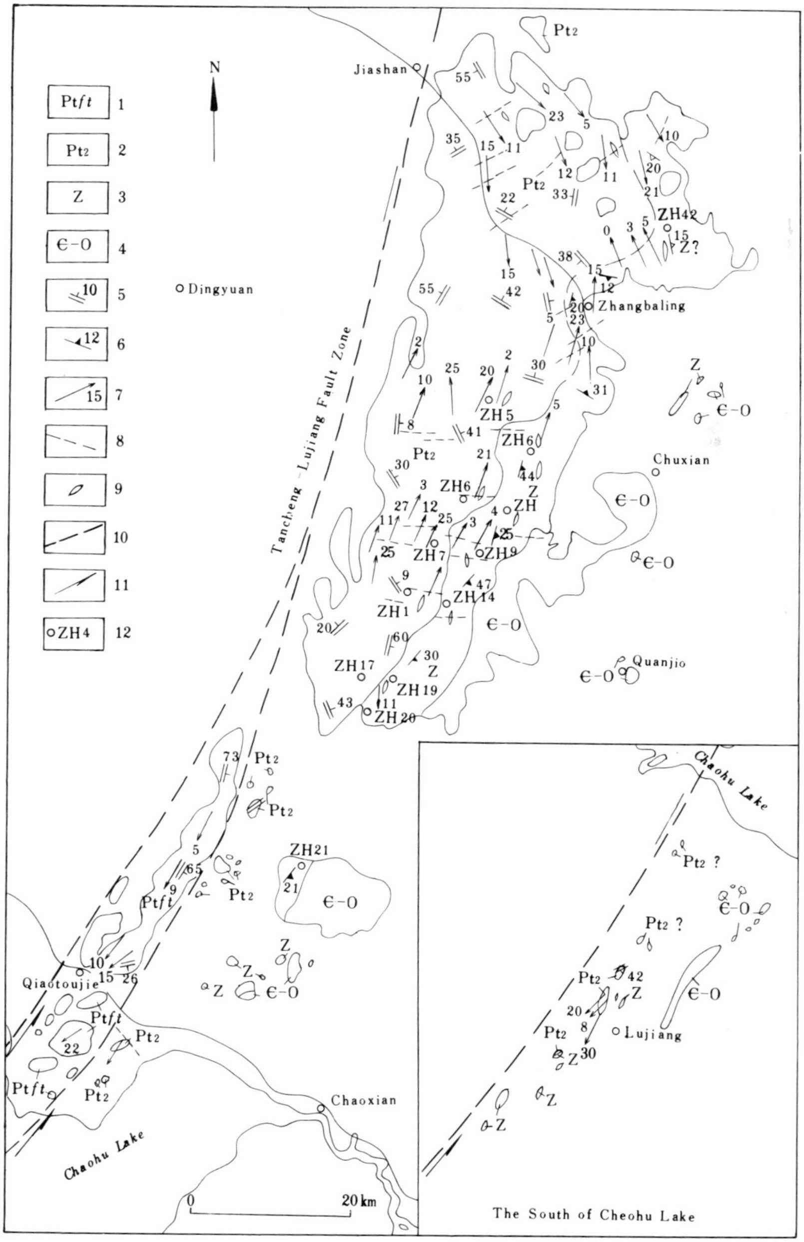

Figure 4

Distribution of foliations in Precambrian system for Zhangbaling belt of southern segment of Tancheng–Lujiang fault zone. Legend: (1) Feidong group, (2) Zhangbaling group, (3) Sinian, (4) Cambro-Ordovician, (5) schistosity, (6) cleavage, (7) lineations, (8) strikes of strain slip cleavage, (9) flattened and elongated pebbles, (10) main fault lines of Tancheng–Lujiang fault zone, (11) directions of displacement, (12) numbered sites of deformed pebbles.

Figure 5
Gash veins cutting schistosity; Zhangbaling group, Zulong-qiao, Chuxian County.

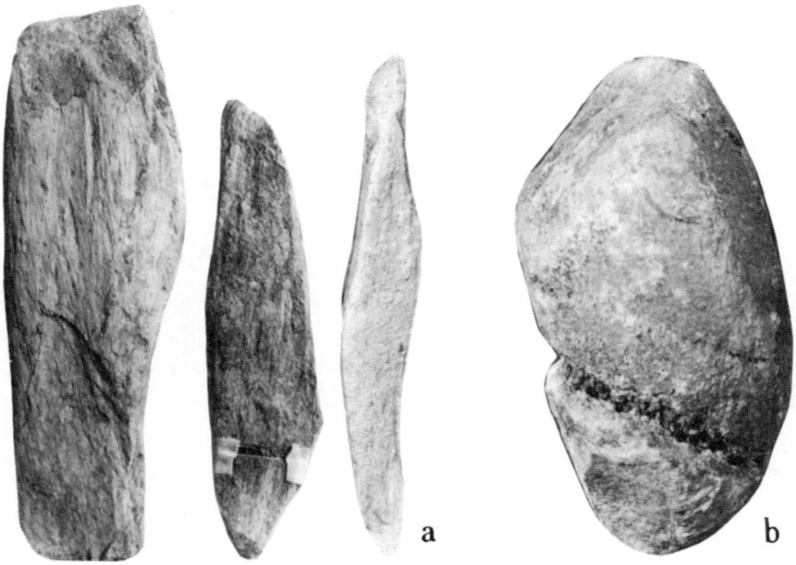

Figure 6
Deformed pebbles; Lower Sinian (Sujiawan formation), Shijie, Chuxian County: (a) fish shape, incompetent pebble with some longitudinal slip lines (0.5 ×); (b) competent pebble with oblique gashes (1 ×).

Figure 7
Fracture cleavage cutting stratifications; Lower Triassic, Wanghu Mount, Chaoxian County.

MERCIER (1980), and HE (1981), they derived a series of parameters related to the regional strain field in this area.

Deformation of pebbles and Strain ellipsoid characteristics

The pebbles studied are of glacial and volcanic origin. The initial shapes of these pebbles are very irregular, but may be taken to be statistically equant. The initial orientations of the long axes of the pebbles are assumed to be random. This assumption is supported by measurements of the axial ratios of weakly deformed pebbles, which are only 2.22/1 to 3.30/1. If there is an initial orientation, the axial ratio must be less than these values. The axial ratios of intensely deformed pebbles in the segment studied are commonly about 20/1 to 30/1. So the initial orientation does not have tectonic significance in that location. The general character of deformation is $a \gg b \gg c$; thus a great number of pebbles become long and flat, in the shape of a fish. Statistically deformed pebbles may be taken to represent the finite strain ellipsoids. The maximum elongation axes (X axes) of these deformed objects are parallel to the stretching lineation and the axial surface of the folds. Except in the later orogeny, the a axes are nearly horizontal. The a–b plane of the deformed pebbles is the maximum flattening plane of a strain ellipsoid, commonly parallel to the metamorphic foliations, axial surface, schistosity, and various cleavages.

The deformation of the flattened and elongated pebbles is neither pure flattening nor pure elongation. These pebbles possibly are the products of differential shearing

Table 1

Columnar section showing characteristics of the deformation of the tectonic levels at different depths in the southern segment of the Tancheng–Lujiang fault zone.

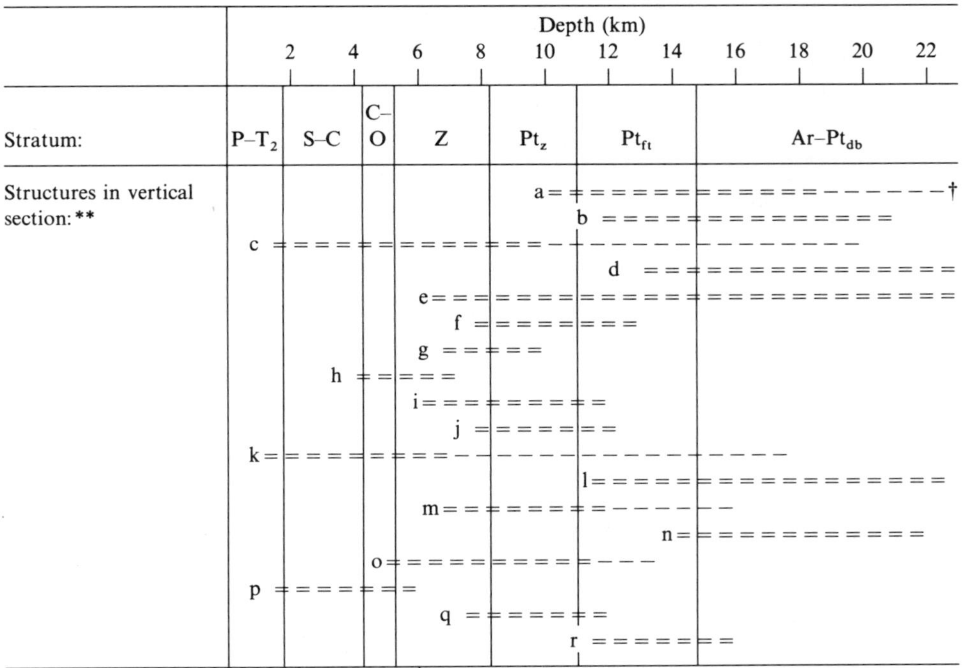

Structure:	Shear structures of fold and fracture	Deformed structures of ductile shear		
Deformation:	Ductile-brittle failure	Paraplastic to plastic flow	Viscous flow	
Metamorphism facies:	Non-metamorph.	Weak meta-morph.	Greenschist facies	Amphibolite facies (retrogressive metamorph.)

* Strata: $P–T_2$, Permian-Middle Triassic; S–C, Silurian-Carboniferous; C–O, Cambrian-Ordovician; Z, Sinian; Pt_z, Zhangbaling group, Pt_{ft}, Feidong group, Pt_{db}, Dabie group.

** Structures: a, mylonite and ultra mulonite; b, pseudotachylyte; c, breccia-cataclasite; d, gneissosity; e, stretching lineations; f, slip schistosity; g, flow cleavage; h, fracture cleavage; i, strain slip cleavage; j, syntectonic gashes; k, jion series, l, plastic boudinage; m, similar boudinage; n, flow structures; o, linear folds; p, paralinear folds; q, kinks; r, present focuses.

† Single dashes represent shallow structures developing at deeper levels owing to later uplift.

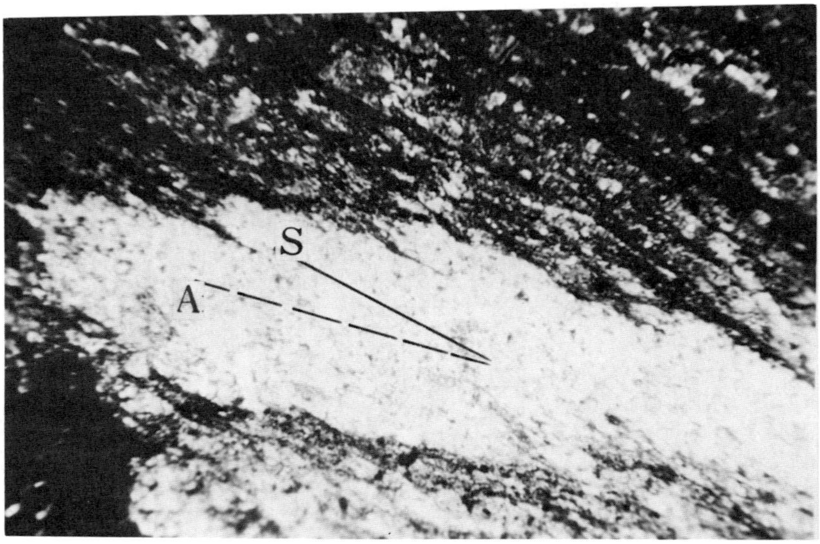

Figure 8a
Photomicrograph (30 ×) showing small angle between stretching axis, A, of elongated pebble and shear plane, S, Sanhejie, Chuxian County.

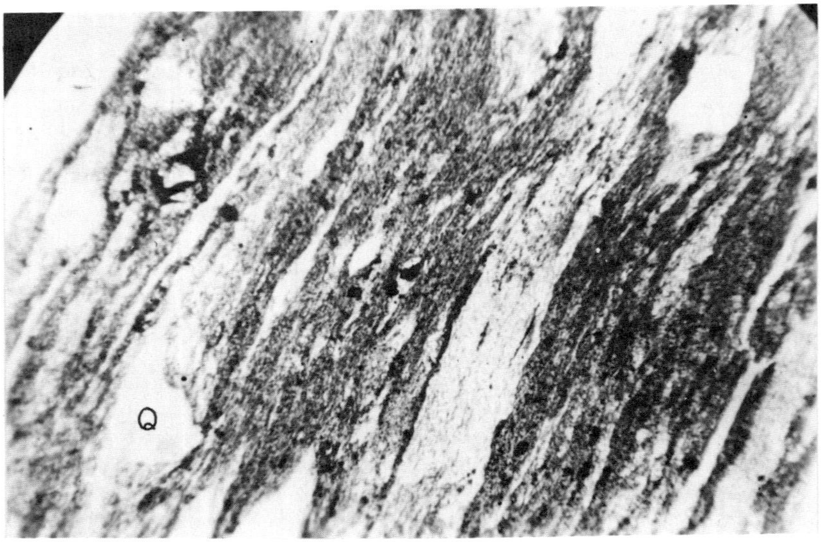

Figure 8b
Photomicrograph (30 ×) showing the elongated small pebbles, in Sujiawan formation. Q, quartz with pressure shadows. From Sanhejie, Quanjio County.

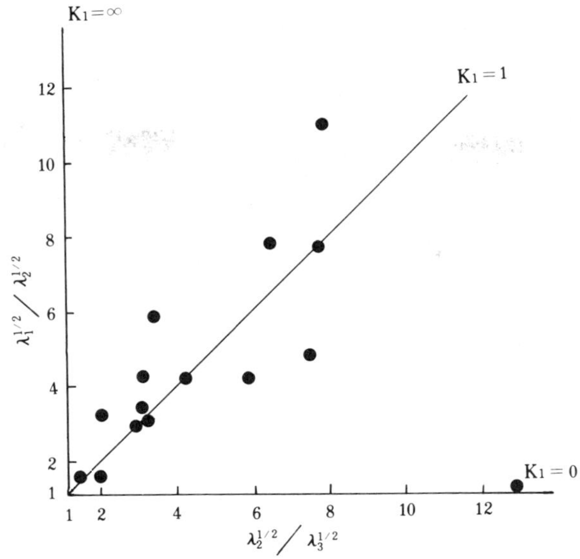

Figure 9
Variation in K_1 (Flinn parameter) for eastern part of southern segment of Tancheng–Lujiang fault zone.

with perpendicular compression and some shear rotation, because there exist both longitudinal slip lines and oblique gashes on their surfaces. As Figure 8 shows there is a small angle between the stretch axis of a pebble and the shear plane—in agreement with the idea of deformation in shear.

Measurements of the deformed pebbles show the following elongations: $e_1 = 3.436$, $e_2 = 0.058$, and $e_3 = -0.713$ (see Table 2). That is, the *a* axis has undergone strong elongation and the *c* axis strong shortening, yet the *b* axis has stayed nearly the same. The Flinn parameters *K*, showing the geometry of a strain ellipsoid, are $K_1 = 1.173$ and $K_2 = 1.053$ over the entire study area; see Figures 9 and 10. Note: $K_1 = [(1 + e_1)/(1 + e_2) - 1]/[(1 + e_2)/(1 + e_3) - 1] = [(\lambda_1^{1/2}/\lambda_2^{1/2}) - 1]/[(\lambda_2^{1/2}/\lambda_3^{1/3}) - 1]$ and $K_2 = (\ln \lambda_1^{1/2} - \ln \lambda_2^{1/2})/(\ln \lambda_2^{1/2} - \ln \lambda_3^{1/2})$, corrected by Ramsey (1967).

That we have $K_1 \approx 1$ suggests that the strain field may be described as simple shear, and $K_2 \approx 1$, that it is a deformation with a constant volume. In other words, in the process of simple shear the *b* axis remains nearly invariant and the strain ellipsoid is flattened and elongated. The strain field around the fault has the character of simple shear. The ellipsoid is of plane-strain type, between the flattening type and the constriction type (Ramsay, 1967). Owing to very intense and continuous shearing, the strain ellipsoid is extremely elongated, and its *X* axis (*a* axis) gradually approaches the direction parallel to the shear and the motion. With the development of the deformation the angle between the schistosity, the plane containing the *X* and *Y* axes (*a* and *b* axes) of the ellipsoid, and the shear zone becomes smaller and

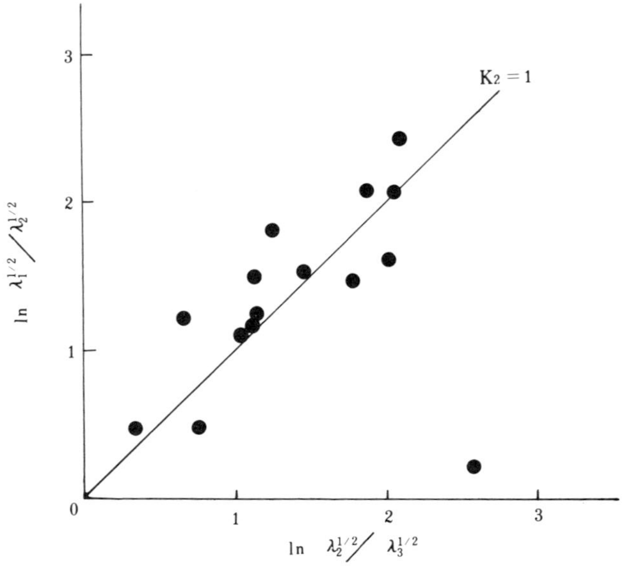

Figure 10
Variation in K_2 for eastern part of southern segment of Tancheng–Lujiang fault zone.

smaller (MATTAUER and MERCIER, 1980). Such progressive strain occurred in this area.

The deformations of the pebbles in the Sujiawan formation and Zhangbaling group are clearly different in different parts of the fault zone. A rapid increase appears in the stretch $\lambda_1^{1/2}$ and the axis ratio a/c toward the center of the Tancheng–Lujiang fault zone (Table 2 and Fig. 11 and 12).

The intense distortion zone on the eastern side of the fault is 40 or 50 km wide. The average strain axis ratio (a/c value) is 27.68/1. At the measurement site nearest to the fault zone center (ZH-5) the maximum strain axis ratio is more than 87.37/1, where the a/c values of several pebbles are 300/1. In the famous Alps area and the Blue Ridge area in America (Mitra, 1979) the maximum a/c values are about 100/1.

Relationship of pebble deformation to nature of rock

The shapes and the axis ratios of deformed pebbles are related not only to the intensity of stress, but also to the mechanical nature of the rock in the pebbles. The authors measured two groups of deformed pebbles composed of different rocks at Sujiawan, Chaoxian County, near the Niao reservoir (Table 3). As expected, quartzite and granite pebbles show competency, resulting in weak flattening and elongation, but the quartz keratophyre tuff is much more deformed. In addition, cement in conglomerates commonly has a lesser competency than that in pebbles; thus the

Table 2

Strain Measurements on Southern Segment of Tancheng–Lujiang Fault Zone

Location	Site (ZH no.)*	F., G.**	No. of meas's.	Av. axis length (mm) a	b	c	Axis ratio (a:b:c) a	b	c	Init. r (mm)	Elongation (mm) E_1 (a/2 − r)	E_2 (b/2 − r)	E_3 (c/2 − r)	Elongation, $e = E/r$ e_1	e_2	e_3	Stretch, $\lambda^{1/2} = e + 1$ $\lambda_1^{1/2}$	$\lambda_2^{1/2}$	$\lambda_3^{1/2}$	$A = \frac{\lambda_1^{1/2}}{\lambda_2^{1/2}}$	$B = \frac{\lambda_2^{1/2}}{\lambda_3^{1/2}}$	K_1,†† $\frac{A}{B}$	ln A	ln B	K_2,†† $\frac{\ln A}{\ln B}$
Zhangbaling area																									
Zhangpoyun	24	S.	183	37.07	12.13	4.20	8.83	2.89	1	6.18	12.36	−0.12	−4.08	2.000	−0.019	−0.660	3.000	0.981	0.340	3.058	2.885	1.092	1.118	1.060	1.055
Zulongqiac	6	S.	198	63.70	8.37	1.11	57.38	7.54	1	4.20	27.65	−0.015	−3.65	6.583	−0.004	−0.869	7.583	0.996	0.131	7.613	7.603	1.002	2.030	2.029	1.000
Zulongqiac	5	Z.	156	119.70	10.70	1.37	87.37	7.81	1	6.03	53.82	−0.679	−5.35	8.925	−0.113	−0.887	9.925	0.887	0.113	11.189	7.850	1.487	2.415	2.060	1.172
Sijie	4	S.	198	53.03	12.06	2.90	18.29	4.16	1	6.11	19.91	−0.08	−4.66	3.259	−0.013	−0.763	4.259	0.987	0.237	4.315	4.165	1.047	1.462	1.427	1.025
Zhougan	9	S.	177	47.39	14.75	4.77	9.94	3.09	1	7.47	16.23	−0.10	−5.09	2.173	−0.013	−0.681	3.173	0.987	0.319	3.215	3.094	1.058	1.168	1.129	1.035
Sijie	3	Z.	175	27.25	4.63	1.35	20.18	3.43	1	2.77	10.86	−0.455	−2.095	3.921	−0.164	−0.756	4.921	0.836	0.244	5.886	3.426	2.014	1.773	1.231	1.440
Zhougan †	7	Z.	218	119.09	98.35	7.65	15.57	12.86	1	21.59	37.95	27.58	−17.77	1.758	1.277	−0.823	2.758	2.277	0.177	1.211	12.864	0.017	0.191	2.554	0.075
Sanhejie	14	S.	168	64.63	15.62	2.60	24.86	6.01	1	6.81	25.51	0.70	−5.51	3.246	0.103	−0.809	4.746	1.103	0.191	4.303	5.775	0.692	1.459	1.754	0.832
Sanhejie	11	Z.	198	47.22	9.36	1.26	37.48	7.43	1	4.12	19.49	0.56	−3.49	4.731	0.136	−0.847	5.731	1.136	0.153	5.045	7.425	0.630	1.618	2.005	0.807
Guanba	20	S.	171	41.61	12.61	6.59	6.31	1.91	1	7.56	13.25	−1.26	−4.27	1.752	−0.167	−0.565	2.752	0.833	0.435	3.303	1.915	1.722	1.195	0.650	1.838
Guanba	17	Z.	198	108.73	13.83	2.16	50.34	6.40	1	7.41	46.96	−0.50	−6.33	6.337	−0.067	−0.854	7.337	0.933	0.146	7.864	6.390	1.273	2.062	1.855	1.112
Sujiawan	21	S.	297	84.06	24.70	8.01	10.49	3.08	1	12.76	29.27	−1.75	−8.75	2.294	−0.032	−0.686	3.294	0.968	0.314	3.403	3.083	1.154	1.225	1.126	1.088
Guanba	19	Zh.	156	19.82	4.56	1.55	12.79	2.94	1	2.60	7.31	−0.28	−1.83	2.812	−0.108	−0.704	3.812	0.892	0.296	4.274	3.014	1.626	1.453	1.103	1.317
Average							27.68	5.35	1					4.59	0.063	−0.817	5.59	1.063	0.173						
Yangtze River vicinity																									
Xintian	25	S.	159	42.19	26.68	19.02	2.22	1.40	1	13.88	7.22	−0.54	−4.37	0.520	−0.039	−0.315	1.520	0.961	0.685	1.582	1.403	1.444	0.459	0.339	1.354
Xintian	26	S.	220	6.515	4.09	1.97	3.30	2.07	1	1.87	1.38	0.17	−0.89	0.738	0.091	−0.476	1.738	1.091	0.524	1.593	2.082	0.548	0.466	0.733	0.636
All areas																									
Average							24.35	4.87	1					3.436	0.058	−0.713	4.436	1.058	0.287			1.173			1.053

* See Figure 4.

** Formation or group: S., Sujiawan formation; Z., Zhangbaling group; Zh., Zhougan formation.

† Not reliable data, owing to poor geological conditions.

†† Flinn parameters.

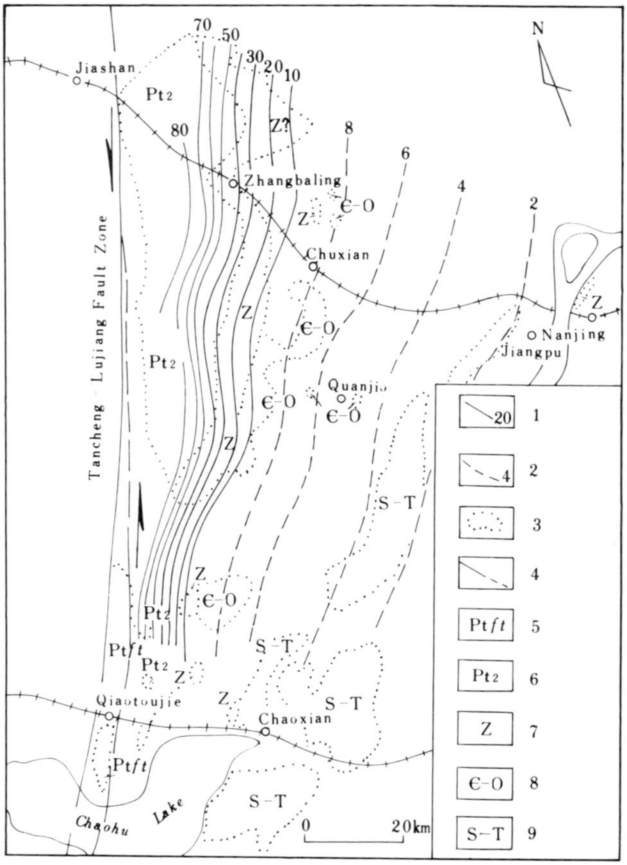

Figure 11
The strain isoline for eastern part of southern segment of Tancheng-Lujiang fault zone, showing variation of axis ratio a/c of deformed pebbles. Legend: (1) main isolines of a/c, (2) subsidiary isolines of a/c, (3) principal geological boundaries, (4) main fault zone, (5) Feidong group, (6) Zhangbaling group, (7) Sinian, (8) Cambro-Ordovician, (9) Silurian-Triassic.

average deformation of pebbles, taken from many measurements, is usually lower than that of whole rock. Different rock beds also exhibit different degrees of deformation; for example, the Zhangbaling group, containing mica schist, and the Sinian, containing clay beds, show the least competency. The more competent interlayers, such as spitite and marble, have been deformed into boudinage structures.

Strain varieties in regional horizontal plane

From measurements of a great number of flattened and elongated bodies in the Zhangbaling group and lower Sinian rocks, the authors have obtained a regional

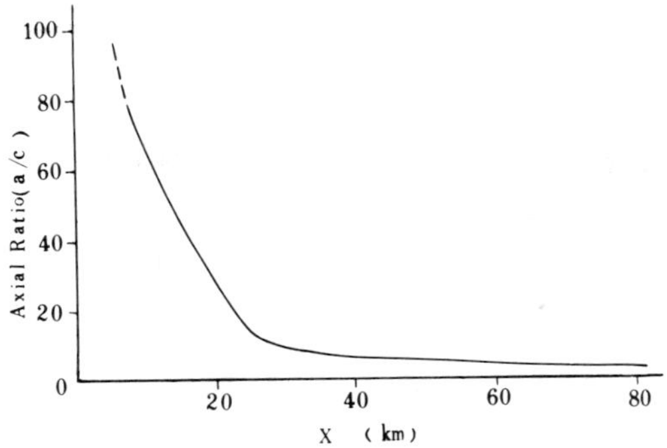

Figure 12
Axial ratio a/c as function of decreasing distance from the Tancheng–Lujiang fault zone.

Table 3
Deformed Pebbles of Various Rocks

Rock type	Axis length (mm) a	b	c	Axial ratio $a:b:c$	Radius r (mm)	Stretch $\lambda_1^{1/2}$	Finn parameter K_1
Tuff of quartz keratophyre	84.06	24.70	8.07	10.49:3.08:1	12.76	3.29	1.16
Quartzite, granite, etc.	128.24	50.50	31.68	4.05:1.59:1	31.00	2.07	2.59

picture of the progressive ductile distortion field on the eastern side of the fault zone. Figure 13 shows different deformations of pebbles in the same bed (Sujiawan formation), both far from the fault center and close to it. The width of the shear distortion zone on the eastern side of the fault zone is about 80 km, where the average a/c value is 24.35/1. At Xintian of Jiangpu County near the Yangtze River, the location farthest from the fault center, the a/c values are only 2.22 to 3.30/1.

In view of these relations it is suggested that the strains show a heterogeneous simple shear and progressive displacement field, which is the first type of displacement field of a ductile shear zone, as suggested by RAMSAY (1980).

Strain varieties in vertical section

The types of strain in a vertical section indicate the nature of strain changes at different depths; that is, the strain varies with depth as a result of temperature and pressure. According to our data (Tables 1 and 4), the differences in strain at different

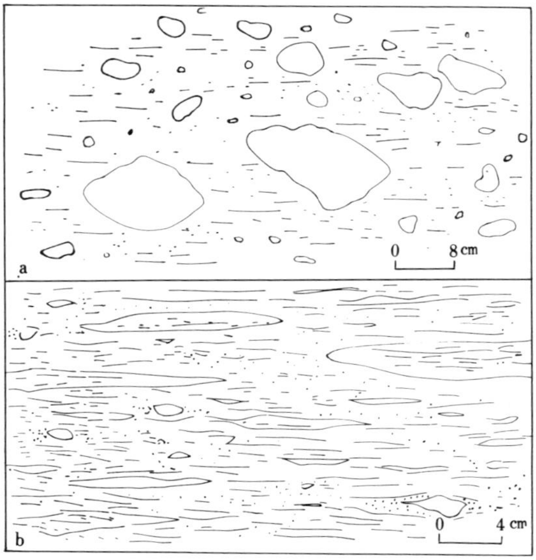

Figure 13
Drawings of deformed pebbles from Sinian (Sujiawan formation): (a) weakly deformed, from Xintian, Jiangpu County; (b) strongly deformed, from Shijie, Chuxian County.

Table 4

Variation of Strain with Depth in Eastern Part of Southern Segment of Tancheng–Lujiang Fault Zone

Stratum	Depth D (km)	Deformed bodies	Axial ratio a/c	$\lambda_1^{1/2}$	K_1
Cambrian	4.65–5.07	tribolite	1.50/1	—	—
Upper Sinian	5.07–6.00	oölite	2.54/1	1.71	—
Lower Sinian					
Sujiawan formation	6.00–7.00	pebbles	12.98/1	3.70	1.222
Zhougan formation	7.00–8.25	pebbles	12.71/1	3.80	1.626
Zhangbaling group	8.25–11.25	volcanic breccia	30.58/1	4.35	1.084

levels are considerable. The tectonic effect causes heterogeneous strain between the upper and lower levels. There is a very strong ductile deformation in the lower levels; for example, the strain ratio is about 30.58/1 in the Zhangbaling group, in which intense shearing, longitudinal stretching, and cross-compressing are found, and the deformations rapidly decrease toward the upper levels.

Calculations of regional deformation

The Zhangbaling belt is intensely deformed on the eastern side of the Tancheng–Lujiang fault zone; the average stretch $\lambda_1^{1/2}$, for $a/2r$, is 5.59, and the elongation e_1 is 4.59, i.e., stretched for 4.59% of the initial length, if it is a homogeneous strain. In the whole 200 km belt, from Hungtse Lake to Caohu Lake, the elongation is 164.23 km, the original length being only 38.77 km. From measurements of a series of syntectonic cross gashes the average extension ratio of these gashes is found to be 4.68%; that is, there is an extension of 9.36 km in a length of 200 km. So the belt has been elongated 173.59 km in all. The elongation e_3 is -0.817, and the average stretch $\lambda_3^{1/2}$, for $c/2r$, is 1/5.45, showing that the intensely deformed belt has been narrowed 222.5 km to the present width of 50 km; thus the original width was 272.5 km. In addition, owing to folding, the shortening also is very considerable. For example, according to measurements of isoclinal folds in the phyllite formation of the lower Sinian, west of Zulongqiao in Chuxian County, the original length of the folded beds was 3.01 m, but the present horizontal distance is only 0.13 m.

Syndeformation metamorphism

During plastic deformation syntectonic metamorphism took place in the fault zone. Like the deformation, the intensity of metamorphism is inversely proportional to the distance from the fault center and directly proportional to the depth.

The intensely deformed belt along the fault, being also the fault metamorphism belt, is 40 to 50 km wide. The metamorphic belt, including the strata of the Zhangbaling group and the Sinian, belongs to the greenschist facies, which may be subdivided into the quartz-feldspar-muscovite-chlorite subfacies and the quartz-albite-epidote-biotite subfacies (after F. J. Dana). In the upper strata the effect of metamorphism rapidly decreases or disappears, for example, weak metamorphism is observed in the upper Sinian, and nonmetamorphism in the Cambro-Ordovician. At the lower levels, such as the Feidong group, retrogressive metamorphism of previous amphibolite-facies rocks resulted in greenschist-facies rocks. On a regional scale the metamorphism clearly diminishes towards the edges of the fault zone. The metamorphic belt of greenschist facies corresponds to a strain ratio of more than 20 and a shear strain γ of more than 2. At the farthest point, near the Yangtze River, the lower Sinian crops out; only weak phyllitization is present, and the pebbles of the Zhangbaling group have not metamorphized.

In the metamorphic belt with greenschist facies the 3T type of polysilic muscovite, glaucophane, jadeite, and other minerals was discovered, which shows that very high differential stresses are present ($\sigma_1 \gg \sigma_2 \gg \sigma_3$). At Qiaotoujie near the fault the quartz grains within the intrusion bodies have undergone plastic deformation, showing that the conditions range from medium pressure to high. At depths greater than 5 km, with pressures on the order of 5 to 7 kbar and temperatures greater than

350 or 400 °C, the greenschist facies appears. The geothermal gradient may get as high as 70 or 79 °C/km, which is obviously higher than the average geothermal gradient in the crust, owing to the frictional heating on the fault. Pseudotachylytes have been found, implying that the fault-related metamorphism took place under conditions of strong dynamic pressure and high temperature.

The syndeformation metamorphism in the greenschist facies of the Zhangbaling belt is similar to other fault metamorphisms in the world, such as the Outer Hebrides thrust belt in the highlands of Scotland; according to SIBSON (1977), the boundary there between the brittle fault and the ductile fault occurs at about 15 km in the Earth's quartz-feldspathic crust, with conditions corresponding to those of greenschist facies. This boundary in the Zhangbaling belt, however, is 5 to 7 km deep, and there begins the ductile deformation and the metamorphism of greenschist facies, perhaps due to strong regional stress and a high geothermal gradient.

Varieties of shear strain

In the southern segment the tectonic orientation of the main fault is about N15°E, and it gradually increases southwards to about N25°E. This major shear zone is associated with many syn-shear tectonics along the eastern side of the fault: schistosity planes, cleavage planes, axial surfaces of folds, etc. Because the diastrophism has always acted on the schistosity in the Zhangbaling group by later folding, we can use the stretching lineation orientation that exists on the schistosity plane as the tectonic strikes. The stretching lineation may be restored to its original horizontal attitude by stereographic projection. In the eastern part of the fault the angle of orientation of the stretching tectonic lines increases gradually from the fault center. In the southernmost part of the study area the orientation angle of the tectonic lines is N60°E, as determined from statistics of the deformed pebbles, and θ', the angle between the stretching lines and the main shear fault plane, is about 45°. In the internal shear-deformed belt near the main fault line angle θ' varies between 0° and 20°; in the place nearest to the center of the fault θ' approaches 0°, and the stretching line is approximately parallel to the fault line. It shows a very strong shearing of horizontal displacement along the fault and progressive strain. If we calculated the value of shear strain γ by using the minimum θ', or $\theta' < 5°$, we can get the maximum value of γ equal to $2/(\tan 2\theta')$, which is more than 11.34. The average value is 6.86 in the intensely deformed zone, which is a greater value than is common in other shear zones, such as the Sihui-Wuchuan fault zone in Guangdong Province, where it is 5.5 (XI, 1982), and the Nordre Stromford shear zone in western Greenland, where it is 6 (BAK et al., 1975). According to calculations of the maximum γ (RAMSAY, 1980), the strain ratio may be more than 130/1 in the study area.

In the northern part of the area studied the orientation of stretching lineation is distorted north-northwest, and near Hungtse Lake it is north-northeast again; see Figure 14. This is a result of another, later sinistral shear zone trending NW-NNW.

Calculation of horizontal displacement

On the basis of the data given above it is suggested that the southern segment of the Tancheng–Lujiang fault zone forms a major ductile shear zone and presents a simple shear deformation. Using the formula suggested by RAMSAY and GRAHAM (1970), $S = \int_0^x \gamma \, dx$, we can calculate the ductile horizontal displacement on the eastern side of this fault zone (Fig. 15). There γ varies from 0.72 to 57.3, and the

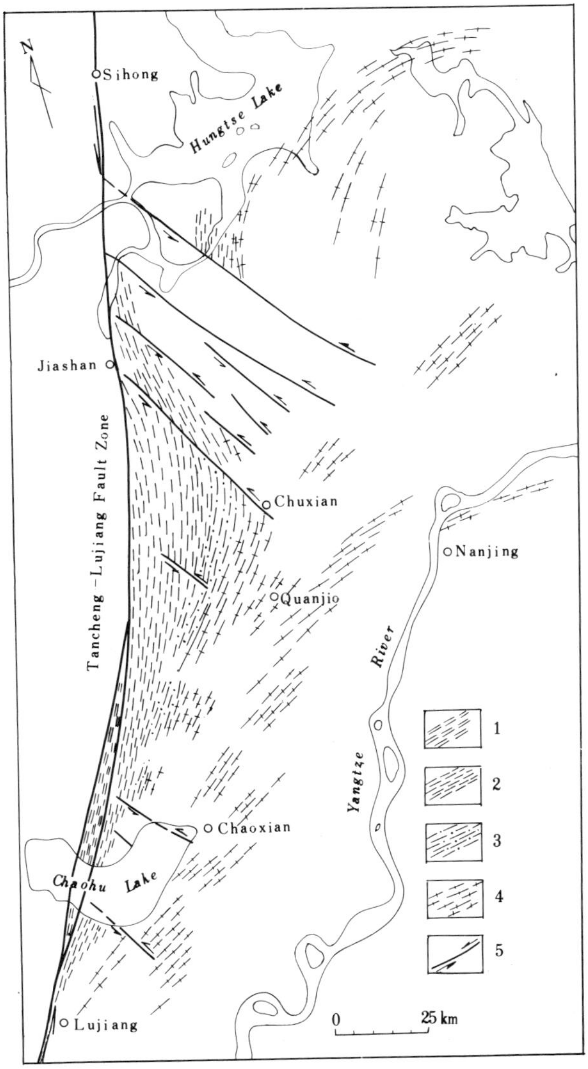

Figure 14
Distributions of lineament over southern segment of Tancheng–Lujiang fault zone. Legend: (1) mineral lineations, (2) stretching lineations, (3) cleavage, (4) axial lines of folds, (5) wrench faults, showing sense of shear.

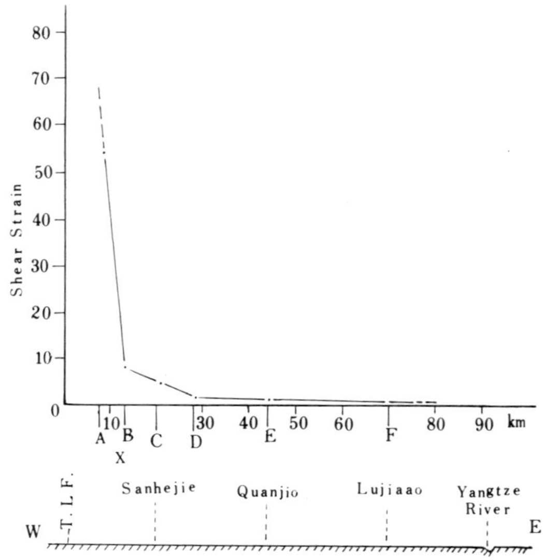

Figure 15
Shear strain curve for section of southern segment of Tancheng–Lujiang fault zone.

horizontal displacement is about 225.27 km. The whole displacement for both sides of the fault may be 450.54 km, which approaches the value obtained from the correlation of geological marks on both sides of the fault. Along this segment, from the Dabie Mountain Uplift on the western side to the Jiaonan Uplift on the eastern side, 450 to 460 km of displacement occurred (after XU, 1980a). From the correlation of geophysical data the displacement is about 400 km (from ZHANG and LI, 1984). A large brittle offset of more than 100 km is noticeable, excepting ductile displacement, so the total displacement of the fault is possibly more than the calculated value given above.

Strain rate and velocity of displacement

In the foregoing calculations strain rate indicates strain in a unit of time (one second). In this study the strain may be indicated by stretch $\lambda_1^{1/2}$, or $a/2r = 5.99$. The age of the strain as found by geological data is the difference between 200 and 180 m.y. B.P., or 20 m.y. (Indo-Sinian orogeny, see later section). The strain rate of stretching is 0.89×10^{-14}/sec. This is similar to that of the famous San Andreas fault zone in North America, which is 10^{-14}/sec. Also according to this calculation, the velocity of shear displacement along the Tancheng–Lujiang fault zone is 1.126 cm/y; the average displacement velocity along the San Andreas fault zone is 1.01 cm/y (CROWELL, 1963), again of similar value.

Age of Ductile Shearing and Tectonic Models of Wrench Shearing

Age of ductile shearing

The foregoing sections have pointed out that the ductile deformation in the zone decreases from lower levels to upper levels and translates into brittle–ductile deformation (see Table 1). The chief formations in which ductile deformation occurs are the Feiding group, the Zhangbaling group, and the Sinian system. Only non-sequence or minor angular unconformity between the Zhangbaling group and the Sinian has been found. The foliation (S_1) and the lineation (L_1) are continuous, and the folds are harmonic between both groups, suggesting that no strong deformation between the Sinian and the Zhangbaling group exists. Although the Feidong group and older formations were deformed during an unspecified earlier episode they were deformed again in the ductile shearing phase. No angular unconformity exists between the Sinian and the upper formations, up to the Middle Triassic series, and the deformation here occurs gradually, which indicates that the Proterozoic to the middle or lower Triassic formations were a single, united tectonic body. The Xiangshan group (J_{1+2}), an Early to Late Cretaceous volcanic rock series with Cretaceous red beds and Yenshanian intrusions that overlie the shear-deformation body unconformably, did not suffer remarkable deformation, and the structures are not harmonic with the deformed tectonic body. So we come to the conclusion that intense ductile shearing along the southern segment of the Tancheng–Lujiang fault zone took place in the Indo-Sinian orogeny.

From the horizontal offset of stratigraphic markers next to the fault zone we know that the major portion of the displacement formed before the late Jurassic or early Cretaceous. If we take the Zhangbaling group as a marker, the offset distance from Lujiang County to east of Taihu County is more than 100 km. Therefore, it is possible that he brittle fracturing climaxed in the early Yenshan stage and that the great Tancheng–Lujiang fault zone developed from then on. It may be concluded that the development of the transcurrent movement of the zone consisted of two stages: ductile shearing in the Indo-Sinian stage and brittle displacement in the early Yenshan stage. In the ductile stage stress was concentrated and significant shearing took place. In the brittle stage stress was released by fracturing and no outstanding ductile deformation took place. As WAKEFIELD (1977) has pointed out, according to the relationship between time and deformation, a ductile shear zone often turns into a brittle shear zone—into a fault. This kind of transition describes the Tancheng–Lujiang fault zone. The transition from ductile deformation to brittle deformation occurred chiefly because the ductile shear zone was uplifted and eroded, because of lower temperature and pressure conditions near the surface (GROCOTT, 1977). Most K–Ar isotopic ages of the markedly deformed metamorphic rock series in the Zhangbaling belt are 230 to 160 m.y. B.P., which was the time that the belt rose and the geothermal temperature fell to 200 °C.

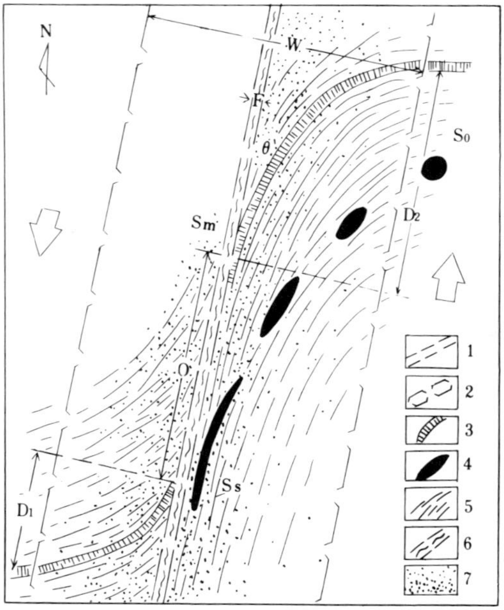

Figure 16
Ductile shear model for Tancheng–Lujiang fault zone. Legend: (1) fault zone, (2) ductile-shear deformed belt, (3) marker beds, (4) flattened and elongated deformed bodies, (5) foliation and lineation structures, (6) mylonite belt, (7) syndeformed metamorphism belts. On the map: (F) width of fault zone, (W) width of shear-deformed zone, (θ') angle between shear fault plane and stretch line, (Sm) foliation mylonite, (Ss) shearing foliation, (S_0) initial foliation, (O) offset, D_1, D_2, horizontal displacements of ductile shear.

Tectonic models of wrench shearing

The southern segment of the fault zone exhibits the features of a major ductile transcurrent fault and is similar to the major Armorican ductile transcurrent fault in Bretagne, France (BERTHE, et al., 1979). Figure 16 shows a ductile shear model for a wrench fault, suggested by those authors, which is characterized by the following.

1. Structural lines beside the fault zone sheared to form curves, which deviate from their original orientation as they approach the fault center. Most of the structural lines approach the center, most of their orientations approach the fault strike, and θ' angles tend to zero. The width of the fault zone is directly proportional to the quantity of shear displacement and inversely proportional to the shear strain (BAK, et al., 1975).

2. The foliation in the zone can indicate its shear direction. Figure 17 shows the left-lateral shearing of the Zhangbaling belt.

3. Deformed pebbles can reflect the nature of the finite-strain ellipsoid. They suggest that the X and Z axes are approximately horizontal and the Y axes

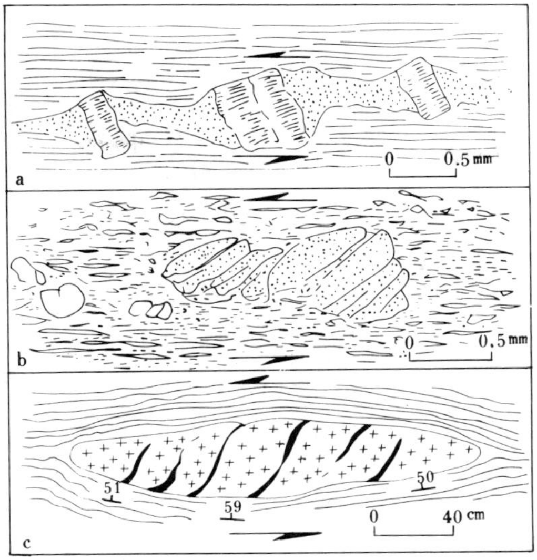

Figure 17
Left-lateral shearing shown by small-scale structures and microstructures: (a) S-shaped porphyritic crystal of feldspar, rotated, with pressure shadows, (b) oblique fractures in clastocrystal in mylonite, (c) oblique gashes in elongated pebble.

approximately vertical. Strain rate rapidly increases towards the center of the fault zone. This exhibits the nature of progressive plane strain. Longitudinal shear extension and cross-shear flattening take place around the center of the zone. The cross-compression gradually transforms into longitudinal compression towards the edges of the zone.

4. The fault zone is characterized by the appearance of abundant mylonites, phyllonites, and syntectonic dynamothermal metamorphism at greenschist facies condition.

5. The direction of activity of the fault zone was consistent, from earlier ductile-shear stages to later brittle-offsetting stages (all left-lateral). The total amount of the fault horizontal displacement includes the quantity of ductile shear displacement ($D_1 + D_2$) and brittle offsetting distance (O); see Figure 17. This model suggests that the features of ductile shear were formed at the middle levels of the crust (5 to 15 km). But later brittle fault deformation was imposed on the ductile deformation zone. See Figure 18.

The authors divide deeper to upper levels of the Tancheng–Lujiang fault zone into four tectonic models of wrench shearing, or four distortion facies. The foregoing discussion pertains to the middle distortion facies. Deep distortion facies (less than 15 to 25 km) produce viscous shear flow deformation; the eastern segment of the Dabie Mountain region in the southern segment of the Tancheng–Lujiang fault zone

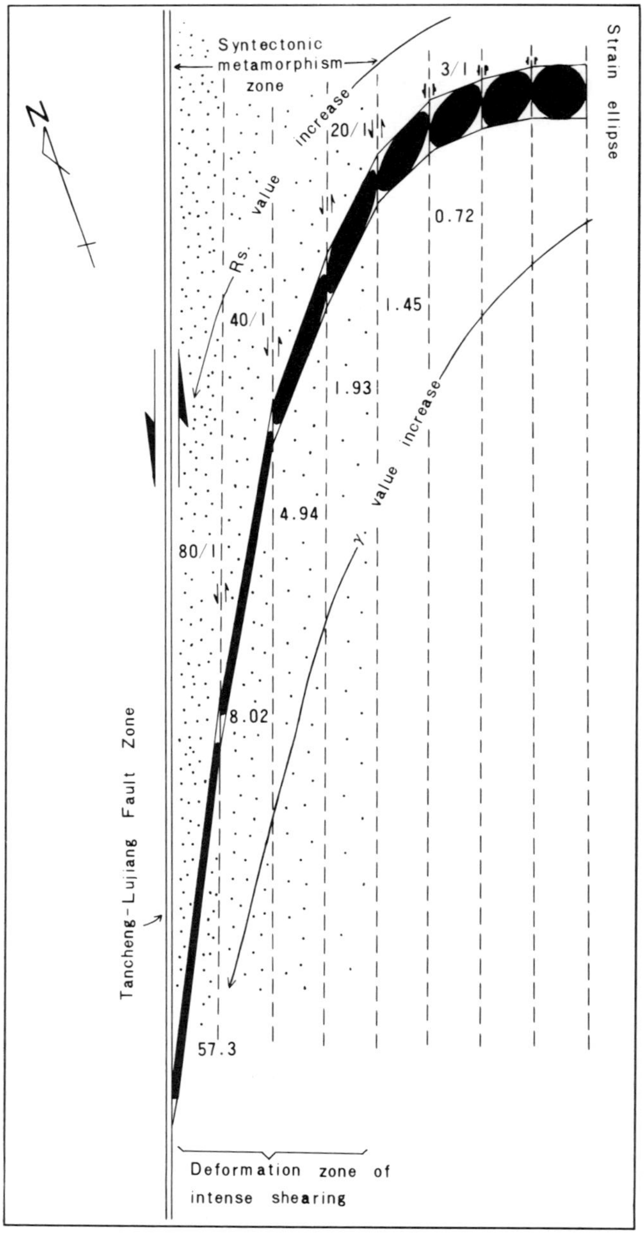

Figure 18
Model showing ductile shear strain on eastern side of southern segment of Tancheng–Lujiang fault zone; strain ratio R_s and shear strain γ, based on many field investigations, but their locations are approximate.

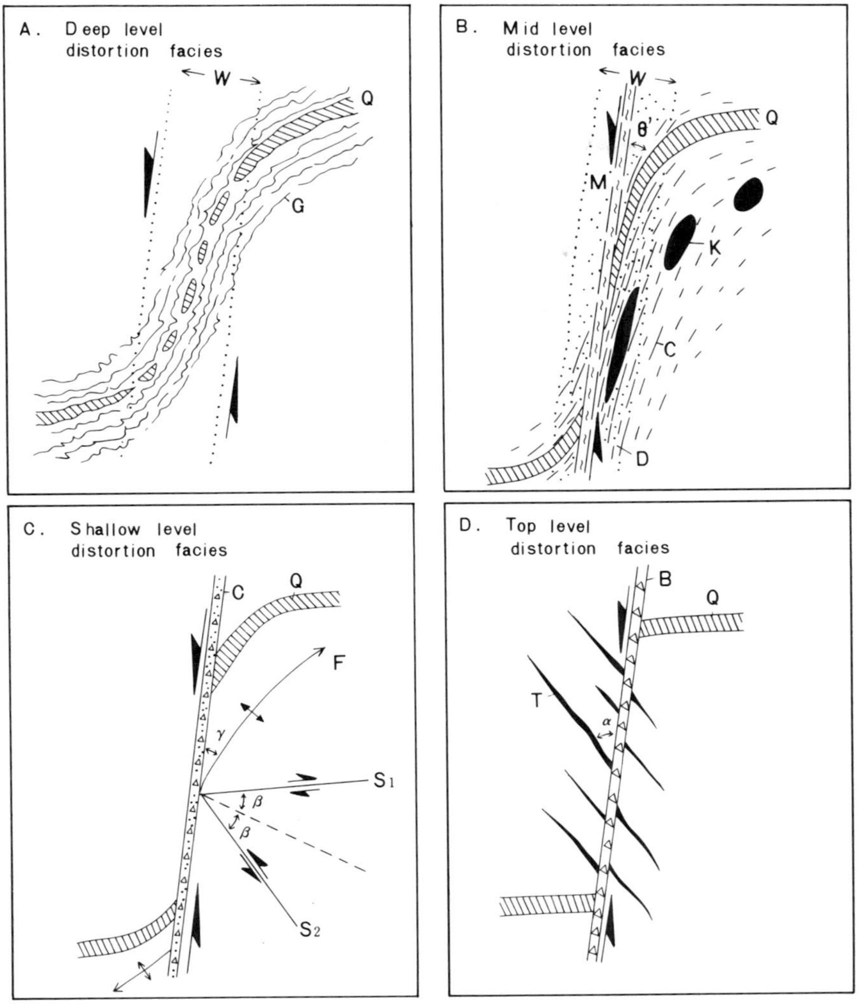

Figure 19
Distortion facies at different tectonic levels in Tancheng–Lujiang fault zone: (G) crystalline schistosity and gneissosity, (W) width of intense shear-deformation zone, (Q) bed indicated, (C in picture B) slipping schistosity and cleavage, (K) deformed bodies of originally similar globular ones, (M) mylonite, (D) fault dynamothermal metamorphism zone, (θ') angle between stretching line and shearing plane, (F) drag folds, (γ) angle between axis trace of drag fold and shear plane (5°–30°), (S_1, S_2) pair of second-order wrench faults, midmost line normal to axis of fold; $\beta = \pm 30°$, (C in picture C) cataclasites, (T) tension fractures or normal fault plane $\pm 45°$), (B) breccias, (α).

is represented by this tectonic model; see Figure 19A. Crystalline schistosity, gneissic schistosity, plastic boudinages, and structural lenticular bodies are found in this deformation facies. In the wider shear zone there are not notable splitting faults. The dominant tectonites are crystalline mylonite schists, mylonite gneisses, and so on; see

Figure 19B. Dynamothermal metamorphism in conditions of amphibolite facies operates at this tectonic level. Shallow distortion facies (0 to 5 km or deeper) present brittle–ductile shear structures; see Figure 19C. At this tectonic level there is a series of diagonal echelon folds (drag folds) and associated secondary compressive and shear faults, which accord with the structural scheme for wrench faults suggested by MOODY and HILL (1956). The shallow distortion facies is characterized by a more narrow fault zone, tectonites of dominant cataclasite, no dynamothermal metamorphism, no penetrative structures, and mylonites occurring locally. The strain axis ratio is generally less than 10/1. This model commonly appears in the lower Yangtze region, the North Anhui, and the North Jiangsu near the Tancheng–Lujiang fault zone. Surface (top) distortion facies (0 to 2 km or deeper) exhibit brittle shear failure; see Figure 19D. Brittle simple offset of pre-existing structures and intense tension fractures oblique to the major fault overprint deformation features at all tectonic levels. Breccias and fault gouges are dominant fault rocks. Surface earthquake structures along some wrench faults are reasonable representatives of this model. The western region of the Shangdong segment in the Tancheng–Lujiang fault zone belongs to this tectonic model.

The coexistence of different structures formed at different levels, such as the ductile shear zone, mylonite zones, and breccia fault gouge zones, occurs because the ductile shear zone uplifted and then suffered erosion in the subsequent stage, and because the nature of faults changes. This conclusion accords with results of research from the Sihui–Wuchuan fault zone in Guangdong. The depths at which the various deformation facies develop obviously are different in different regions and different tectonic settings. Strong compressive-shear stress is the condition that produces ductile deformation at the lower levels of the Tancheng–Lujiang fault zone (SIBSON, 1977; SERMAN, 1980; ZHANG, 1982). The shape of the shear zone changes from the upper levels to the lower. In the research on the West Greenland shear zone, BAK (1975) pointed out that the shape of the zone is a wedge pointed upwards. From the north to the south of the southern segment of the Tancheng–Lujiang fault zone, as the tectonic levels gradually get lower, the width of the zone corresponding decreases, which suggests that the shape of the fault zone in the vertical direction is a wedge.

Just as there is a difference in deformation rates at different levels in the Tancheng–Lujiang fault zone, there is differential extension and compression between upper levels and lower levels. As a result, the effect of displacement movement on the zone is to show smaller longitudinal shear elongation and larger brittle offsetting at shallow levels, extensive ductile deformation increasing and brittle offsetting decreasing at middle levels, and shear viscous flow with no brittle offsetting at deep levels. So there must exist décollements between different tectonic levels in the fault zone.

Intense deformation of terrains next to major ductile shear zones results in intense extension, compression, and rotation of pre-existing structures and geological bodies, besides the décollements between upper levels and lower levels. This must effect the original appearance of the paleotectonics and paleogeography and the pre-existing sedimentary facies zone.

Research is continuing in the Tancheng–Lujiang fault zone in eastern China. With the use of minor plastic deformation structures and, especially, measurements of many deformed pebbles the authors attempt to probe into the evolution of the regional strain field in time and in space, and they tentatively define four shear tectonic models for wrench faults at different tectonic levels.

Acknowledgments

We are grateful to Prof. X. Zhu of the Chinese Marine Geological Surveys, Prof. X. Y. Ma of the Institute of Geology, Chinese State Seimological Bureau, and Prof. S. X. He of the Central-South Institute of Mining and Metallurgy, for their enthusiastic concern and help. We also thank H. L. Song, Z. P. Zhu, G. Zhu, and W. S. Tong for their beneficial help.

REFERENCES

BAK, J., KORSTGÅRD, J. and SØRENSEN, K. (1975), *A major shear zone within the Nagssugtoqidian of west Greenland.* Tectonophys. *27*, 191–209.

BERTHE, D., CHOUKROUNE, P. and JEGOUZO, P. (1979), *Orthogneiss: The example of the South American shear zone.* J. Struct. Geol. *1*, No. 1, 31–42.

BILLINGS, M. P., *Structural Geology.* Prentice-Hall, New Jersey, 1972, p. 606.

CROWELL, J. C. (1963), *The San-Andreas fault system.* Trans. Amer. Geophys. Union, *44*, No. 2.

DURNEY, D. W. and RAMSAY, J. G., 'Increment strains measured by syntectonic crystal growths,' in *Gravity and Tectonics.* John Wiley, 1973, p. 67–96.

GROCOTT, J. (1977), *The relationship between Precambrian shear belts and modern fault systems.* J. Geol. Soc. *133*, No. 3, 257–262.

HE, S. X. (1981), *Strain analyses of deformed rocks.* Special paper, Central-South Institute of Mining and Metallurgy, p. 130.

HILLS, E., *Element of Structural Geology.* Methuen, 1972, p. 483.

HOBBS, B. E., MEANS, W. D. and WILLIAMS, P. F., *An Outline of Structural Geology.* John Wiley, 1976, p. 340.

MATTAUER, M. and MERCIER, J. L. (1980), *Microtectonique et grande tectonique.* Mem. H. ser. Soc. geol. de France, N. 10.

MEANS, W. D., *Stress and Strain.* Springer-Verlag, New York, McGraw-Hill, 1976, p. 310.

MITRA, G. (1979), *Ductile deformation zones in Blue Ridge basement rocks and estimation of finite strains.* Geol. Soc. Amer. Bull. *90*, 935–951.

MOODY, J. D. and HILL, M. J. (1956), *Wrench fault tectonics.* Geol. Soc. Amer. Bull. *67*, 1207–1246.

RAMSAY, J. G., *Folding and Fracturing of Rocks.* McGraw-Hill, New York, 1967, p. 568.

RAMSAY, J. G. (1980), *Shear zone geometry: A review.* J. Struct. Geol. *2*, No. 1/2, 83–89.

RAMSAY, J. G. and GRAHAM, R. H. (1970), *Strain variation in shear belt.* Can. J. Earth Sci. *7*, No. 3, 786–813.

SERMAN, S. E. (1980), *The physical laws of fracture developing in the earth's crust* (from the Russian, translation in Chinese), 'Earth quake'. Publ. House, Beijing, p. 143.

SIBSON, R. H. (1977), *Fault rocks and fault mechanisms.* J. Geol. Soc., *133*, Part 3, 191–213.

WAKEFIELD, J. (1977), *Mylonitization in the Lethakane shear zone, eastern Botswana.* J. Geol. Soc. *133*, 263–375.

XI, X. S. (1982), *The structure features and development history of the ductile shear zones in Yongning region, Yangchun County, Guangdong province.* J. Central-South Inst. Mining Metall. No. 4, Dec., 135–142 (in Chinese).

XU, J. W. (1980a), *The horizontal displacement of Tancheng–Lujiang fault zone and its geological significance*, among scientific papers on geology for international exchange. Publ. House of Geol. Beijing, (1), 129–142 (in Chinese with English abstr.).

XU, J. W. (1980b), *The great left-lateral horizontal displacement of Tancheng–Lujiang fault zone, eastern China.* J. Hefei Polytech. Univ. No. 1, March, 1–26 (in Chinese with English abstr.).

XU, J. W. (1984), 'The Tancheng–Lujiang wrench fault system,' in *Collection of Structural Geology.* No. 3. Publ. House of Geol. Beijing, p. 18–33 (in Chinese with English abstr.).

XU, J. W., CUI, K. R., LIU, Q., TONG, W. X. and ZHU, G. (1985), *Mesozoic sinistral transcurrent faulting along the continent margin in eastern Asia.* Marine geol. Quatern. Geol. 5, No. 2, 51–64 (in Chinese with English abstr.).

ZHANG, Z. T. (1982), *The plastic fault and the plastic deformation zone of fault.* Bull. Xi'an Inst. Geol. Min. Res., Chinese Acad. Geol. Sci., No. 4, 94–101 (in Chinese).

ZHANG, Y. X. and LI, L. L. (1984), 'The giant strike slip along Tancheng–Lujiang fault zone and its influence on the nearby structures,' in 'Collection of Structural Geology,' No. 3. Publ. House of Geol. Beijing, p. 1–8 (in Chinese).

(Received 4th September 1985, revised 20th March 1986, accepted 23rd March 1986)

PAGEOPH, Vol. 124, Nos. 1/2 (1986) 0033-4533/86/020365-09\$1.50+0.20/0

Fault Dimensions, Displacements and Growth

JUAN WATTERSON[1]

Abstract—Maximum total displacement (D) is plotted against fault or thrust width (W) for 65 faults, thrusts, and groups of faults from a variety of geological environments. Displacements range from 0.4 m to 40 km and widths from 150 m to 630 km, and there is a near linear relationship between D and W^2. The required compatibility strains (e_s) in rocks adjacent to these faults increases linearly with W and with $\sqrt{D}$ and ranges from $e_s = 2 \times 10^{-4}$ to $e_s = 3 \times 10^{-1}$. These are permanent ductile strains, which compare with values of $e_s = 2 \times 10^{-5}$ for the elastic strains imposed during single slip earthquake events, which are characterised by a linear relationship between slip (u) and W.

The data are consistent with a simple growth model for faults and thrusts, in which the slip in successive events increases by increments of constant size, and which predicts a relationship between displacement and width of the form $D = cW^2$. Incorporation of constant ductile strain rate into the model shows that the repeat time for slip events remains constant throughout the life of a fault, while the displacement rate increases with time. An internally consistent model with $e_s = 2 \times 10^{-5}$, giving repeat times of 160 years and instantaneous displacement rates of 0.02 cm/yr, 0.2 cm/yr, and 2.0 cm/yr when total displacement is 1 m, 100 m, and 10 km, and slip increasing by 0.5 mm with each event, gives a good approximation of the data. The model is also applicable to stable sliding, the slip rate varying with ductile strain rate and with W^2.

Key words: Fault, displacement, seismic slip, strain, strain rate, displacement rate.

Introduction

The problem addressed concerns the relationship between fault dimensions and the total displacement accrued through numerous slip events or stable sliding, and the consequent ductile and elastic strains required to maintain compatibility in rocks adjacent to the fault. The nomenclature used is illustrated in Figure 1, which shows the essential geometry of an ideal simple normal fault which does not intersect a free surface, i.e. a blind fault. The slip surface is taken to be an elliptical plane with the slip vector parallel to the short axis of the ellipse; the dimension of the slip surface parallel to the slip vector is the length L. The dimension normal to the slip vector within the slip plane is the fault width W. Displacement varies from a maximum at

[1] Department of Geological Sciences, Liverpool University, P.O. Box 147, Liverpool L69 3BX, United Kingdom.

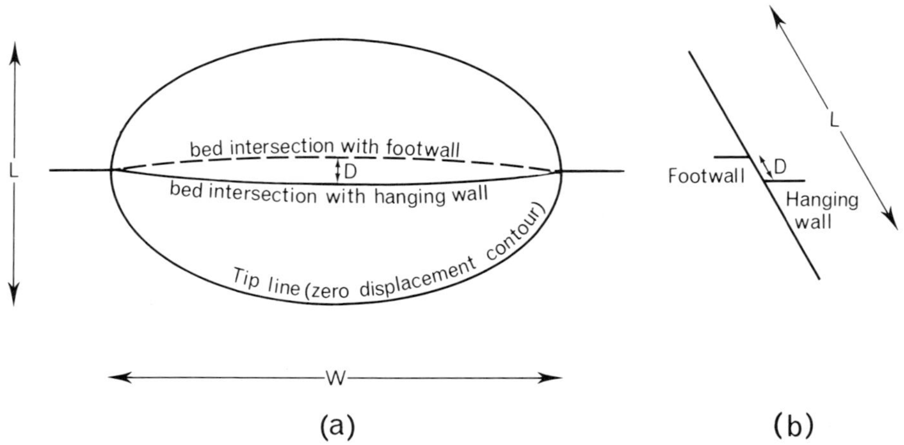

Figure 1
Essential geometry of an idealised simple normal fault with elliptical tip line, or zero displacement contour: (a) viewed along normal to slip surface; (b) side elevation. Disposition of an originally horizontal horizon on the fault plane is shown for footwall and hanging wall. L is length (parallel to slip direction), W is width, and D is maximum displacement.

the centre of the elliptical surface to zero at the edge of the slip surface which, following thrust terminology (BOYER and ELLIOTT, 1982), is an elliptical tip line. The fault is assumed to grow by radial extension of the tip line with no migration of the point of maximum displacement.

The maximum compatibility strains required by this geometry are, for total fault displacement D, length L, and width W, as follows.

1. In rocks adjacent to the fault surface and lying along the minor axis of the ellipse (parallel to the slip direction), an average linear strain $e_s = D/L$: this is the strain arising from the edge component of the fault dislocation, referred to by MURAOKA and KAMATA (1983) as the slip-parallel strain.

2. In rocks adjacent to the fault surface and initially lying along the major axis of the ellipse (perpendicular to the slip direction), an average shear strain $\gamma = D/W$: this is the strain arising from the screw component of the fault dislocation. For comparative purposes a more useful quantity is the linear strain of the long axis of the rotational strain ellipse e_s corresponding to the shear strain (RAMSAY and HUBER, 1983), which for the comparatively small strains ($\gamma < 0.2$) involved in faulting is $e_s = \gamma/2$, or $e_s = D/2W$.

The average strains can thus be simply expressed in terms of the maximum displacement and the principal dimensions of the fault surface. A more rigorous definition of the strains is not possible without knowledge of the distribution of displacements across and normal to the fault surface.

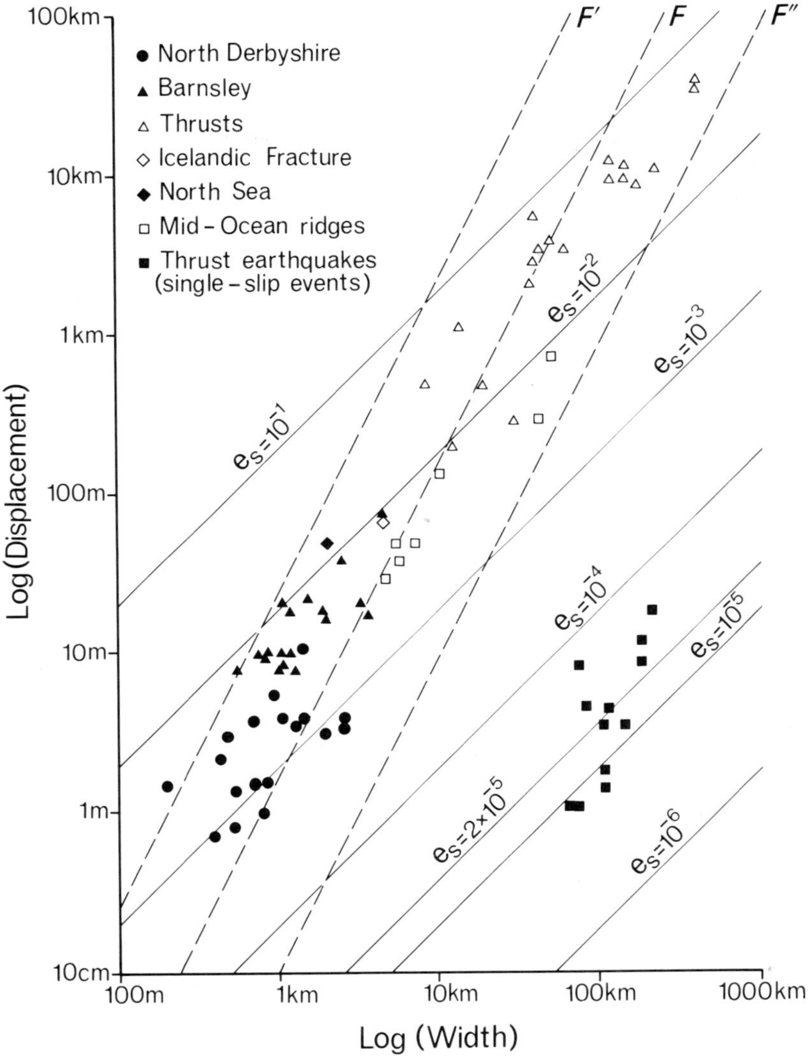

Figure 2

Width versus maximum displacement for faults, thrusts, and groups of faults (coal field faults from Barnsley and N. Derbyshire areas distinguished) and slip versus width for seismic events; e_s is linear strain required for compatibility, and F, F', and F'' are model growth curves. See text for explanation.

Fault and thrust data

The data available are sufficient only for examining the relationship between total displacement and fault width, and this is shown in Figure 2 for a range of faults and thrusts from different geological environments and covering a wide range of scale. The systematic distribution is in part an artifact of the log–log scale required by the range of size. Sources for the data are given below.

1. Mine data are for normal faults encountered during deep mining operations in British coal fields. Where several worked coal seams intersect the same fault at different levels and the displacement on the fault has been recorded at several places in each seam, sufficient data points may exist for contours of displacement to be drawn over the whole, or a substantial part, of the fault surface (RIPPON, 1985). This allows both the maximum displacement and the principal dimensions of the fault to be identified. Single fault surfaces generally approximate the geometry shown diagrammatically in Figure 1a, i.e., elliptical tip lines with displacement parallel to the short axis of the ellipse. The data plotted are derived either from faults where the complete fault plane has been defined or from faults for which sufficient information is available for definition of the maximum displacement and the half-width.

2. Thrust data are from ELLIOTT (1976) for thrusts in low-grade rocks in the foot hills and front ranges of the Canadian Rockies. The thrusts are not blind, and the assumption is made that the maximum widths and maximum displacements are those at the leading edge.

3. A single data point is included for a normal fault scarp in Iceland (GUDMUNDSSON, 1980), for which the same assumptions are made as for the thrusts.

4. Seven data points refer to normal fault scarps from several ocean floor areas (SEARLE, 1983; LAUGHTON and SEARLE, 1979) adjacent to spreading centres: the original data are of mean values for length and height of several scarps within each area. The plotted data may therefore underestimate the maximum displacements on these faults. The same assumptions are made as in Paragraphs 2 and 3.

5. A single data point represents two faults of similar size and displacement reconstructed from high-resolution seismic-reflection data; the faults are located in the North Sea.

Also plotted in Figure 2 are contours of the corresponding linear strains e_s. Not included in the plotted data are two faults of lengths 1.3 and 2.5 m, with displacements of 13.5 and 40.0 cm (MURAOKA and KAMATA, 1983). The displacements on these faults are an order of magnitude greater than would be expected if the linear trend in Figure 2 were to be extended to smaller scales, and differences between length and width are unlikely to reduce the anomaly significantly. These faults occur in Quaternary lacustrine sediments, which may have very different mechanical properties to the host rocks of the other faults for which data are available.

Interpretation

The immediate conclusions to be drawn from the data are as follows.

1. The strains required to accommodate these faults and thrusts, just less than $e_s = 0.1$, or 10% strain in the case of the major structures, are much too great to be

accommodated elastically and must therefore represent permanent ductile strains. In this context 'ductile' means strain in which material continuity is maintained, or in which discontinuities are on a very small scale relative to that of the strained volume.

2. Strain e_s increases with the size of the fault or thrust.

3. The relationship between total displacement and fault width is not linear.

A linear relationship between fault length and the mean slip u for individual large earthquake events has been demonstrated (SCHOLZ, 1982) for both strike slip and thrust movements, the characteristic values of e^1 being 1.25×10^{-5} for strike-slip faults and 2×10^{-5} for thrusts. The width of strike-slip faults is limited by the thickness of the seismogenic layer, and no relationship between width and seismic slip is expected or found (SCHOLZ, op. cit.). This limitation does not apply to deep thrust displacements, and the slips and fault widths for twelve such events listed by SCHOLZ (op. cit.) are plotted in Figure 2. The stress drop of small earthquakes is known to be nearly constant (HANKS, 1977) and independent of fault dimensions; as the stress drop is proportional to mean slip per fault radius, a linear relationship between u and W is expected, *pace* length per slip for large earthquakes. A slip per width ratio of 4×10^{-5} for single slip events, corresponding to $e_s = 2 \times 10^{-5}$, will be taken as the representative value. This is the strain change or strain drop (KANAMORI and ANDERSON, 1975). The earthquake data are not exactly comparable, since the values of total displacement and of e_s for the inactive faults refer to the maximum value of displacement on a fault, whereas the values for single slip earthquake events refers to mean values over the slip surface; as the mean slip is likely to exceed 0.8 of the maximum value (ESHELBY, 1957), no correction is necessary for present purposes.

From the data, and the assumed knowledge of the representative value of e_s associated with single slip events, an estimate can be made of the size of the final slip event on each of the faults for which the total displacement is known; e.g., a fault of width 100 km with a total displacement of 10 km will have had a final slip of 4 m, given the assumptions discussed previously. This allows some constraints to be placed on the manner in which faults grow and on the number of slip events which are required to achieve the total displacement on a seismic fault. Some absolute limitations are immediately apparent. If a fault is the product of only a single slip event, i.e. $D = u$, then it would plot somewhere along the line corresponding to $e_s = 2 \times 10^{-5}$; this is clearly not the case for any of the faults for which data are available. If every slip event on a fault were of the same size, then, given a linear relation between u and W, the fault would not increase in width, and the growth curve of the fault would parallel the y axis ($\log D$) of Figure 2; such a relationship would require complete relaxation of all elastic compatibility strains between successive slip events, and there would be no systematic relationship between width and total displacement in a sample of faults. This is not the case for the sample available. On the other hand, if no relaxation of elastic compatibility strains took place between successive slip events, the established linear relationship between u and W for earthquake events would not obtain. The real case must lie between these extremes, and it requires relaxation between successive seismic faulting events of a proportion of the elastic strain, which is replaced by an equivalent permanent strain. It is evident that in the general case a fault will increase in width with time and that individual displacements of successive slip events must also increase.

Fault growth model

It might be expected that a model for the increase of successive slip events would require a complicated growth law, but this seems not to be the case. A model reasonably consistent with the data is one in which the amount of slip in individual events increases in simple arithmetic progression. Where successive slips increase by a constant increment,

$$A = u/2 \quad \text{and } D = N \,.\, A, \quad \text{so } D = N \,.\, u/2.$$
$$u = N \,.\, k, \quad \text{so } N = u/k, \quad \text{and therefore } D = u^2/2k.$$
$$\text{But } u \propto W, \quad \text{so, for constant } k, D = c \,.\, W^2, \quad \text{or } \log D = c + 2 \log W.$$

Here D is total displacement, u is last slip, W is fault width, A is average slip, N is number of slip events, and k is the increment by which successive slip events increase, and c is a constant.

A fault growing according to this model will have a growth curve which plots as a straight line on a log–log plot with a slope of 63.5 ($\tan^{-1} = 2$). The position of the growth curve is determined by the values of k and u/W. A representative growth curve F, Figure 2, corresponds to values of $k = 0.5$ mm and $u/W = 4 \times 10^{-5}$. Most of the data are included in the range bounded by the two lines F' and F'', which correspond to $k = 0.03$ and $k = 8$ mm respectively for $u/W = 4 \times 10^{-5}$. Alternatively, F' and F'' correspond to a u/W of 8×10^{-5} and 5×10^{-6} respectively at constant $k = 0.5$ mm.

If $D = cW^2$, the value of the constant c is 1.6×10^{-3} for line F, 2.5×10^{-2} for F', and 1×10^{-4} for F''. Given the range in type and scale of the faults and thrusts in the data set, and the range of material properties of the rocks in which they occur, the required range of values for the constant and for either k or u/W is sufficiently small for the model to be a potentially useful approximation.

The conclusions drawn here conflict in some respects with the assumptions of Williams and Chapman (1983), who accept the popular concept of a 'ductile bead' at the propagating tip line of a fault. This concept requires that accommodation strains are imposed only adjacent to the current tip line, thereafter remaining unmodified during subsequent evolution and growth of the fault. This requirement is a consequence of the unstated assumptions (op. cit.) that (a) total maximum displacement is directly proportional to fault dimension (length or width), and that (b) the displacement gradient from fault tip to point of maximum displacement is linear.

Table 1

Characteristics of Fault Growth with Constant Values $e_s = 5 \times 10^{-15}$ sec^{-1}, $u/W = 4 \times 10^{-5}$, and u Incrementing by 0.5 mm with each Slip Event

Total maximum displacement (km)	0.001	0.01	0.1	1.0	10	50
Width (km)	1.6	5	15.8	25	158	173
Time (yr $\times 10^3$)	11	33	107	327	1070	2356
Average displacement rate (cm/yr)	0.01	0.03	0.1	0.32	1.0	2.19
Instantaneous displacement rate (cm/yr)	0.02	0.06	0.2	0.63	1.98	4.3
Last slip (cm)	3.16	10	31.6	100	316	692
Number of slip events	63	200	633	2000	6329	14451
Repeat time (yr)	160	160	160	160	160	160

The first of these assumptions is inconsistent with the data given here, and the second assumption is doubtful, given the radial distribution of slip required by the ESHELBY (1957) model for single slip events. The increase in strains with fault dimensions demonstrated in Figure 2 strongly suggests that rocks adjacent to the whole slip surface undergo strain changes with each increment of slip, rather than just those adjacent to the propagating tip line, although this cannot be demonstrated conclusively, given that e_s is defined as the mean strain.

Rates of strain and of displacement

Taking the growth curve F as representative, the simple growth model can be used in examining the relationship between rates of strain, rates of displacement, and repeat times for seismic events. During the growth of a fault the most likely constant factor is the linear ductile strain rate, and this may in fact be the parameter which determines all other rates.

Taking $\dot{e}_s = 5 \times 10^{-15}$ sec^{-1} as the constant ductile strain rate, the other changes during growth of a fault or thrust to a total displacement of 50 km are listed in Table 1. This shows the model to predict constant repeat times for seismic events throughout the life of an individual fault, regardless of size attained, and to predict an increase in both average and instantaneous displacement rates with increase in size of the fault. The value of 5×10^{-15} was chosen for $\dot{e}_s$ in order to give a reasonable repeat time of 160 years, but it is within the range expected within the seismogenic layer. The ductile strain rates expected below the seismogenic layer range from 10^{-12} sec^{-1} to 10^{-14} sec^{-1} and are likely to decrease by approximately one order of magnitude for each 100° reduction in temperature (see data for quartz in WHITE, 1976, Fig. 7).

Although the model has been expressed in terms of seismic, i.e. intermittent, fault displacement, it is also applicable to faults which are characterised by stable sliding.

If stable sliding takes place with constant elastic strain, i.e. plastic yielding *ss*, the ductile strain rate is constant and the relationships between total displacement, width, displacement rates, and linear ductile strain rate are the same as for seismic faults.

The model is directly applicable only to growing faults in which each slip event, or stable sliding, occurs over the entire fault surface, and in which the fault surface is extended either with each seismic event or continuously during stable sliding. It may prove possible to adapt the model to slip events that affect only a part of an existing fault surface, but it cannot be applied to kinematically controlled faults, where the displacement rate on a single slip surface is determined directly by plate motions; such faults in any case terminate in transfer structures rather than tip lines, and length and width as defined here cannot be measured.

Acknowledgments

Much of the coal field fault data plotted in Figure 2 was provided by John Rippon and Jim Barnett of the British National Coal Board; I am grateful to these colleagues and to John Mortimer and John Walsh of the Liverpool Fault Analysis Group for advice and assistance, and to Britoil plc for access to seismic-reflection data.

This work is supported by Grant GR/4719 from the British Natural Environment Research Council and contracts from the National Coal Board (YCE.330/19572) and the U.S. Geological Survey Earthquake Hazards Reduction Program (14-08-0001-22072). Publication is with the permission of the National Coal Board.

References

Boyer, S. E. and Elliott, D. (1982), *Thrust systems*. Amer. Assoc. Petr. Geol. *66*, 9, 1196–1230.

Elliott, D. (1976), *The energy balance and deformation mechanisms of thrust sheets*. Phil. Trans. R. Soc. London A *283*, 289–312.

Eshelby, J. D. (1957), *The determination of the elastic field of an ellipsoidal inclusion and related problems*. Proc. Roy. Soc. London A *241*, 376–396.

Gudmundsson, A. (1980), *The Vogur Fissure Swarm, Reykjanes Peninsula, SW Iceland*. Jokull *30*, 43–64.

Hanks, T. C. (1977), *Earthquake stress drops, ambient tectonic stress, and stresses that drive plate motions*. Pure Appl. Geophys. *115*, 441–458.

Kanamori, H. and Anderson, D. L. (1975), *Theoretical basis of some empirical relation in seismology*. Bull. Seism. Soc. Am. *65*, 1073–1095.

Laughton, A. S. and Searle, R. C. (1979), *Tectonic Processes of Slow Spreading Ridges, in Deep Drilling Results in the Atlantic Ocean: Ocean Crust*. M. Talwani, C. G. Harrison, and D. E. Hayes (eds.), Maurice Ewing Ser., Vol. 2, AGU, Washington, D.C.

Muraoka, H. and Kamata, H. (1983), *Displacement distribution along minor fault traces*. J. Struct. Geol. *5*, 483–495.

Ramsay, J. G. and Huber, M. I. (1983), *The techniques of modern structural geology. Vol. 1, Strain Analysis*. Academic Press, London, 307 pp.

Rippon, J. H. (1985), *Contoured patterns of the throw and hade of normal faults in the coal measures (Westphalian) of north-east Derbyshire.* Proc. Yorkshire Geol. Soc. *45*, 147–161.

Scholz, C. H. (1982), *Scaling laws for large earthquakes: Consequences for physical models.* Bull. Seism. Soc. Am. *72*, 1–14.

Searle, R. C. (1983), *Gloria survey of the East Pacific Rise near 3.5° S: Tectonic and volcanic characteristics of a fast spreading mid-ocean rise.* Tectonophys. *101*, 319–344.

White, S. (1976), *The effects of strain on the microstructures, fabrics, and deformation mechanisms in quartzites.* Phil. Trans. R. Soc. London A *283*, 289–312.

Williams, G. and Chapman, T. (1983), *Strains developed in the hanging walls of thrusts due to their slip/propagation rates: A dislocation model.* J. Struct. Geol. *5*, 563–571.

(Received 1st November 1985, revised 2nd February 1986, accepted 18th February 1986)